T0189740

Fractional Dynamic Calculus and Fractional
Dynamic Equations on Time Scales

Svetlin G. Georgiev

Fractional Dynamic Calculus and Fractional Dynamic Equations on Time Scales

 Springer

Svetlin G. Georgiev
Faculty of Mathematics and Informatics
Sofia University St Kliment Ohridski
Sofia, Bulgaria

ISBN 978-3-030-08892-7   ISBN 978-3-319-73954-0   (eBook)
https://doi.org/10.1007/978-3-319-73954-0

Printed on acid-free paper

This Springer imprint is published by the registered company Springer International Publishing AG part
of Springer Nature.
The registered company address is: Gewerbestrasse 11, 6330 Cham, Switzerland

Preface

Fractional differential equation theory is an important branch of mathematics that includes continuous fractional differential equations and discrete fractional difference equations. The theory of fractional differential equations has gained considerable popularity and importance over the past three decades or so. Many applications in numerous seemingly diverse and widespread fields of science and engineering have been found. The theory indeed provides several potentially useful tools for solving differential and integral equations and various other problems involving special functions of mathematical physics as well as their extensions and generalizations in one and more variables. To unify differential equations and difference equations, Stefan Hilger proposed the time scale and built the relevant basic theories. This book encompasses recent developments in fractional calculus and fractional differential equations on time scales. This book is intended for use in the field of fractional calculus and fractional dynamic equations on time scales. It is also suitable for graduate courses in the above fields. The book contains nine chapters, which are pedagogically organized. This book is specifically designed for those without extensive mathematical background who wish to understand fractional calculus and fractional dynamic equations on time scales.

The basic definitions of forward and backward jump operators are due to Hilger. In Chapter 1, examples of jump operators on some time scales are presented. The graininess function, which is the distance from a point to the closed point on the right, is introduced in this chapter. The definitions of delta derivative and delta integral are given, and some of their properties are deduced. Chapter 2 introduces the Laplace transform on time scales. Its main properties are deduced, and conditions are given on the class of functions that have a transform. An inversion formula for the transform is given. Chapter 3 deals with convolution on time scales. Using an initial value problem containing a dynamic version of the transport equation, the delay (or shift) of a function defined on a time scale is introduced, and the delay in turn is used to introduce the convolution of two functions defined on the time scale. Some elementary properties of the delay and the convolution are given, and the main convolution theorem is proved. Chapter 4 is concerned with the Riemann–Liouville fractional Δ-integral and the Riemann–Liouville fractional Δ-derivative. Some of the properties of the Δ-power function and Riemann–Liouville fractional Δ-integral and derivative are deduced and proved. In Chapter 5, a Cauchy-type problem with Riemann–Liouville fractional Δ-derivative is considered. The existence and uniqueness of the solution are proved along with the dependence of the solution on the initial data. Riemann–Liouville fractional dynamic equations with constant coefficients are investigated in Chapter 6. In Chapter 7, the Caputo fractional Δ-derivative on time scales is introduced, and some of its properties are deduced. In Chapter 8, the existence and uniqueness of the solution of the Cauchy-type problem with the Caputo fractional Δ-derivative are proved, and the dependence of the solution upon the initial value is investigated. Chapter 9 is devoted to Caputo fractional dynamic equations with constant coefficients.

The aim of this book is to present a clear and well-organized treatment of the concepts behind the development of the relevant mathematics and solution techniques. The material of this book is presented in a highly readable, mathematically solid format. Many practical problems are illustrated, displaying a wide variety of solution techniques.

Paris, France   Svetlin G. Georgiev
October 2017

Contents

Chapter 1
Elements of Time Scale Calculus

This chapter is devoted to a brief exposition of the time scale calculus that provides the framework for the study of fractional dynamic calculus and fractional dynamic equations on time scales. A detailed discussion of the time scale calculus is beyond the scope of this book, and for this reason, the author limits himself to outlining a minimal set of properties that will be needed in what follows. The presentation in this chapter follows the books [5] and [2]. A deep and thorough investigation of the time scale calculus, as well as a discussion of the available literature on this subject, can be found in those books.

1.1 Forward and Backward Jump Operators, Graininess Function

Definition 1.1. A time scale is an arbitrary nonempty closed subset of real numbers.

We will denote a time scale by the symbol $\mathbb{T}$.
We suppose that a time scale $\mathbb{T}$ has the topology that it inherits from the real numbers with standard topology.

Example 1.1. $[1, 2]$, $\mathbb{R}$, $\mathbb{N}$,

$$\left\{-2, \quad -1, \quad -\frac{1}{2}, \quad 0, \quad \frac{1}{4}, \quad \frac{1}{3}, \quad 2, \quad 3, \quad 6\right\}$$

are time scales.

Example 1.2. $[a, b)$, $(a, b]$, (a, b), $(a, b] \bigcup \{2b - a\}$ are not time scales.

© Springer International Publishing AG, part of Springer Nature 2018
S. G. Georgiev, *Fractional Dynamic Calculus and Fractional Dynamic Equations on Time Scales*, https://doi.org/10.1007/978-3-319-73954-0_1

Definition 1.2. For $t \in \mathbb{T}$ we define the forward jump operator $\sigma : \mathbb{T} \to \mathbb{T}$ as follows:

$$\sigma(t) = \inf\{s \in \mathbb{T} : s > t\}.$$

We note that $\sigma(t) \geq t$ for all $t \in \mathbb{T}$.

Definition 1.3. For $t \in \mathbb{T}$ we define the backward jump operator $\rho : \mathbb{T} \to \mathbb{T}$ by

$$\rho(t) = \sup\{s \in \mathbb{T} : s < t\}.$$

We note that $\rho(t) \leq t$ for all $t \in \mathbb{T}$.

Definition 1.4. We set

$$\inf \emptyset = \sup \mathbb{T}, \quad \sup \emptyset = \inf \mathbb{T}.$$

Definition 1.5. For $t \in \mathbb{T}$ we have the following cases.

1. If $\sigma(t) > t$, then we say that t is right-scattered.
2. If $t < \sup \mathbb{T}$ and $\sigma(t) = t$, then we say that t is right-dense.
3. If $\rho(t) < t$, then we say that t is left-scattered.
4. If $t > \inf \mathbb{T}$ and $\rho(t) = t$, then we say that t is left-dense.
5. If t is left-scattered and right-scattered at the same time, then we say that t is isolated.
6. If t is left-dense and right-dense at the same time, then we say that t is dense.

Example 1.3. Let $\mathbb{T} = \{\sqrt{3n+2} : n \in \mathbb{N}\}$. If $t = \sqrt{3n+2}$ for some $n \in \mathbb{N}$, then $n = \dfrac{t^2 - 2}{3}$ and

$$\sigma(t) = \inf\{l \in \mathbb{N} : \sqrt{3l+2} > \sqrt{3n+2}\} = \sqrt{3n+5} = \sqrt{t^2+3} \quad \text{for } n \in \mathbb{N},$$
$$\rho(t) = \sup\{l \in \mathbb{N} : \sqrt{3l+2} < \sqrt{3n+2}\} = \sqrt{3n-1} = \sqrt{t^2-3} \quad \text{for } n \in \mathbb{N}, n \geq 2.$$

For $n = 1$ we have

$$\rho(\sqrt{5}) = \sup \emptyset = \inf \mathbb{T} = \sqrt{5}.$$

Since

$$\sqrt{t^2 - 3} < t < \sqrt{t^2 + 3} \qquad \text{for} \qquad n \geq 2,$$

we conclude that every point $\sqrt{3n+2}$, $n \in \mathbb{N}$, $n \geq 2$, is right-scattered and left-scattered, i.e., every point $\sqrt{3n+2}$, $n \in \mathbb{N}$, $n \geq 2$, is isolated.

Because

$$\sqrt{5} = \rho(\sqrt{5}) < \sigma(\sqrt{5}) = \sqrt{8},$$

we have that the point $\sqrt{5}$ is right-scattered.

Example 1.4. Let $\mathbb{T} = \left\{ \dfrac{2}{2n+1} : n \in \mathbb{N} \right\} \cup \{0\}$ and $t \in \mathbb{T}$ be arbitrarily chosen.

1. $t = \dfrac{2}{3}$.   Then

$$\sigma\left(\frac{2}{3}\right) = \inf \emptyset = \sup \mathbb{T} = \frac{2}{3},$$

$$\rho\left(\frac{2}{3}\right) = \sup\left\{ \frac{2}{2l+1}, 0 : \frac{2}{2l+1}, 0 < \frac{2}{3}, l \in \mathbb{N} \right\} = \frac{2}{5} < \frac{2}{3},$$

i.e., $\dfrac{2}{3}$ is left-scattered.

2. $t = \dfrac{2}{2n+1}$, $n \in \mathbb{N}$, $n \geq 2$.   Then

$$\sigma\left(\frac{2}{2n+1}\right) = \inf\left\{ \frac{2}{2l+1} : \frac{2}{2l+1} > \frac{2}{2n+1}, l \in \mathbb{N} \right\}$$

$$= \frac{2}{2(n-1)+1} > \frac{2}{2n+1},$$

$$\rho\left(\frac{2}{2n+1}\right) = \sup\left\{ \frac{2}{2l+1}, 0 : \frac{2}{2l+1}, 0 < \frac{2}{2n+1}, l \in \mathbb{N} \right\}$$

$$= \frac{2}{2(n+1)+1} < \frac{2}{2n+1}.$$

Therefore, all points $\dfrac{2}{2n+1}$, $n \in \mathbb{N}$, $n \geq 2$, are right-scattered and left-scattered, i.e., all points $\dfrac{2}{2n+1}$, $n \in \mathbb{N}$, $n \geq 2$, are isolated.

3. $t = 0$.   Then

$$\sigma(0) = \inf\{s \in \mathbb{T} : s > 0\} = 0,$$

$$\rho(0) = \sup\{x \in \mathbb{T} : s < 0\} = \sup \emptyset = \inf \mathbb{T} = 0.$$

Example 1.5. Let $\mathbb{T} = \left\{ \dfrac{n}{7} : n \in \mathbb{N}_0 \right\}$ and $t = \dfrac{n}{7}$, $n \in \mathbb{N}_0$, be arbitrarily chosen.

1. $n \in \mathbb{N}$.   Then

$$\sigma\left(\frac{n}{7}\right) = \inf\left\{\frac{l}{7}, 0 : \frac{l}{7}, 0 > \frac{n}{7}, l \in \mathbb{N}_0\right\} = \frac{n+1}{7} > \frac{n}{7},$$

$$\rho\left(\frac{n}{7}\right) = \sup\left\{\frac{l}{7}, 0 : \frac{l}{7}, 0 < \frac{n}{7}, l \in \mathbb{N}_0\right\} = \frac{n-1}{7} < \frac{n}{7}.$$

Therefore, all points $t = \dfrac{n}{7}$, $n \in \mathbb{N}$, are right-scattered and left-scattered, i.e., all points $t = \dfrac{n}{7}$, $n \in \mathbb{N}$, are isolated.

2. $n = 0$.   Then

$$\sigma(0) = \inf\left\{\frac{l}{7}, 0 : \frac{l}{7}, 0 > 0, l \in \mathbb{N}_0\right\} = \frac{1}{7} > 0,$$

$$\rho(0) = \sup\left\{\frac{l}{7} : \frac{l}{7}, 0 < 0, l \in \mathbb{N}_0\right\} = \sup\emptyset = \inf\mathbb{T} = 0,$$

i.e., $t = 0$ is right-scattered.

Exercise 1.1. Classify each point $t \in \mathbb{T} = \{\sqrt[7]{5n - 2} : n \in \mathbb{N}_0\}$ as left-dense, left-scattered, right-dense, or right-scattered.

Answer. The points $\sqrt[7]{5n - 2}$, $n \in \mathbb{N}$, are isolated; the point $\sqrt[7]{-2}$ is right-scattered.

Definition 1.6. The numbers

$$H_0 = 0, \quad H_n = \sum_{k=1}^{n} \frac{1}{k}, \quad n \in \mathbb{N},$$

will be called harmonic numbers.

Exercise 1.2. Let

$$\mathbb{H} = \{H_n : n \in \mathbb{N}_0\}.$$

Prove that $h\mathbb{H}$ is a time scale for all $h > 0$. Find $\sigma(t)$ and $\rho(t)$ for $t \in \mathbb{H}$.

Answer. $\sigma(H_n) = H_{n+1}$, $\quad \rho(H_n) = H_{n-1}$, $\quad n \in \mathbb{N}$, $\quad \rho(H_0) = H_0$.

Definition 1.7. The graininess function $\mu : \mathbb{T} \to [0, \infty)$ is defined by

$$\mu(t) = \sigma(t) - t.$$

Example 1.6. Let $\mathbb{T} = \{7^n : n \in \mathbb{N}\}$. Let also $t = 7^n \in \mathbb{T}$ for some $n \in \mathbb{N}$. Then

$$\sigma(t) = \inf\left\{7^l : 7^l > 7^n, l \in \mathbb{N}\right\} = 7^{n+1} = 7t.$$

Hence,

$$\mu(t) = \sigma(t) - t = 7t - t = 6t \quad \text{or} \quad \mu\left(7^n\right) = 6\left(7^n\right), \quad n \in \mathbb{N}.$$

Exercise 1.3. Let $\mathbb{T} = \left\{\sqrt[5]{n+4} : n \in \mathbb{N}_0\right\}$. Find $\mu(t), t \in \mathbb{T}$.

Answer. $\mu\left(\sqrt[5]{n+4}\right) = \sqrt[5]{n+5} - \sqrt[5]{n+4}, n \in \mathbb{N}_0$.

Definition 1.8. If $f : \mathbb{T} \to \mathbb{R}$ is a function, then we define the function $f^\sigma : \mathbb{T} \to \mathbb{R}$ by

$$f^\sigma(t) = f(\sigma(t)) \quad \text{for any} \quad t \in \mathbb{T}, \quad \text{i.e.,} \quad f^\sigma = f \circ \sigma.$$

Example 1.7. Let $\mathbb{T} = 3^{\mathbb{N}_0}$,

$$f(t) = t^3 - 7t, \quad t \in \mathbb{T}.$$

We will find

$$f^\sigma(t), \quad t \in \mathbb{T}.$$

Here

$$\sigma(t) = 3t, \quad t \in \mathbb{T}.$$

Then

$$\begin{aligned}
f^\sigma(t) &= (\sigma(t))^3 - 7\sigma(t) \\
&= (3t)^3 - 7(3t) \\
&= 27t^3 - 21t, \quad t \in \mathbb{T}.
\end{aligned}$$

Example 1.8. Let $\mathbb{T} = \mathbb{N}_0^2$ and

$$f(t) = \frac{t+1}{t+2}, \quad t \in \mathbb{T}.$$

We will find

$$f^\sigma(t), \quad t \in \mathbb{T}.$$

Here

$$\sigma(t) = \left(\sqrt{t}+1\right)^2, \quad t \in \mathbb{T}.$$

Then

$$f^\sigma(t) = \frac{\left(\sqrt{t}+1\right)^2 + 1}{\left(\sqrt{t}+1\right)^2 + 2}$$

$$= \frac{t + 2\sqrt{t} + 2}{t + 2\sqrt{t} + 3}, \quad t \in \mathbb{T}.$$

Example 1.9. Let $\mathbb{T} = 2^{\mathbb{N}_0}$ and

$$f(t) = t^2 - t + 1, \quad t \in \mathbb{T}.$$

We will find

$$f^\sigma(t), \quad t \in \mathbb{T}.$$

Here

$$\sigma(t) = 2t, \quad t \in \mathbb{T}.$$

Then

$$f^\sigma(t) = (\sigma(t))^2 - \sigma(t) + 1$$

$$= (2t)^2 - 2t + 1$$

$$= 4t^2 - 2t + 1, \quad t \in \mathbb{T}.$$

Exercise 1.4. Let $\mathbb{T} = 5^{\mathbb{N}_0}$ and

$$f(t) = \frac{t^2 + 3}{t^2 + 7}, \quad t \in \mathbb{T}.$$

Find $f^\sigma(t), t \in \mathbb{T}$.

Answer.

$$f^\sigma(t) = \frac{25t^2 + 3}{25t^2 + 7}, \quad t \in \mathbb{T}.$$

Definition 1.9. We define the set

$$\mathbb{T}^\kappa = \begin{cases} \mathbb{T} \backslash (\rho(\sup \mathbb{T}), \sup \mathbb{T}] & \text{if} \quad \sup \mathbb{T} < \infty, \\ \mathbb{T} & \text{otherwise.} \end{cases}$$

Example 1.10. Let $\mathbb{T} = \left\{ \dfrac{1}{2n+1} : n \in \mathbb{N} \right\} \cup \{0\}$. Then $\sup \mathbb{T} = \dfrac{1}{3}$ and

$$\rho\left(\frac{1}{3}\right) = \sup\left\{ \frac{1}{2l+1}, 0 : \frac{1}{2l+1}, 0 < 1, l \in \mathbb{N} \right\} = \frac{1}{5}.$$

Therefore,

$$\mathbb{T}^\kappa = \mathbb{T}\backslash\left(\frac{1}{5}, \frac{1}{3}\right] = \left\{ \frac{1}{2n+1} : n \in \mathbb{N}, n \geq 2 \right\} \cup \{0\}.$$

Example 1.11. Let $\mathbb{T} = \{7n^2 + n - 10 : n \in \mathbb{N}\}$. Then $\sup \mathbb{T} = \infty$ and $\mathbb{T}^\kappa = \mathbb{T}$.

Example 1.12. Let $\mathbb{T} = \left\{ \dfrac{11}{n^2+3} : n \in \mathbb{N} \right\} \cup \{0\}$. Then $\sup \mathbb{T} = \dfrac{11}{4} < \infty$,

$$\rho\left(\frac{11}{4}\right) = \sup\left\{ \frac{11}{l^2+3}, 0 : \frac{11}{l^2+3}, 0 < \frac{11}{4}, l \in \mathbb{N} \right\} = \frac{11}{7}.$$

Hence,

$$\mathbb{T}^\kappa = \mathbb{T}\backslash\left(\frac{11}{7}, \frac{11}{4}\right] = \left\{ \frac{11}{n^2+3} : n \geq 2 \right\} \cup \{0\}.$$

Definition 1.10. We assume that $a \leq b$. We define the interval $[a, b]$ in $\mathbb{T}$ by

$$[a, b] = \{t \in \mathbb{T} : a \leq t \leq b\}.$$

Open intervals, half-open intervals, and so on are defined accordingly.

Example 1.13. Let $[a, b]$ be an interval in $\mathbb{T}$ and let b be a left-dense point, $b < \infty$. Then $\sup[a, b] = b$, and since b is a left-dense point, we have that $\rho(b) = b$. Hence,

$$[a, b]^\kappa = [a, b]\backslash(b, b] = [a, b]\backslash\emptyset = [a, b].$$

Example 1.14. Let $[a, b]$ be an interval in $\mathbb{T}$, and b a left-scattered point, $b < \infty$. Then $\sup[a, b] = b$, and since b is a left-scattered point, we have that $\rho(b) < b$. We assume that there is $c \in (\rho(b), b]$, $c \in \mathbb{T}$. Then $\rho(b) < c \leq b$, which is a contradiction. Therefore,

$$[a, b]^\kappa = [a, b]\backslash(\rho(b), b] = [a, b).$$

Exercise 1.5. Let $\mathbb{T} = \left\{ \dfrac{4}{2n^2+n+10} : n \in \mathbb{N} \right\} \cup \{0\}$. Find $\mathbb{T}^\kappa$.

Answer. $\left\{ \dfrac{4}{2n^2+n+10} : n \in \mathbb{N}, n \geq 2 \right\} \cup \{0\}$.

1.2 Differentiation

Definition 1.11. Assume that $f : \mathbb{T} \to \mathbb{R}$ is a function and let $t \in \mathbb{T}^{\kappa}$. We define $f^{\Delta}(t)$ to be the number, if it exists, defined as follows: for every $\epsilon > 0$ there is a neighborhood U of t, $U = (t - \delta, t + \delta) \cap \mathbb{T}$ for some $\delta > 0$, such that

$$|f(\sigma(t)) - f(s) - f^{\Delta}(t)(\sigma(t) - s)| \le \epsilon |\sigma(t) - s| \quad \text{for all} \quad s \in U, \quad s \ne \sigma(t).$$

We call $f^{\Delta}(t)$ the delta or Hilger derivative of f at t.

We say that f is delta or Hilger differentiable, or differentiable for short, in T^{κ} if $f^{\Delta}(t)$ exists for all $t \in \mathbb{T}^{\kappa}$. The function $f^{\Delta} : \mathbb{T} \to \mathbb{R}$ is said to be the delta derivative or Hilger derivative, or simply the derivative, of f in T^{κ}.

Remark 1.1. If $\mathbb{T} = \mathbb{R}$, then the delta derivative coincides with the classical derivative.

Theorem 1.1. *The delta derivative is well defined.*

Proof. Let $t \in \mathbb{T}^{\kappa}$ and $f_i^{\Delta}(t)$, $i = 1, 2$, be such that

$$|f(\sigma(t)) - f(s) - f_1^{\Delta}(t)(\sigma(t) - s)| \le \frac{\epsilon}{2}|\sigma(t) - s|,$$

$$|f(\sigma(t)) - f(s) - f_2^{\Delta}(t)(\sigma(t) - s)| \le \frac{\epsilon}{2}|\sigma(t) - s|,$$

for all $\epsilon > 0$ and all s belonging to a neighborhood U of t, $U = (t - \delta, t + \delta) \cap \mathbb{T}$, for some $\delta > 0$, $s \ne \sigma(t)$. Hence,

$$|f_1^{\Delta}(t) - f_2^{\Delta}(t)| = \left| f_1^{\Delta}(t) - \frac{f(\sigma(t)) - f(s)}{\sigma(t) - s} + \frac{f(\sigma(t)) - f(s)}{\sigma(t) - s} - f_2^{\Delta}(t) \right|$$

$$\le \left| f_1^{\Delta}(t) - \frac{f(\sigma(t)) - f(s)}{\sigma(t) - s} \right| + \left| \frac{f(\sigma(t)) - f(s)}{\sigma(t) - s} - f_2^{\Delta}(t) \right|$$

$$= \frac{|f(\sigma(t)) - f(s) - f_1^{\Delta}(t)(\sigma(t) - s)|}{|\sigma(t) - s|} + \frac{|f(\sigma(t)) - f(s) - f_2^{\Delta}(t)(\sigma(t) - s)|}{|\sigma(t) - s|}$$

$$\le \frac{\epsilon}{2} + \frac{\epsilon}{2}$$

$$= \epsilon.$$

Since $\epsilon > 0$ was chosen arbitrarily, we conclude that

$$f_1^{\Delta}(t) = f_2^{\Delta}(t),$$

which completes the proof.

Remark 1.2. Let us assume that $\sup \mathbb{T} < \infty$ and $f^{\Delta}(t)$ is defined at a point $t \in \mathbb{T} \backslash \mathbb{T}^{\kappa}$ with the same definition as given in Definition 1.11. Then the unique point $t \in \mathbb{T} \backslash \mathbb{T}^{\kappa}$ is $\sup \mathbb{T}$.

Hence, for all $\epsilon > 0$ there is a neighborhood $U = (t - \delta, t + \delta) \cap (\mathbb{T} \backslash \mathbb{T}^{\kappa})$ for some $\delta > 0$, such that

$$f(\sigma(t)) = f(s) = f(\sigma(\sup \mathbb{T})) = f(\sup \mathbb{T}), \quad s \in U, \quad s \neq \sigma(t).$$

Therefore, for all $\alpha \in \mathbb{R}$ and $s \in U$ we have

$$|f(\sigma(t)) - f(s) - \alpha(\sigma(t) - s)| = |f(\sup \mathbb{T}) - f(\sup \mathbb{T}) - \alpha(\sup \mathbb{T} - \sup \mathbb{T})|$$
$$\leq \epsilon |\sigma(t) - s|,$$

i.e., every $\alpha \in \mathbb{R}$ is the delta derivative of f at the point $t \in \mathbb{T} \backslash \mathbb{T}^{\kappa}$.

Example 1.15. Let $f(t) = \alpha \in \mathbb{R}$. We will prove that $f^{\Delta}(t) = 0$ for all $t \in \mathbb{T}^{\kappa}$. Indeed, for $t \in \mathbb{T}^{\kappa}$ and for all $\epsilon > 0$ there exists $\delta > 0$ such that $s \in (t - \delta, t + \delta) \cap \mathbb{T}, s \neq \sigma(t)$, implies

$$|f(\sigma(t)) - f(s) - 0(\sigma(t) - s)| = |\alpha - \alpha|$$
$$\leq \epsilon |\sigma(t) - s|.$$

Example 1.16. Let $f(t) = 2t, t \in \mathbb{T}$. We will prove that $f^{\Delta}(t) = 2$ for all $t \in \mathbb{T}^{\kappa}$. Indeed, for $t \in \mathbb{T}^{\kappa}$ and for all $\epsilon > 0$ there exists $\delta > 0$ such that $s \in (t - \delta, t + \delta) \cap \mathbb{T}$, $s \neq \sigma(t)$, implies

$$|f(\sigma(t)) - f(s) - 2(\sigma(t) - s)| = 2|\sigma(t) - s - (\sigma(t) - s)|$$
$$\leq \epsilon |\sigma(t) - s|.$$

Example 1.17. Let $f(t) = 4t^2, t \in \mathbb{T}$. We will prove that $f^{\Delta}(t) = 4\sigma(t) + 4t$, $t \in \mathbb{T}^{\kappa}$.
Indeed, for $t \in \mathbb{T}^{\kappa}$ and for all $\epsilon > 0$ and for $s \in \left(t - \dfrac{\epsilon}{8}, t + \dfrac{\epsilon}{8}\right) \cap \mathbb{T}, s \neq \sigma(t)$, we have $|t - s| < \dfrac{\epsilon}{4}$ and

$$|f(\sigma(t)) - f(s) - 4(\sigma(t) + t)(\sigma(t) - s)| = 4\left|(\sigma(t))^2 - s^2 - (\sigma(t) + t)(\sigma(t) - s)\right|$$
$$= 4|(\sigma(t) - s)(\sigma(t) + s)$$
$$-(\sigma(t) + t)(\sigma(t) - s)|$$
$$= 4|\sigma(t) - s||t - s|$$
$$\leq \epsilon |\sigma(t) - s|.$$

Exercise 1.6. Let $f(t) = \sqrt[3]{t}$, $t \in \mathbb{T}$, $t > 0$. Prove that

$$f^{\Delta}(t) = \frac{1}{\sqrt[3]{t^2} + \sqrt[3]{t\sigma(t)} + \sqrt[3]{(\sigma(t))^2}} \quad \text{for } t \in \mathbb{T}^{\kappa}, t > 0.$$

Exercise 1.7. Let $f(t) = t^4$, $t \in \mathbb{T}$. Prove that $f^{\Delta}(t) = (\sigma(t))^3 + t(\sigma(t))^2 + t^2\sigma(t)$ $+ t^3$ for $t \in \mathbb{T}^{\kappa}$.

Theorem 1.2. *Assume $f : \mathbb{T} \to \mathbb{R}$ is a function and let $t \in \mathbb{T}^{\kappa}$. Then we have the following.*

1. If f is differentiable at t, then f is continuous at t.
2. If f is continuous at t and t is right-scattered, then f is differentiable at t with

$$f^{\Delta}(t) = \frac{f(\sigma(t)) - f(t)}{\mu(t)}.$$

3. If t is right-dense, then f is differentiable iff the limit

$$\lim_{s \to t} \frac{f(t) - f(s)}{t - s}$$

exists as a finite number. In this case,

$$f^{\Delta}(t) = \lim_{s \to t} \frac{f(t) - f(s)}{t - s}.$$

4. If f is differentiable at t, then

$$f(\sigma(t)) = f(t) + \mu(t)f^{\Delta}(t).$$

Example 1.18. Let $\mathbb{T} = \left\{ \dfrac{1}{2n+1} : n \in \mathbb{N}_0 \right\} \cup \{0\}$, $f(t) = \sigma(t)$, $t \in \mathbb{T}$. We will find $f^{\Delta}(t)$, $t \in \mathbb{T}$. For $t \in \mathbb{T}$, $t = \dfrac{1}{2n+1}$, $n = \dfrac{1-t}{2t}$, $n \geq 1$, we have

$$\sigma(t) = \inf\left\{ \frac{1}{2l+1}, 0 : \frac{1}{2l+1}, 0 > \frac{1}{2n+1}, l \in \mathbb{N}_0 \right\} = \frac{1}{2n-1}$$

$$= \frac{1}{2\frac{1-t}{2t} - 1} = \frac{t}{1 - 2t} > t,$$

i.e., every point $t = \dfrac{1}{2n+1}$, $n \geq 1$, is right-scattered. At these points,

$$f^{\Delta}(t) = \frac{f(\sigma(t)) - f(t)}{\sigma(t) - t} = \frac{\sigma(\sigma(t)) - \sigma(t)}{\sigma(t) - t} = 2\frac{(\sigma(t))^2}{(1 - 2\sigma(t))(\sigma(t) - t)}$$

$$= 2\frac{\left(\frac{t}{1-2t}\right)^2}{\left(1 - \frac{2t}{1-2t}\right)\left(\frac{t}{1-2t} - t\right)} = 2\frac{\frac{t^2}{(1-2t)^2}}{\frac{1-4t}{1-2t}\frac{2t^2}{1-2t}} = 2\frac{t^2}{2t^2(1 - 4t)} = \frac{1}{1 - 4t}.$$

Let $n = 0$, i.e., $t = 1$. Then

$$\sigma(1) = \inf\left\{\frac{1}{2l+1}, 0 : \frac{1}{2l+1}, 0 > 1, l \in \mathbb{N}_0\right\} = \inf\varnothing = \sup\mathbb{T} = 1,$$

i.e., $t = 1$ is a right-dense point. Also,

$$\lim_{s\to 1}\frac{f(1) - f(s)}{1 - s} = \lim_{s\to 1}\frac{\sigma(1) - \sigma(s)}{1 - s} = \lim_{s\to 1}\frac{1 - \frac{s}{1-2s}}{1 - s}$$

$$= \lim_{s\to 1}\frac{1 - 3s}{(1 - s)(1 - 2s)} = +\infty.$$

Therefore, $\sigma'(1)$ doesn't exist.

Let now $t = 0$. Then

$$\sigma(0) = \inf\left\{\frac{1}{2l+1}, 0 : \frac{1}{2l+1}, 0 > 0, l \in \mathbb{N}_0\right\} = 0.$$

Consequently, $t = 0$ is right-dense. Also,

$$\lim_{h\to 0}\frac{\sigma(h) - \sigma(0)}{h} = \lim_{h\to 0}\frac{\frac{h}{1-2h} - 0}{h} = \lim_{h\to 0}\frac{1}{1 - 2h} = 1.$$

Therefore, $\sigma'(0) = 1$.

Example 1.19. Let $\mathbb{T} = \{n^2 : n \in \mathbb{N}_0\}$, $f(t) = t^2$, $g(t) = \sigma(t)$, $t \in \mathbb{T}$. We will find $f^\Delta(t)$ and $g^\Delta(t)$ for $t \in \mathbb{T}^\kappa$. For $t \in \mathbb{T}^\kappa$, $t = n^2$, $n = \sqrt{t}$, $n \in \mathbb{N}_0$, we have

$$\sigma(t) = \inf\{l^2 : l^2 > n^2, l \in \mathbb{N}_0\} = (n + 1)^2 = (\sqrt{t} + 1)^2 > t.$$

Therefore, all points of $\mathbb{T}$ are right-scattered. We note that $f(t)$ and $g(t)$ are continuous functions in $\mathbb{T}$. Hence,

$$f^\Delta(t) = \frac{f(\sigma(t)) - f(t)}{\sigma(t) - t} = \frac{(\sigma(t))^2 - t^2}{\sigma(t) - t} = \sigma(t) + t$$

$$= \left(\sqrt{t} + 1\right)^2 + t = t + 2\sqrt{t} + 1 + t = 1 + 2\sqrt{t} + 2t,$$

$$g^\Delta(t) = \frac{g(\sigma(t)) - g(t)}{\sigma(t) - t} = \frac{\sigma(\sigma(t)) - \sigma(t)}{\sigma(t) - t}$$

$$= \frac{(\sqrt{\sigma(t)} + 1)^2 - \sigma(t)}{\sigma(t) - t} = \frac{\sigma(t) + 2\sqrt{\sigma(t)} + 1 - \sigma(t)}{\sigma(t) - t}$$

$$= \frac{1 + 2\sqrt{\sigma(t)}}{\sigma(t) - t} = \frac{1 + 2(\sqrt{t} + 1)}{(\sqrt{t} + 1)^2 - t} = \frac{3 + 2\sqrt{t}}{1 + 2\sqrt{t}}.$$

Example 1.20. Let $\mathbb{T} = \{\sqrt[4]{2n+1} : n \in \mathbb{N}_0\}$, $f(t) = t^4$, $t \in \mathbb{T}$. We will find $f^\Delta(t)$, $t \in \mathbb{T}$. For $t \in \mathbb{T}$, $t = \sqrt[4]{2n+1}$, $n = \dfrac{t^4-1}{2}$, $n \in \mathbb{N}_0$, we have

$$\sigma(t) = \inf\{\sqrt[4]{2l+1} : \sqrt[4]{2l+1} > \sqrt[4]{2n+1}, l \in \mathbb{N}_0\} = \sqrt[4]{2n+3}$$
$$= \sqrt[4]{t^4+2} > t.$$

Therefore, every point of $\mathbb{T}$ is right-scattered. We note that the function $f(t)$ is continuous in $\mathbb{T}$. Hence,

$$f^\Delta(t) = \frac{f(\sigma(t)) - f(t)}{\sigma(t) - t} = \frac{(\sigma(t))^4 - t^4}{\sigma(t) - t}$$
$$= (\sigma(t))^3 + t(\sigma(t))^2 + t^2\sigma(t) + t^3$$
$$= \sqrt[4]{(t^4+2)^3} + t^2\sqrt[4]{t^4+2} + t\sqrt{t^4+2} + t^3.$$

Exercise 1.8. Let $\mathbb{T} = \{\sqrt[5]{n+1} : n \in \mathbb{N}_0\}$, $f(t) = t + t^3$, $t \in \mathbb{T}$. Find $f^\Delta(t)$, $t \in \mathbb{T}^\kappa$.

Answer. $1 + \sqrt[5]{(t^5+1)^2} + t\sqrt[5]{t^5+1} + t^2$.

Theorem 1.3. *Assume $f, g : \mathbb{T} \to \mathbb{R}$ are differentiable at $t \in \mathbb{T}^\kappa$. Then the following hold:*

1. *The sum $f + g : \mathbb{T} \to \mathbb{R}$ is differentiable at t with*

$$(f+g)^\Delta(t) = f^\Delta(t) + g^\Delta(t).$$

2. *For every constant α, $\alpha f : \mathbb{T} \to \mathbb{R}$ is differentiable at t with*

$$(\alpha f)^\Delta(t) = \alpha f^\Delta(t).$$

3. *If $f(t)f(\sigma(t)) \neq 0$, we have that $\dfrac{1}{f} : \mathbb{T} \to \mathbb{R}$ is differentiable at t and*

$$\left(\frac{1}{f}\right)^\Delta(t) = -\frac{f^\Delta(t)}{f(t)f(\sigma(t))}.$$

4. *If $g(t)g(\sigma(t)) \neq 0$, we have that $\dfrac{f}{g} : \mathbb{T} \to \mathbb{R}$ is differentiable at t with*

$$\left(\frac{f}{g}\right)^\Delta(t) = \frac{f^\Delta(t)g(t) - f(t)g^\Delta(t)}{g(t)g(\sigma(t))}.$$

5. *The product $fg : \mathbb{T} \to \mathbb{R}$ is differentiable at t with*

$$(fg)^\Delta(t) = f^\Delta(t)g(t) + f(\sigma(t))g^\Delta(t) = f(t)g^\Delta(t) + f^\Delta(t)g(\sigma(t)).$$

Example 1.21. Let $f, g, h : \mathbb{T} \to \mathbb{R}$ be differentiable at $t \in \mathbb{T}^\kappa$. Then

$$(fgh)^\Delta(t) = ((fg)h)^\Delta(t) = (fg)^\Delta(t)h(t) + (fg)(\sigma(t))h^\Delta(t)$$
$$= (f^\Delta(t)g(t) + f(\sigma(t))g^\Delta(t))h(t) + f^\sigma(t)g^\sigma(t)h^\Delta(t)$$
$$= f^\Delta(t)g(t)h(t) + f^\sigma(t)g^\Delta(t)h(t) + f^\sigma(t)g^\sigma(t)h^\Delta(t).$$

Example 1.22. Let $f : \mathbb{T} \to \mathbb{R}$ be differentiable at $t \in \mathbb{T}^\kappa$. Then

$$\left(f^2\right)^\Delta(t) = (ff)^\Delta(t) = f^\Delta(t)f(t) + f(\sigma(t))f^\Delta(t) = f^\Delta(t)(f^\sigma(t) + f(t)).$$

Also,

$$(f^3)^\Delta(t) = (ff^2)^\Delta(t) = f^\Delta(t)(f(t))^2 + f(\sigma(t))(f^2)^\Delta(t)$$
$$= f^\Delta(t)(f(t))^2 + f^\sigma(t)f^\Delta(t)(f^\sigma(t) + f(t))$$
$$= f^\Delta(t)(f(t))^2 + f(t)f^\sigma(t) + (f^\sigma(t))^2.$$

We assume that

$$(f^n)^\Delta(t) = (f(t))^k \sum_{k=0}^{n-1} (f(t))^k \left(f^\sigma(t)\right)^{n-1-k}$$

for some $n \in \mathbb{N}$.
We will prove that

$$(f^{n+1})^\Delta(t) = f^\Delta(t) \sum_{k=0}^{n} (f(t))^k \left(f^\sigma(t)\right)^{n-k}.$$

Indeed,

$$(f^{n+1})^\Delta(t) = (ff^n)^\Delta(t) = f^\Delta(t)(f(t))^n + f^\sigma(t)(f^n)^\Delta(t)$$
$$= f^\Delta(t)(f(t))^n + f^\Delta(t)(f(t))^{n-1} + (f(t))^{n-2}f^\sigma(t)$$
$$+ \cdots + f(t)\left(f^\sigma(t)\right)^{n-2} + \left(f^\sigma(t)\right)^{n-1}f^\sigma(t)$$
$$= f^\Delta(t)\left((f(t))^n + (f(t))^{n-1}f^\sigma(t) + (f(t))^{n-2}\left(f^\sigma(t)\right)^2 + \cdots + \left(f^\sigma(t)\right)^n\right)$$
$$= f^\Delta(t) \sum_{k=0}^{n} (f(t))^k \left(f^\sigma(t)\right)^{n-k}.$$

Example 1.23. Now we consider $f(t) = (t - a)^m$ for $a \in \mathbb{R}$ and $m \in \mathbb{N}$. We set $h(t) = (t - a)$. Then $h^\Delta(t) = 1$. From this and the previous exercise, we get

$$f^\Delta(t) = h^\Delta(t) \sum_{k=0}^{m-1} (h(t))^k \left(h^\sigma(t)\right)^{m-1-k}$$

$$= \sum_{k=0}^{m-1} (t - a)^k (\sigma(t) - a)^{m-1-k}.$$

Let now $g(t) = \dfrac{1}{f(t)}$. Then

$$g^\Delta(t) = -\frac{f^\Delta(t)}{f(\sigma(t))f(t)},$$

whereupon

$$g^\Delta(t) = -\frac{1}{(\sigma(t) - a)^m (t - a)^m} \sum_{k=0}^{m-1} (t - a)^k (\sigma(t) - a)^{m-1-k}$$

$$= -\sum_{k=0}^{m-1} \frac{1}{(t - a)^{m-k}} \frac{1}{(\sigma(t) - a)^{k+1}}.$$

Definition 1.12. Let $f : \mathbb{T} \to \mathbb{R}$ and $t \in \left(T^\kappa\right)^\kappa = \mathbb{T}^{\kappa^2}$. We define the second derivative of f at t by

$$f^{\Delta^2}(t) = \left(f^\Delta\right)^\Delta(t),$$

provided it exists. Similarly we define higher-order derivatives $f^{\Delta^n} : \mathbb{T}^{\kappa^n} \to \mathbb{R}$.

Theorem 1.4 (*Leibniz Formula*). *Let $S_k^{(n)}$ be the set consisting of all possible strings of length n containing σ exactly k times and Δ $n - k$ times. If*

$$f^\Lambda \quad \text{exists} \quad \text{for} \quad \text{all} \quad \Lambda \in S_k^{(n)},$$

then

$$(fg)^{\Delta^n} = \sum_{k=0}^{n} \left(\sum_{\Lambda \in S_k^{(n)}} f^\Lambda \right) g^{\Delta^k}.$$

Example 1.24. Let μ be differentiable at $t \in \mathbb{T}^\kappa$ and suppose that t is right-scattered. Then

$$f^{\Delta\sigma}(t) = \left(f^\Delta\right)^\sigma(t) = \left(\frac{f(\sigma(t)) - f(t)}{\sigma(t) - t}\right)^\sigma = \frac{f(\sigma(\sigma(t))) - f(\sigma(t))}{\sigma(\sigma(t)) - \sigma(t)}$$

$$= \frac{f(\sigma(\sigma(t))) - f(\sigma(t))}{\sigma(t) - t} \frac{1}{\frac{\sigma(\sigma(t)) - \sigma(t)}{\sigma(t) - t}}$$

$$= \left(f^\sigma\right)^\Delta(t)\frac{1}{\sigma^\Delta(t)} = f^{\sigma\Delta}(t)\frac{1}{1 + \mu^\Delta(t)},$$

i.e.,

$$f^{\sigma\Delta}(t) = (1 + \mu^\Delta(t))f^{\Delta\sigma}(t).$$

Also,

$$f^{\sigma\sigma\Delta}(t) = \left(f^\sigma\right)^{\sigma\Delta}(t) = (1 + \mu^\Delta(t))\left(f^{\sigma\Delta}\right)^\sigma(t) = (1 + \mu^\Delta(t))f^{\sigma\Delta\sigma}(t),$$

$$f^{\sigma\Delta\sigma}(t) = \left(f^\sigma\right)^{\Delta\sigma}(t) = \left(\left(f^\sigma\right)^\Delta\right)^\sigma(t)$$

$$= \left((1 + \mu^\Delta(t))(f^\Delta)^\sigma(t)\right)^\sigma = \left(1 + \mu^{\Delta\sigma}(t)\right)f^{\Delta\sigma\sigma}(t).$$

Theorem 1.5 (*Chain Rule*). *Assume that $g : \mathbb{R} \to \mathbb{R}$ is continuous, $g : \mathbb{T} \to \mathbb{R}$ is delta differentiable on $\mathbb{T}^\kappa$, and $f : \mathbb{R} \to \mathbb{R}$ is continuously differentiable. Then there exists $c \in [t, \sigma(t)]$ with*

$$(f \circ g)^\Delta(t) = f'(g(c))g^\Delta(t).$$

Example 1.25. Let $\mathbb{T} = \mathbb{Z}$, $f(t) = t^3 + 1$, $g(t) = t^2$. We have that $g : \mathbb{R} \to \mathbb{R}$ is continuous, $g : \mathbb{T} \to \mathbb{R}$ is delta differentiable on $\mathbb{T}^\kappa$, $f : \mathbb{R} \to \mathbb{R}$ is continuously differentiable, $\sigma(t) = t + 1$. Then

$$g^\Delta(t) = \sigma(t) + t,$$

$$(f \circ g)^\Delta(1) = f'(g(c))g^\Delta(1) = 3(g(c))^2(\sigma(1) + 1) = 9c^4. \tag{1.1}$$

Here $c \in [1, \sigma[1]] = [1, 2]$.
Also,

$$f \circ g(t) = f(g(t)) = (g(t))^3 + 1 = t^6 + 1,$$

$$(f \circ g)^\Delta(t) = (\sigma(t))^5 + t(\sigma(t))^4 + t^2(\sigma(t))^3 + t^3(\sigma(t))^2 + t^4\sigma(t) + t^5,$$

$$(f \circ g)^\Delta(1) = (\sigma(1))^5 + (\sigma(1))^4 + (\sigma(1))^3 + (\sigma(1))^2 + \sigma(1) + 1 = 63.$$

From this and (1.1), we get

$$63 = 9c^4 \quad or \quad c^4 = 7 \quad or \quad c = \sqrt[4]{7} \in [1, 2].$$

Example 1.26. Let $\mathbb{T} = \{2^n : n \in \mathbb{N}_0\}$, $f(t) = t + 2$, $g(t) = t^2 - 1$. We note that $g : \mathbb{T} \to \mathbb{R}$ is delta differentiable, $g : \mathbb{R} \to \mathbb{R}$ is continuous, and $f : \mathbb{R} \to \mathbb{R}$ is continuously differentiable.
For $t \in \mathbb{T}$, $t = 2^n$, $n \in \mathbb{N}_0$, $n = \log_2 t$, we have

$$\sigma(t) = \inf \left\{ 2^l : 2^l > 2^n, l \in \mathbb{N}_0 \right\} = 2^{n+1} = 2t > t.$$

Therefore, all points of $\mathbb{T}$ are right-scattered. Since $\sup \mathbb{T} = \infty$, we have that $\mathbb{T}^\kappa = \mathbb{T}$. Also, for $t \in \mathbb{T}$, we have

$$(f \circ g)(t) = f(g(t)) = g(t) + 2 = t^2 - 1 + 2 = t^2 + 1,$$
$$(f \circ g)^\Delta(t) = \sigma(t) + t = 2t + t = 3t.$$

Hence,

$$(f \circ g)^\Delta(2) = 6. \tag{1.2}$$

Now, using Theorem 1.5, we get that there is $c \in [2, \sigma(2)] = [2, 4]$ such that

$$(f \circ g)^\Delta(2) = f'(g(c))g^\Delta(2) = g^\Delta(2) = \sigma(2) + 2 = 4 + 2 = 6. \tag{1.3}$$

From (1.2) and (1.3), we find that for every $c \in [2, 4]$, we have

$$(f \circ g)^\Delta(2) = f'(g(c))g^\Delta(2).$$

Example 1.27. Let $\mathbb{T} = \left\{ 3^{n^2} : n \in \mathbb{N}_0 \right\}$, $f(t) = t^2 + 1$, $g(t) = t^3$. We note that $g : \mathbb{R} \to \mathbb{R}$ is continuous, $g : \mathbb{T} \to \mathbb{R}$ is delta differentiable, and $f : \mathbb{R} \to \mathbb{R}$ is continuously differentiable.
For $t \in \mathbb{T}$, $t = 3^{n^2}$, $n \in \mathbb{N}_0$, $n = (\log_3 t)^{\frac{1}{2}}$, we have

$$\sigma(t) = \inf \left\{ 3^{l^2} : 3^{l^2} > 3^{n^2}, l \in \mathbb{N}_0 \right\} = 3^{(n+1)^2}$$

$$= 3 \cdot 3^{n^2} \cdot 3^{2n} = 3t3^{2(\log_3 t)^{\frac{1}{2}}} > t.$$

Consequently, all points of $\mathbb{T}$ are right-scattered. Also, $\sup \mathbb{T} = \infty$. Then $\mathbb{T}^\kappa = \mathbb{T}$. Hence for $t \in \mathbb{T}$, we have

$$(f \circ g)(t) = f(g(t)) = (g(t))^2 + 1 = t^6 + 1,$$

$$(f \circ g)^{\Delta}(t) = (\sigma(t))^5 + t(\sigma(t))^4 + t^2(\sigma(t))^3$$
$$+ t^3(\sigma(t))^2 + t^4\sigma(t) + t^5,$$

$$(f \circ g)^{\Delta}(1) = (\sigma(1))^5 + (\sigma(1))^4 + (\sigma(1))^3 + (\sigma(1))^2 + \sigma(1) + 1$$
$$= 3^5 + 3^4 + 3^3 + 3^2 + 3 + 1 = 364.$$

(1.4)

From Theorem 1.5, it follows that there exists $c \in [1, \sigma(1)] = [1, 3]$ such that

$$(f \circ g)^{\Delta}(1) = f'(g(c))g^{\Delta}(1) = 2g(c)g^{\Delta}(1) = 2c^3 g^{\Delta}(1). \qquad (1.5)$$

Because all points of $\mathbb{T}$ are right-scattered, we have

$$g^{\Delta}(1) = (\sigma(1))^2 + \sigma(1) + 1 = 9 + 3 + 1 = 13.$$

From this and (1.5), we obtain

$$(f \circ g)^{\Delta}(1) = 26c^3.$$

From the last equation and (1.4), we obtain

$$364 = 26c^3, \quad or \quad c^3 = \frac{364}{26} = 14, \quad or \quad c = \sqrt[3]{14}.$$

Exercise 1.9. Let $\mathbb{T} = \mathbb{Z}$, $f(t) = t^2 + 2t + 1$, $g(t) = t^2 - 3t$. Find a constant $c \in [1, \sigma(1)]$ such that

$$(f \circ g)^{\Delta}(1) = f'(g(c))g^{\Delta}(1).$$

Answer. $\forall c \in [1, 2]$.

Theorem 1.6 (*Chain Rule*). *Assume that $v : \mathbb{T} \to \mathbb{R}$ is strictly increasing and $\tilde{\mathbb{T}} = v(\mathbb{T})$ is a time scale. Let $w : \tilde{\mathbb{T}} \to \mathbb{R}$. If $v^{\Delta}(t)$ and $w^{\tilde{\Delta}}(v(t))$ exist for $t \in \mathbb{T}^{\kappa}$, then*

$$(w \circ v)^{\Delta} = (w^{\tilde{\Delta}} \circ v)v^{\Delta}.$$

Example 1.28. Let $\mathbb{T} = \left\{2^{2n} : n \in \mathbb{N}_0\right\}$, $v(t) = t^2$, $w(t) = t^2 + 1$. Then $v : \mathbb{T} \to \mathbb{R}$ is strictly increasing and $\tilde{\mathbb{T}} = v(\mathbb{T}) = \left\{2^{4n} : n \in \mathbb{N}_0\right\}$ is a time scale. For $t \in \mathbb{T}$, $t = 2^{2n}$, $n \in \mathbb{N}_0$, we have

$$\sigma(t) = \inf\left\{2^{2l} : 2^{2l} > 2^{2n}, l \in \mathbb{N}_0\right\} = 2^{2n+2} = 4t,$$

$$v^{\Delta}(t) = \sigma(t) + t = 5t.$$

For $t \in \tilde{\mathbb{T}}, t = 2^{4n}, n \in \mathbb{N}_0$, we have

$$\tilde{\sigma}(t) = \inf \left\{ 2^{4l} : 2^{4l} > 2^{4n}, l \in \mathbb{N}_0 \right\} = 2^{4n+4} = 16t.$$

Also, for $t \in \mathbb{T}$, we have

$$(w \circ v)(t) = w(v(t)) = (v(t))^2 + 1 = t^4 + 1,$$
$$(w \circ v)^\Delta(t) = (\sigma(t))^3 + t(\sigma(t))^2 + t^2\sigma(t) + t^3$$

$$= 64t^3 + 16t^3 + 4t^3 + t^3 = 85t^3,$$
$$w^{\tilde{\Delta}} \circ v(t) = \tilde{\sigma}(v(t)) + v(t) = 16v(t) + v(t) = 17v(t) = 17t^2,$$
$$\left(w^{\tilde{\Delta}} \circ v(t) \right) v^\Delta(t) = 17t^2(5t) = 85t^3.$$

Consequently,

$$(w \circ v)^\Delta(t) = (w^{\tilde{\Delta}} \circ v(t))v^\Delta(t), \quad t \in \mathbb{T}^\kappa.$$

Example 1.29. Let $\mathbb{T} = \{n + 1 : n \in \mathbb{N}_0\}$, $v(t) = t^2$, $w(t) = t$. Then $v : \mathbb{T} \to \mathbb{R}$ is strictly increasing and $\tilde{\mathbb{T}} = \{(n + 1)^2 : n \in \mathbb{N}_0\}$ is a time scale.
For $t \in \mathbb{T}, t = n + 1, n \in \mathbb{N}_0$, we have

$$\sigma(t) = \inf\{l + 1 : l + 1 > n + 1, l \in \mathbb{N}_0\} = n + 2 = t + 1,$$
$$v^\Delta(t) = \sigma(t) + t = t + 1 + t = 2t + 1.$$

For $t \in \tilde{\mathbb{T}}, t = (n + 1)^2, n \in \mathbb{N}_0$, we have

$$\tilde{\sigma}(t) = \{(l + 1)^2 : (l + 1)^2 > (n + 1)^2, l \in \mathbb{N}_0\} = (n + 2)^2$$
$$= (n + 1)^2 + 2(n + 1) + 1 = t + 2\sqrt{t} + 1.$$

Hence for $t \in \mathbb{T}$, we get

$$(w^{\tilde{\Delta}} \circ v)(t) = 1, \quad (w^{\tilde{\Delta}} \circ v)(t)v^\Delta(t) = 1(2t + 1) = 2t + 1,$$
$$w \circ v(t) = v(t) = t^2, \quad (w \circ v)^\Delta(t) = \sigma(t) + t = 2t + 1.$$

Consequently,

$$(w \circ v)^\Delta(t) = (w^{\tilde{\Delta}} \circ v(t))v^\Delta(t), \quad t \in \mathbb{T}^\kappa.$$

Example 1.30. Let $\mathbb{T} = \{2^n : n \in \mathbb{N}_0\}$, $v(t) = t$, $w(t) = t^2$. Then $v : \mathbb{T} \to \mathbb{R}$ is strictly increasing, $v(\mathbb{T}) = \mathbb{T}$.
For $t \in \mathbb{T}$, $t = 2^n$, $n \in \mathbb{N}_0$, we have

$$\sigma(t) = \inf\{2^l : 2^l > 2^n, l \in \mathbb{N}_0\} = 2^{n+1} = 2t, \quad v^\Delta(t) = 1,$$

$$(w \circ v)(t) = w(v(t)) = (v(t))^2 = t^2,$$

$$(w \circ v)^\Delta(t) = \sigma(t) + t = 2t + t = 3t,$$

$$(w^\Delta \circ v)(t) = \sigma(v(t)) + v(t) = 2v(t) + v(t) = 3v(t) = 3t,$$

$$(w^\Delta \circ v)(t)v^\Delta(t) = 3t.$$

Consequently,

$$(w \circ v)^\Delta(t) = (w^{\tilde{\Delta}} \circ v(t))v^\Delta(t), \quad t \in \mathbb{T}^\kappa.$$

Exercise 1.10. Let $\mathbb{T} = \{2^{3n} : n \in \mathbb{N}_0\}$, $v(t) = t^2$, $w(t) = t$. Prove that

$$(w \circ v)^\Delta(t) = (w^{\tilde{\Delta}} \circ v(t))v^\Delta(t), \quad t \in \mathbb{T}^\kappa.$$

Theorem 1.7 (*Derivative of the Inverse*). *Assume that* $v : \mathbb{T} \to \mathbb{R}$ *is strictly increasing and* $\tilde{\mathbb{T}} := v(\mathbb{T})$ *is a time scale. Then*

$$(v^{-1})^{\tilde{\Delta}} \circ v(t) = \frac{1}{v^\Delta(t)}$$

for all $t \in \mathbb{T}^\kappa$ *such that* $v^\Delta(t) \neq 0$.

Example 1.31. Let $\mathbb{T} = \mathbb{N}$, $v(t) = t^2 + 1$. Then $\sigma(t) = t + 1$, $v : \mathbb{T} \to \mathbb{R}$ is strictly increasing, and

$$v^\Delta(t) = \sigma(t) + t = 2t + 1.$$

Hence,

$$\left(v^{-1}\right)^{\tilde{\Delta}} \circ v(t) = \frac{1}{v^\Delta(t)} = \frac{1}{2t + 1}.$$

Example 1.32. Let $\mathbb{T} = \{n + 3 : n \in \mathbb{N}_0\}$, $v(t) = t^2$. Then $v : \mathbb{T} \to \mathbb{R}$ is strictly increasing, $\sigma(t) = t + 1$,

$$v^\Delta(t) = \sigma(t) + t = 2t + 1.$$

Hence,

$$\left(v^{-1}\right)^{\tilde{\Delta}} \circ v(t) = \frac{1}{v^{\Delta}(t)} = \frac{1}{2t+1}.$$

Example 1.33. Let $\mathbb{T} = \left\{2^{n^2} : n \in \mathbb{N}_0\right\}$, $v(t) = t^3$. Then $v : \mathbb{T} \to \mathbb{R}$ is strictly increasing, and for $t \in \mathbb{T}$, $t = 2^{n^2}$, $n \in \mathbb{N}_0$, $n = \left(\log_2 t\right)^{\frac{1}{2}}$, we have

$$\sigma(t) = \inf\left\{2^{l^2} : 2^{l^2} > 2^{n^2}, l \in \mathbb{N}_0\right\} = 2^{(n+1)^2}$$

$$= 2^{n^2} 2^{2n+1} = t2^{2(\log_2 t)^{\frac{1}{2}}+1}.$$

Then

$$v^{\Delta}(t) = (\sigma(t))^2 + t\sigma(t) + t^2 = t^2 2^{4(\log_2 t)^{\frac{1}{2}}+2} + t^2 2^{2(\log_2 t)^{\frac{1}{2}}+1} + t^2.$$

Hence,

$$\left(v^{-1}\right)^{\tilde{\Delta}} \circ v(t) = \frac{1}{t^2 2^{4(\log_2 t)^{\frac{1}{2}}+2} + t^2 2^{2(\log_2 t)^{\frac{1}{2}}+1} + t^2}.$$

Exercise 1.11. Let $\mathbb{T} = \{n+5 : n \in \mathbb{N}_0\}$, $v(t) = t^2 + t$. Find $\left(v^{-1}\right)^{\tilde{\Delta}} \circ v(t)$.

Answer. $\dfrac{1}{2t+2}$.

1.3 Mean Value Theorems

Let $\mathbb{T}$ be a time scale and $a, b \in \mathbb{T}$, $a < b$, Let $f : \mathbb{T} \to \mathbb{R}$ be a function.

Theorem 1.8. *Suppose that f has delta derivative at each point of $[a, b]$. If $f(a) = f(b)$, then there exist points $\xi_1, \xi_2 \in [a, b]$ such that*

$$f^{\Delta}(\xi_2) \leq 0 \leq f^{\Delta}(\xi_1).$$

Proof. Since f is delta differentiable at each point of $[a, b]$, it follows that f is continuous on $[a, b]$. Therefore, there exist $\xi_1, \xi_2 \in [a, b]$ such that

$$m = \min_{[a,b]} f(t) = f(\xi_1), \quad M = \max_{[a,b]} f(t) = f(\xi_2).$$

Because $f(a) = f(b)$, we assume that $\xi_1, \xi_2 \in [a, b)$.

1. Let $\sigma(\xi_1) > \xi_1$. Then

$$f^\Delta(\xi_1) = \frac{f(\sigma(\xi_1)) - f(\xi_1)}{\sigma(\xi_1) - \xi_1} \geq 0.$$

2. Let $\sigma(\xi_1) = \xi_1$. Then

$$f^\Delta(\xi_1) = \lim_{t \to \xi_1} \frac{f(\xi_1) - f(t)}{\xi_1 - t} \geq 0.$$

3. Let $\sigma(\xi_2) > \xi_2$. Then

$$f^\Delta(\xi_2) = \frac{f(\sigma(\xi_2)) - f(\xi_2)}{\sigma(\xi_2) - \xi_2} \leq 0.$$

4. Let $\sigma(\xi_2) = \xi_2$. Then

$$f^\Delta(\xi_2) = \lim_{t \to \xi_2} \frac{f(\xi_2) - f(t)}{\xi_2 - t} \leq 0.$$

This completes the proof.

Theorem 1.9. *If f is delta differentiable at t_0, then*

$$f(\sigma(t)) = f(t_0) + \left(f^\Delta(t_0) + E(t)\right)(\sigma(t) - t_0), \tag{1.6}$$

where $E(t)$ is defined in a neighborhood of t_0 and

$$\lim_{t \longrightarrow t_0} E(t) = E(t_0) = 0.$$

Proof. Define

$$E(t) = \begin{cases} \frac{f(\sigma(t)) - f(t_0)}{\sigma(t) - t_0} - f^\Delta(t_0), & t \in \mathbb{T}, \quad t \neq t_0, \\ 0, & t = t_0. \end{cases} \tag{1.7}$$

Solving (1.7) for $f(\sigma(t))$ yields (1.6) if $t \neq t_0$.
Let $t = t_0$. Then

1. $\sigma(t_0) > t_0$. Then (1.6) is obvious.
2. $\sigma(t_0) = t_0$. Then (1.6) is obvious.

This completes the proof.

Theorem 1.10. *Let f be delta differentiable at t_0. If $f^\Delta(t_0) > (<)0$, then there exists $\delta > 0$ such that*

$$f(\sigma(t)) \geq (\leq)f(t_0) \quad \text{for} \quad \forall t \in (t_0, t_0 + \delta)$$

and

$$f(\sigma(t)) \le (\ge) f(t_0) \quad \text{for} \quad \forall t \in (t_0 - \delta, t_0).$$

Proof. Using (1.6), we have for $t \ne t_0$ that

$$\frac{f(\sigma(t)) - f(t_0)}{\sigma(t) - t_0} = f^\Delta(t_0) + E(t). \qquad (1.8)$$

Let $\delta > 0$ be chosen such that $|E(t)| \le f^\Delta(t_0)$ for all $t \in (t_0 - \delta, t_0 + \delta)$. Such a $\delta > 0$ exists because $\lim_{t \longrightarrow t_0} E(t) = 0$. Hence for all $t \in (t_0 - \delta, t_0 + \delta)$, we have

$$f^\Delta(t_0) + E(t) \ge 0.$$

If $t \in (t_0, t_0 + \delta)$, then $\sigma(t) \ge t_0$, and from (1.8) we obtain

$$\frac{f(\sigma(t)) - f(t_0)}{\sigma(t) - t_0} \ge 0,$$

i.e., $f(\sigma(t)) \ge f(t_0)$.
If $t \in (t_0 - \delta, t_0)$, then $t \le \sigma(t) \le t_0$, and from (1.8), we get $f(\sigma(t)) \le f(t_0)$. This completes the proof.

Theorem 1.11 (*Mean Value Theorem*). *Suppose that f is continuous on $[a, b]$ and has a delta derivative at each point of $[a, b)$. Then there exist $\xi_1, \xi_2 \in [a, b)$ such that*

$$f^\Delta(\xi_1)(b - a) \le f(b) - f(a) \le f^\Delta(\xi_2)(b - a). \qquad (1.9)$$

Proof. Consider the function ϕ defined on $[a, b]$ by

$$\phi(t) = f(t) - f(a) - \frac{f(b) - f(a)}{b - a}(t - a).$$

Then ϕ is continuous on $[a, b]$ and has a delta derivative at each point of $[a, b)$. Also, $\phi(a) = \phi(b) = 0$. Hence, there exist $\xi_1, \xi_2 \in [a, b)$ such that

$$\phi^\Delta(\xi_1) \le 0 \le \phi^\Delta(\xi_2),$$

or

$$f^\Delta(\xi_1) - \frac{f(b) - f(a)}{b - a} \le 0 \le f^\Delta(\xi_2) - \frac{f(b) - f(a)}{b - a},$$

whereupon we get (1.9). This completes the proof.

Corollary 1.1. *Let f be a continuous function on* $[a, b]$ *that has a delta derivative at each point of* $[a, b)$. *If* $f^\Delta(t) = 0$ *for all* $t \in [a, b)$, *then f is a constant function on* $[a, b]$.

Proof. For every $t \in [a, b]$, using (1.9), we have that there exist $\xi_1, \xi_2 \in [a, b)$ such that

$$0 = f^\Delta(\xi_1)(t - a) \le f(t) - f(a) \le f^\Delta(\xi_2)(t - a) = 0,$$

i.e., $f(t) = f(a)$. This completes the proof.

Corollary 1.2. *Let f be a continuous function on* $[a, b]$ *that has a delta derivative at each point of* $[a, b)$. *Then f is increasing, decreasing, nondecreasing, or nonincreasing on* $[a, b]$ *according to whether* $f^\Delta(t) > 0$, $f^\Delta(t) < 0$, $f^\Delta(t) \ge 0$, *or* $f^\Delta(t) \le 0$ *for all* $t \in [a, b)$, *respectively.*

Proof. 1. Let $f^\Delta(t) > 0$ for all $t \in [a, b]$. Then for all $t_1, t_2 \in [a, b]$, $t_1 < t_2$, there exists $\xi_1 \in (t_1, t_2)$ such that

$$f(t_1) - f(t_2) \le f^\Delta(\xi_1)(t_1 - t_2) < 0,$$

i.e., $f(t_1) < f(t_2)$.
2. Let $f^\Delta(t) < 0$ for all $t \in [a, b]$. Then for $t_1, t_2 \in [a, b]$, $t_1 < t_2$, there exists $\xi_1 \in (t_1, t_2)$ such that

$$f(t_1) - f(t_2) \ge f^\Delta(\xi_1)(t_1 - t_2) > 0,$$

i.e., $f(t_1) > f(t_2)$.
The cases $f^\Delta(t) \ge 0$ and $f^\Delta(t) \le 0$ we leave to the reader as an exercise. This completes the proof.

1.4 Integration

Definition 1.13. A function $f : \mathbb{T} \to \mathbb{R}$ is called regulated if its right-sided limits exist (and are finite) at all right-dense points in $\mathbb{T}$ and its left-sided limits exist (and are finite) at all left-dense points in $\mathbb{T}$.

Example 1.34. Let $\mathbb{T} = \mathbb{N}$ and

$$f(t) = \frac{t^2}{t - 1}, \quad g(t) = \frac{t}{t + 1}, \quad t \in \mathbb{T}.$$

We note that all points of $\mathbb{T}$ are right-scattered. The points $t \in \mathbb{T}$, $t \neq 1$, are left-scattered. The point $t = 1$ is left-dense. Also, $\lim\limits_{t \to 1-} f(t)$ is not finite and $\lim\limits_{t \to 1-} g(t)$ exists and is finite. Therefore, the function f is not regulated and the function g is regulated.

Example 1.35. Let $\mathbb{T} = \mathbb{R}$ and

$$f(t) = \begin{cases} 0 & \text{for } t = 0 \\ \frac{1}{t} & \text{for } t \in \mathbb{R}\backslash\{0\}. \end{cases}$$

We have that all points of $\mathbb{T}$ are dense, and $\lim\limits_{t \to 0-} f(t)$ and $\lim\limits_{t \to 0+} f(t)$ are not finite. Therefore, the function f is not regulated.

Exercise 1.12. Let $\mathbb{T} = \mathbb{R}$ and

$$f(t) = \begin{cases} 11 & \text{for } t = 1, \\ \frac{1}{t-1} & \text{for } t \in \mathbb{R}\backslash\{1\}. \end{cases}$$

Determine whether f is regulated.

Answer. It is not.

Definition 1.14. A continuous function $f : \mathbb{T} \to \mathbb{R}$ is called predifferentiable with region of differentiation D if

1. $D \subset \mathbb{T}^\kappa$,
2. $\mathbb{T}^\kappa \backslash D$ is countable and contains no right-scattered elements of $\mathbb{T}$,
3. f is differentiable at each $t \in D$.

Example 1.36. Let $\quad \mathbb{T} = P_{a,b} = \bigcup\limits_{k=0}^{\infty} [k(a+b), k(a+b) + a] \quad$ for $\quad a > b > 0$, $f : \mathbb{T} \to \mathbb{R}$, be defined by

$$f(t) = \begin{cases} 0 & \text{if } t \in \bigcup_{k=0}^{\infty}[k(a+b), k(a+b)+b], \\ t - (a+b)k - b & \text{if } t \in [(a+b)k+b, (a+b)k+a]. \end{cases}$$

Then f is predifferentiable with $D = T\backslash \bigcup\limits_{k=0}^{\infty}\{(a+b)k + b\}$.

Example 1.37. Let $\mathbb{T} = \mathbb{R}$ and

$$f(t) = \begin{cases} 0 & \text{if } t = 3, \\ \frac{1}{t-3} & \text{if } \mathbb{R}\backslash\{3\}. \end{cases}$$

Since $f : \mathbb{T} \to \mathbb{R}$ is not continuous, it follows that f is not predifferentiable.

Example 1.38. Let $\mathbb{T} = \mathbb{N}_0 \bigcup \left\{1 - \frac{1}{n} : n \in \mathbb{N}\right\}$ and

$$f(t) = \begin{cases} 0 & \text{if } t \in \mathbb{N} \\ t & \text{otherwise.} \end{cases}$$

Then f is predifferentiable with $D = \mathbb{T}\backslash\{1\}$.

Exercise 1.13. Let $\mathbb{T} = \mathbb{R}$ and

$$f(t) = \begin{cases} 0 & \text{if} \quad t = -3, \\ \frac{1}{t+3} & \text{if} \quad t \in \mathbb{R} \backslash \{-3\}. \end{cases}$$

Check whether $f : \mathbb{T} \to \mathbb{R}$ is predifferentiable, and if it is, find the region of differentiation.

Answer. It is not predifferentiable.

Definition 1.15. A function $f : \mathbb{T} \to \mathbb{R}$ is called rd-continuous if it is continuous at right-dense points in $\mathbb{T}$ and its left-sided limits exist (and are finite) at left-dense points in $\mathbb{T}$. The set of rd-continuous functions $f : \mathbb{T} \to \mathbb{R}$ will be denoted by $\mathscr{C}_{rd}(\mathbb{T})$.
The set of functions $f : \mathbb{T} \to \mathbb{R}$ that are differentiable with rd-continuous derivative is denoted by $\mathscr{C}^1_{rd}(\mathbb{T})$.

Some results concerning rd-continuous and regulated functions are contained in the following theorem. Since its statements follow directly from the definitions, we leave its proof to the reader.

Theorem 1.12. *Assume $f : \mathbb{T} \to \mathbb{R}$.*

1. *If f is continuous, then f is rd-continuous.*
2. *If f is rd-continuous, then f is regulated.*
3. *The jump operator σ is rd-continuous.*
4. *If f is regulated or rd-continuous, then so is f^σ.*
5. *Assume that f is continuous. If $g : \mathbb{T} \to \mathbb{R}$ is regulated or rd-continuous, then $f \circ g$ has that property.*

Theorem 1.13. *Every regulated function on a compact interval is bounded.*

Proof. Assume that $f : [a, b] \to \mathbb{R}$, $[a, b] \subset \mathbb{T}$, is unbounded. Then for each $n \in \mathbb{N}$ there exists $t_n \in \mathbb{T}$ such that $|f(t_n)| > n$. Because $\{t_n\}_{n \in \mathbb{N}} \subset [a, b]$, there exists a subsequence $\{t_{n_k}\}_{k \in \mathbb{N}} \subset \{t_n\}_{n \in \mathbb{N}}$ such that

$$\lim_{k \to \infty} t_{n_k} = t_0.$$

Since $\mathbb{T}$ is closed, we have that $t_0 \in \mathbb{T}$. Also, t_0 is a dense point. Using that f is regulated, we get

$$\left| \lim_{k \to \infty} f(t_{n_k}) = f(t_0) \right| \neq \infty,$$

which is a contradiction. This completes the proof.

Theorem 1.14 (Induction Principle). *Let $t_0 \in \mathbb{T}$ and assume that*

$$\{S(t) : t \in [t_0, \infty)\}$$

is a family of statements satisfying the following conditions:

 (i) $S(t_0)$ *is true.*
 (ii) *If* $t \in [t_0, \infty)$ *is right-scattered and* $S(t)$ *is true, then* $S(\sigma(t))$ *is true.*
 (iii) *If* $t \in [t_0, \infty)$ *is right-dense and* $S(t)$ *is true, then there is a neighborhood U of* t *such that* $S(s)$ *is true for all* $s \in U \cap (t, \infty)$.
 (iv) *If* $t \in (t_0, \infty)$ *is left-dense and* $S(s)$ *is true for* $s \in [t_0, t)$, *then* $S(t)$ *is true.*

Then $S(t)$ *is true for all* $t \in [t_0, \infty)$.

Proof. Let

$$S^* = \{t \in [t_0, \infty) : S(t) \quad is \quad not \quad true\}.$$

We assume that $S^* \neq \emptyset$. Let $\inf S^* = t^*$. Because $\mathbb{T}$ is closed, we have that $t^* \in \mathbb{T}$.

1. If $t^* = t_0$, then $S(t^*)$ is true.
2. If $t^* \neq t_0$ and $t^* = \rho(t^*)$, using (iv), we get that $S(t^*)$ is true.
3. If $t^* \neq t_0$ and $\rho(t^*) < t^*$, then $\rho(t^*)$ is right-scattered. Since $S(\rho(t^*))$ is true, we get that $S(t^*)$ is true.

Consequently, $t^* \notin S^*$.
If we suppose that t^* is right-scattered, then using that $S(t^*)$ is true and (ii), we conclude that $S(\sigma(t^*))$ is true, which is a contradiction. From the definition of t^* it follows that $t^* \neq \max \mathbb{T}$. Since t^* is not right-scattered and $t^* \neq \max \mathbb{T}$, we obtain that t^* is right-dense. Because $S(t^*)$ is true, using (iii), there exists a neighborhood U of t^* such that $S(s)$ is true for all $s \in U \bigcap (t^*, \infty)$, which is a contradiction.
Consequently, $S^* = \emptyset$. This completes the proof.

Theorem 1.15 (*Dual Version of Induction Principle*). *Let* $t_0 \in \mathbb{T}$ *and assume that*

$$\{S(t) : t \in (-\infty, t_0]\}$$

is a family of statements satisfying the following conditions:

 (i) $S(t_0)$ *is true.*
 (ii) *If* $t \in (-\infty, t_0]$ *is left-scattered and* $S(t)$ *is true, then* $S(\rho(t))$ *is true.*
 (iii) *If* $t \in (-\infty, t_0]$ *is left-dense and* $S(t)$ *is true, then there is a neighborhood U of* t *such that* $S(s)$ *is true for all* $s \in U \cap (-\infty, t)$.
 (iv) *If* $t \in (-\infty, t_0)$ *is right-dense and* $S(s)$ *is true for* $s \in (t, t_0)$, *then* $S(t)$ *is true.*

Then $S(t)$ *is true for all* $t \in (-\infty, t_0]$.

Theorem 1.16. *Let f and g be real-valued functions defined on* $\mathbb{T}$, *both prediffer-entiable with region D. Then*

$$|f^{\Delta}(t)| \leq |g^{\Delta}(t)| \quad for \quad all \quad t \in D$$

implies

$$|f(s) - f(r)| \leq g(s) - g(r) \quad \text{for all} \quad r, s \in \mathbb{T}, \quad r \leq s. \tag{1.10}$$

Proof. Let $r, s \in \mathbb{T}$ with $r \leq s$. Let also

$$[r, s) \backslash D = \{t_n : n \in \mathbb{N}\}.$$

We take $\epsilon > 0$. We consider the statements

$$S(t) : |f(t) - f(r)| \leq g(t) - g(r) + \epsilon \left(t - r + \sum_{t_n < t} 2^{-n} \right)$$

for $t \in [r, s]$.

We will prove, using the induction principle, that $S(t)$ is true for all $t \in [r, s]$.

1. $S(r) : 0 \leq \epsilon \sum_{t_n < t} 2^{-n}$ is true.

2. Let $t \in [r, s]$ be right-scattered and assume that $S(t)$ holds. Then for $t \in D$ we have

$$\begin{aligned}
|f(\sigma(t)) - f(r)| &= |f(t) + \mu(t)f^{\Delta}(t) - f(r)| \\
&\leq \mu(t)|f^{\Delta}(t)| + |f(t) - f(r)| \\
&\leq \mu(t)|f^{\Delta}(t)| + g(t) - g(r) + \epsilon \left(t - r + \sum_{t_n < t} 2^{-n} \right) \\
&\leq \mu(t)g^{\Delta}(t) + g(t) - g(r) + \epsilon \left(t - r + \sum_{t_n < t} 2^{-n} \right) \\
&= g(\sigma(t)) - g(r) + \epsilon \left(t - r + \sum_{t_n < t} 2^{-n} \right) \quad (t < \sigma(t)) \\
&< g(\sigma(t)) - g(r) + \epsilon \left(\sigma(t) - r + \sum_{t_n < \sigma(t)} 2^{-n} \right),
\end{aligned}$$

i.e., $S(\sigma(t))$ holds.

3. Let $t \in [r, s)$ and suppose that t is right-dense.

   Case 1. $t \in D$. Then f and g are differentiable at t. Then there exists a neighborhood U of t such that

$$|f(t) - f(\tau) - f^{\Delta}(t)(t - \tau)| \leq \frac{\epsilon}{2}|t - \tau|$$

   for all $\tau \in U$, and

$$|g(t) - g(\tau) - g^{\Delta}(t)(t - \tau)| \leq \frac{\epsilon}{2}|t - \tau|$$

for all $\tau \in U$. Thus

$$|f(t) - f(\tau)| \le \left(|f^{\Delta}(t)| + \frac{\epsilon}{2} \right) |t - \tau|$$

for all $\tau \in U$, and

$$g(\tau) - g(t) + g^{\Delta}(t)(t - \tau) \ge -\frac{\epsilon}{2}|t - \tau|$$

for all $\tau \in U$ or

$$g(\tau) - g(t) - g^{\Delta}(t)(\tau - t) \ge -\frac{\epsilon}{2}|t - \tau|$$

for all $\tau \in U$.
Hence, for all $\tau \in U \cap (t, \infty)$,

$$
\begin{aligned}
|f(\tau) - f(r)| &= |f(\tau) - f(t) + f(t) - f(r)| \\
&\le |f(\tau) - f(t)| + |f(t) - f(r)| \\
&\le \left(|f^{\Delta}(t)| + \tfrac{\epsilon}{2} \right) |t - \tau| + g(t) - g(r) \\
&\quad + \epsilon \left(t - r + \sum_{t_n < t} 2^{-n} \right) \\
&\le \left(g^{\Delta}(t) + \tfrac{\epsilon}{2} \right) |t - \tau| + g(t) - g(r) \\
&\quad + \epsilon \left(t - r + \sum_{t_n < t} 2^{-n} \right) \\
&= g^{\Delta}(t)(\tau - t) + \tfrac{\epsilon}{2}(\tau - t) + g(t) - g(r) \\
&\quad + \epsilon \left(t - r + \sum_{t_n < t} 2^{-n} \right) \\
&\le g(\tau) - g(t) + \tfrac{\epsilon}{2}|t - \tau| + \tfrac{\epsilon}{2}(\tau - t) + g(t) - g(r) \\
&\quad + \epsilon \left(t - r + \sum_{t_n < t} 2^{-n} \right) \\
&= g(\tau) - g(r) + \epsilon(\tau - t) + \epsilon \left(t - r + \sum_{t_n < t} 2^{-n} \right) \\
&= g(\tau) - g(r) + \epsilon \left(\tau - r + \sum_{t_n < t} 2^{-n} \right),
\end{aligned}
$$

so $S(\tau)$ follows for all $\tau \in U \cap (t, \infty)$.

Case 2. $t \notin D$. Then $t = t_m$ for some $m \in \mathbb{N}$. Since f and g are predifferentiable, they are continuous. Therefore, there exists a neighborhood U of t such that

$$|f(\tau) - f(t)| \le \frac{\epsilon}{2} 2^{-m} \quad \text{for all} \quad \tau \in U$$

and

$$|g(\tau) - g(t)| \leq \frac{\epsilon}{2} 2^{-m} \quad \text{for all} \quad \tau \in U.$$

Therefore,

$$g(\tau) - g(t) \geq -\frac{\epsilon}{2} 2^{-m} \quad \text{for all} \quad \tau \in U.$$

Consequently,

$$
\begin{aligned}
|f(\tau) - f(r)| &= |f(\tau) - f(t) + f(t) - f(r)| \\
&\leq |f(\tau) - f(t)| + |f(t) - f(r)| \\
&\leq \frac{\epsilon}{2} 2^{-m} + g(t) - g(r) + \epsilon \left(t - r + \sum_{t_n < t} 2^{-n} \right) \\
&\leq \frac{\epsilon}{2} 2^{-m} + g(\tau) + \frac{\epsilon}{2} 2^{-m} - g(r) + \epsilon \left(t - r + \sum_{t_n < t} 2^{-n} \right) \\
&= \epsilon 2^{-m} + g(\tau) - g(r) + \epsilon \left(\tau - r + \sum_{t_n < t} 2^{-n} \right) \\
&\leq \epsilon 2^{-m} + g(\tau) - g(r) + \epsilon \left(\tau - r + \sum_{t_n < \tau} 2^{-n} \right),
\end{aligned}
$$

so $S(\tau)$ follows for all $\tau \in U \cap (t, \infty)$.

4. Let t be left-dense and suppose that $S(\tau)$ is true for $\tau < t$. Then

$$
\begin{aligned}
\lim_{\tau \to t-} |f(\tau) - f(r)| &\leq \lim_{\tau \to t-} \left\{ g(\tau) - g(r) + \epsilon \left(\tau - r + \sum_{t_n < \tau} 2^{-n} \right) \right\} \\
&\leq \lim_{\tau \to t-} \left\{ g(\tau) - g(r) + \epsilon \left(\tau - r + \sum_{t_n < t} 2^{-n} \right) \right\}
\end{aligned}
$$

implies $S(t)$, since f and g are continuous at t.

From this and the induction principle, it follows that $S(t)$ is true for all $t \in [r, s]$. Consequently, (1.10) holds for all $r \leq s$, $r, s \in \mathbb{T}$. This completes the proof.

Theorem 1.17. *Suppose $f : \mathbb{T} \to \mathbb{R}$ is predifferentiable with D. If U is a compact interval with endpoints $r, s \in \mathbb{T}$, then*

$$|f(s) - f(r)| \leq \left\{ \sup_{t \in U^\kappa \cap D} |f^\Delta(t)| \right\} |s - r|.$$

Proof. Without loss of generality, we suppose that $r \leq s$. We set

$$g(t) := \left\{ \sup_{t \in U^\kappa \cap D} |f^\Delta(t)| \right\} (t - r), \quad t \in \mathbb{T}.$$

Then

$$g^\Delta(t) = \left\{ \sup_{t \in U^k \cap D} |f^\Delta(t)| \right\} \geq |f^\Delta(t)|$$

for all $t \in D \cap [r, s]^\kappa$.

From this and Theorem 1.16, it follows that

$$|f(t) - f(r)| \leq g(t) - g(r) \quad \text{for} \quad \text{all} \quad t \in [r, s],$$

whereupon

$$|f(s) - f(r)| \leq g(s) - g(r) = g(s) = \left\{ \sup_{t \in U^\kappa \cap D} |f^\Delta(t)| \right\} (s - r).$$

This completes the proof.

Theorem 1.18. *Let f be predifferentiable with region D. If $f^\Delta(t) = 0$ for all $t \in D$, then f is a constant function.*

Proof. From Theorem 1.17 it follows that for all $r, s \in \mathbb{T}$,

$$|f(s) - f(r)| \leq \left\{ \sup_{t \in U^\kappa \cap D} |f^\Delta(t)| \right\} |s - r| = 0,$$

i.e., $f(s) = f(r)$. Therefore, f is a constant function. This completes the proof.

Theorem 1.19. *Let f and g be predifferentiable with region D and $f^\Delta(t) = g^\Delta(t)$ for all $t \in D$. Then*

$$g(t) = f(t) + C \quad \text{for} \quad \text{all} \quad t \in \mathbb{T},$$

where C is a constant.

Proof. Let $h(t) = f(t) - g(t)$, $t \in \mathbb{T}$. Then

$$h^\Delta(t) = f^\Delta(t) - g^\Delta(t) = 0 \quad \text{for} \quad \text{all} \quad t \in D.$$

From this and Theorem 1.18, it follows that $h(t)$ is a constant function. This completes the proof.

Theorem 1.20. *Suppose* $f_n : \mathbb{T} \to \mathbb{R}$ *is predifferentiable with region D for each* $n \in \mathbb{N}$. *Assume that for each* $t \in \mathbb{T}^\kappa$ *there exists a compact interval* $U(t)$ *such that the sequence* $\{f_n^\Delta\}_{n\in\mathbb{N}}$ *converges uniformly on* $U(t) \cap D$.

(i) *If* $\{f_n\}_{n\in\mathbb{N}}$ *converges at some* $t_0 \in U(t)$ *for some* $t \in \mathbb{T}^\kappa$, *then it converges uniformly on* $U(t)$.

(ii) *If* $\{f_n\}_{n\in\mathbb{N}}$ *converges at some* $t_0 \in \mathbb{T}$, *then it converges uniformly on* $U(t)$ *for all* $t \in \mathbb{T}^\kappa$.

(iii) *The limit mapping* $f = \lim\limits_{n\longrightarrow\infty} f_n$ *is predifferentiable with D, and we have*

$$f^\Delta(t) = \lim_{n\longrightarrow\infty} f_n^\Delta(t) \quad \text{for} \quad \text{all} \quad t \in D.$$

Proof. (i) Since $\{f_n^\Delta\}_{n\in\mathbb{N}}$ converges uniformly on $U(t) \cap D$, there exists $N \in \mathbb{N}$ such that

$$\sup_{s\in U(t)\cap D} |(f_m - f_n)^\Delta(s)|$$

is finite for all $m, n \in \mathbb{N}$. Let $m, n \geq N$ and $r \in U(t)$. Then

$$|f_n(r) - f_m(r)| = |f_n(r) - f_m(r) - (f_n(t_0) - f_m(t_0)) + (f_n(t_0) - f_m(t_0))|$$
$$\leq |f_n(t_0) - f_m(t_0)| + \left\{\sup_{s\in U(t)\cap D} |(f_n - f_m)^\Delta(s)|\right\} |r - t_0|.$$

Hence, $\{f_n\}_{n\in\mathbb{N}}$ converges uniformly on $U(t)$, i.e., $\{f_n\}_{n\in\mathbb{N}}$ is a locally uniformly convergent sequence.

(ii) Suppose that $\{f_n(t_0)\}_{n\in\mathbb{N}}$ converges for some $t_0 \in \mathbb{T}$. Suppose that

$$S(t) : \{f_n(t)\}_{n\in\mathbb{N}}$$

converges.

1. $S(t_0) : \{f_n(t_0)\}$ converges is true.
2. Let t be right-scattered and assume that $S(t)$ holds. Then

$$f_n(\sigma(t)) = f_n(t) + \mu(t)f_n^\Delta(t)$$

converges by the assumption, i.e., $S(\sigma(t))$ holds.
3. Let t be right-dense and assume that $S(t)$ holds. Then by (i), $\{f_n\}_{n\in\mathbb{N}}$ converges on $U(t)$, and so $S(r)$ holds for all $r \in U(t) \cap (t, \infty)$.
4. Let t be left-dense and suppose that $S(r)$ holds for all $t_0 \leq r < t$. Since $U(t) \cap [t_0, t) \neq \emptyset$, using again part (i), we have that $\{f_n\}_{n\in\mathbb{N}}$ converges on $U(t)$; in particular, $S(t)$ is true.

Consequently, $S(t)$ is true for all $t \in [t_0, \infty)$. Using the dual version of the induction principle for the negative direction, we have that $S(t)$ is also true for all $t \in (-\infty, t_0]$. (We note that the first part of this has already been shown; the second part follows by $f_n(\rho(t)) = f_n(t) - \mu(\rho(t))f_n^\Delta(\rho(t))$, and the third and fourth parts follow again by (i).)

(iii) Let $t \in D$. Without loss of generality we can assume that $\sigma(t) \in U(t)$. We take $\epsilon > 0$ arbitrarily. Using (i), we see that there exists $N \in \mathbb{N}$ such that

$$|(f_n - f_m)(r) - (f_n - f_m)(\sigma(t))| \le \left\{\sup_{s \in U(t) \cap D} |(f_n - f_m)^\Delta(s)|\right\} |r - \sigma(t)|$$

for all $r \in U(t)$ and all $m, n \ge N$. Since $\{f_n^\Delta\}_{n \in \mathbb{N}}$ converges uniformly on $U(t) \cap D$, there exists $N_1 \ge N$ such that

$$\sup_{s \in U(t) \cap D} |(f_n - f_m)^\Delta(s)| \le \frac{\epsilon}{3} \quad \text{for all} \quad m, n \ge N_1.$$

Hence,

$$|(f_n - f_m)(r) - (f_n - f_m)(\sigma(t))| \le \frac{\epsilon}{3}|r - \sigma(t)|$$

for all $r \in U(t)$ and for all $m, n \ge N_1$. Now, letting $m \longrightarrow \infty$, we have that

$$|(f_n - f)(r) - (f_n - f)(\sigma(t))| \le \frac{\epsilon}{3}|r - \sigma(t)|$$

for all $r \in U(t)$ and all $n \ge N_1$. Let

$$g = \lim_{n \longrightarrow \infty} f_n^\Delta.$$

Then there exists $M \ge N_1$ such that

$$|f_M^\Delta(t) - g(t)| \le \frac{\epsilon}{3},$$

and since f_M is differentiable at t, there also exists a neighborhood W of t with

$$|f_M(\sigma(t)) - f_M(r) - f_M^\Delta(t)(\sigma(t) - r)| \le \frac{\epsilon}{3}|\sigma(t) - r|$$

for all $r \in W$.

From this we see that for all $r \in U(t) \cap W$, we have

$$
\begin{aligned}
|f(\sigma(t)) - f(r) - g(t)|\sigma(t) - r|| &\le |(f_M - f)((\sigma(t)) - (f_M - f)(r)| \\
&\quad + |f_M^\Delta(t) - g(t)||\sigma(t) - r| \\
&\quad + |f_M(\sigma(t)) - f_M(r) - f_M^\Delta(t)|\sigma(t) - r|| \\
&\le \frac{\epsilon}{3}|\sigma(t) - r| + \frac{\epsilon}{3}|\sigma(t) - r| \\
&\quad + \frac{\epsilon}{3}|\sigma(t) - r| \\
&= \epsilon|\sigma(t) - r|.
\end{aligned}
$$

Consequently, f is differentiable at t with $f^\Delta(t) = g(t)$. This completes the proof.

Theorem 1.21. *Let $t_0 \in \mathbb{T}$, $x_0 \in \mathbb{R}$, and let $f : \mathbb{T}^\kappa \to \mathbb{R}$ be a given regulated map. Then there exists exactly one predifferentiable function F satisfying*

$$F^\Delta(t) = f(t) \quad \text{for} \quad \text{all} \quad t \in D, \quad F(t_0) = x_0.$$

Proof. Let $n \in \mathbb{N}$ and

$$S(t) : \begin{cases} \text{there} & \text{exists} & \text{a} & \text{predifferentiable} & (F_{nt}, D_{nt}), \\ F_{nt} : [t_0, t] \to \mathbb{R} & \text{with} & F_{nt}(t_0) = x_0 & \text{and} \\ |F_{nt}^\Delta(s) - f(s)| \leq \frac{1}{n} & \text{for} & s \in D_{nt}. \end{cases}$$

1. $t = t_0$. Then $D_{nt_0} = \emptyset$ and $F_{nt_0}(t_0) = x_0$. Then the statement $S(t_0)$ follows.
2. Let t be right-scattered and suppose that $S(t)$ is true. Define

$$D_{n\sigma(t)} = D_{nt} \cup \{t\}$$

and $F_{n\sigma(t)}$ on $[t_0, \sigma(t)]$ by

$$F_{n\sigma(t)}(s) = \begin{cases} F_{nt}(s) & \text{if} \quad s \in [t_0, t], \\ F_{nt}(t) + \mu(t)f(t) & \text{if} \quad s = \sigma(t). \end{cases}$$

Then

$$F_{n\sigma(t)}(t_0) = F_{nt}(t_0) = x_0,$$

$$|F_{n\sigma(t)}^\Delta(s) - f(s)| = |F_{nt}^\Delta(s) - f(s)| \leq \frac{1}{n} \quad \text{for} \quad s \in D_{nt},$$

and

$$\begin{aligned} |F_{n\sigma(t)}^\Delta(t) - f(t)| &= \left| \frac{F_{n\sigma(t)}(\sigma(t)) - F_{n\sigma(t)}(t)}{\mu(t)} - f(t) \right| \\ &= \left| \frac{F_{nt}(t) + \mu(t)f(t) - F_{n\sigma(t)}(t)}{\mu(t)} - f(t) \right| \\ &= \left| \frac{F_{nt}(t) + \mu(t)f(t) - F_{nt}(t)}{\mu(t)} - f(t) \right| \\ &= \left| \frac{\mu(t)f(t)}{\mu(t)} - f(t) \right| = 0 \leq \frac{1}{n}, \end{aligned}$$

and therefore, the statement $S(\sigma(t))$ is valid.
3. Suppose t is right-dense and $S(t)$ is true. Since t is right-dense and $f(t)$ is regulated,

$$f(t^+) = \lim_{s \to t, s > t} f(s) \quad \text{exists}.$$

Hence, there is a neighborhood U of t with

$$|f(s) - f(t^+)| \leq \frac{1}{n} \quad \text{for all} \quad s \in U \cap (t, \infty). \tag{1.11}$$

Let $r \in U \cap (t, \infty)$. Define

$$D_{nr} = [D_{nt} \setminus \{t\}] \cup [t, r]^\kappa$$

and F_{nr} on $[t_0, r]$ by

$$F_{nr}(s) = \begin{cases} F_{nt}(s) & \text{if} \quad s \in [t_0, t], \\ F_{nt}(t) + f(t^+)(s - t) & \text{if} \quad s \in (t, r]. \end{cases}$$

Then F_{nr} is continuous at t and hence on $[t_0, r]$. Also, F_{nr} is differentiable on $(t, r]^\kappa$ with

$$F_{nr}^\Delta(s) = f(t^+) \quad \text{for all} \quad s \in (t, r]^\kappa.$$

Hence F_{nr} is predifferentiable on $[t_0, t)$. Since t is right-dense, we have that F_{nr} is predifferentiable with D_{nr}. From this and (1.11), we also have

$$|F_{nr}^\Delta(s) - f(s)| \leq \frac{1}{n} \quad \text{for all} \quad s \in D_{nr}.$$

Therefore, the statement $S(r)$ is true for all $r \in U \cap (t, \infty)$.

4. Now we suppose that t is left-dense and the statement $S(r)$ is true for $r < t$. Since $f(t)$ is regulated, it follows that

$$f(t^-) = \lim_{s \to t, s < t} f(s) \quad \text{exists.} \tag{1.12}$$

Hence, there exists a neighborhood U of t with

$$|f(s) - f(t^-)| \leq \frac{1}{n} \quad \text{for all} \quad s \in U \cap (-\infty, t).$$

Fix some $r \in U \cap (-\infty, t)$ and define

$$D_{nt} = \begin{cases} D_{nr} \cup (r, t) & \text{if} \quad r \quad \text{is} \quad \text{right-dense,} \\ D_{nr} \cup [r, t) & \text{if} \quad r \quad \text{is} \quad \text{right-scattered} \end{cases}$$

and F_{nt} on $[t_0, t]$ by

$$F_{nt}(s) = \begin{cases} F_{nr}(s) & \text{if} \quad s \in (t_0, r] \\ F_{nr}(t) + f(t^-)(s - r) & \text{if} \quad s \in (r, t]. \end{cases}$$

We note that F_{nt} is continuous at r and hence in $[t_0, t]$. Since

$$F_{nt}^{\Delta}(s) = f(t^-) \quad \text{for} \quad \text{all} \quad s \in (r, t],$$

it follows that F_{nt} is predifferentiable with D_{nt} and

$$|F_{nt}^{\Delta}(s) - f(s)| \leq \frac{1}{n} \quad \text{for} \quad \text{all} \quad s \in D_{nt}.$$

Hence the statement $S(t)$ holds.

By the induction principle, $S(t)$ is true for all $t \geq t_0$, $t \in \mathbb{T}$. Similarly, we can show that $S(t)$ is valid for $t \leq t_0$. Hence, F_n is predifferentiable with D_n, $F_n(t_0) = x_0$, and

$$|F_n^{\Delta}(t) - f(t)| \leq \frac{1}{n} \quad \text{for} \quad \text{all} \quad t \in D_n.$$

Now let

$$F = \lim_{n \longrightarrow \infty} F_n \quad \text{and} \quad D = \bigcap_{n \in \mathbb{N}} D_n.$$

Then $F(t_0) = x_0$, F is predifferentiable on D, and using Theorem 1.20, we obtain

$$F^{\Delta}(t) = \lim_{n \longrightarrow \infty} F_n^{\Delta}(t) = f(t) \quad \text{for} \quad \text{all} \quad t \in D.$$

This completes the proof.

Definition 1.16. Assume that $f : \mathbb{T} \to \mathbb{R}$ is a regulated function. A function F whose existence is guaranteed by Theorem 1.21 is called a pre-antiderivative of f. We define the indefinite integral of a regulated function f by

$$\int f(t) \Delta t = F(t) + c,$$

where c is an arbitrary constant and F is a pre-antiderivative of f. We define the Cauchy integral by

$$\int_{\tau}^{s} f(t) \Delta t = F(s) - F(\tau) \quad \text{for} \quad \text{all} \quad \tau, s \in \mathbb{T}.$$

A function $F : \mathbb{T} \to \mathbb{R}$ is called an antiderivative of $f : \mathbb{T} \to \mathbb{R}$ if

$$F^{\Delta}(t) = f(t) \quad \text{holds} \quad \text{for} \quad \text{all} \quad t \in \mathbb{T}^{\kappa}.$$

Example 1.39. Let $\mathbb{T} = \mathbb{Z}$. Then $\sigma(t) = t + 1$, $t \in \mathbb{T}$. Let also $f(t) = 3t^2 + 5t + 2$, $g(t) = t^3 + t^2$, $t \in \mathbb{T}$. Since

$$g^\Delta(t) = (\sigma(t))^2 + t\sigma(t) + t^2 + \sigma(t) + t$$
$$= (t+1)^2 + t(t+1) + t^2 + t + 1 + t$$
$$= t^2 + 2t + 1 + t^2 + t + t^2 + 2t + 1$$
$$= 3t^2 + 5t + 2,$$

we conclude that

$$\int (3t^2 + 5t + 2)\Delta t = t^3 + t^2 + c.$$

Example 1.40. Let $\mathbb{T} = 2^{\mathbb{N}}$, $f : \mathbb{T} \to \mathbb{R}$ be defined by $f(t) = 2\sin\dfrac{t}{2}\cos\dfrac{3t}{2}, t \in \mathbb{T}$. Let also $g(t) = \sin t, t \in \mathbb{T}$. In this case, we have that $\sigma(t) = 2t$. Since

$$g^\Delta(t) = \frac{\sin\sigma(t) - \sin t}{\sigma(t) - t}$$
$$= \frac{\sin(2t) - \sin t}{t}$$
$$= \frac{2}{t}\sin\tfrac{t}{2}\cos\tfrac{3t}{2},$$

we get

$$\int \frac{2}{t}\sin\frac{t}{2}\cos\frac{3t}{2}\Delta t = \sin t + c.$$

Example 1.41. Let $\mathbb{T} = \mathbb{N}_0^2$, $f : \mathbb{T} \to \mathbb{R}$ be defined by $f(t) = \dfrac{1}{1+2\sqrt{t}}\log\dfrac{(\sqrt{t}+1)^2}{t}$, $t \in \mathbb{T}$. Let also $g(t) = \log t, t \in \mathbb{T}$, Since $\sigma(t) = (\sqrt{t}+1)^2$ and

$$g^\Delta(t) = \frac{\log\sigma(t) - \log t}{\sigma(t) - t}$$
$$= \frac{\log(\sqrt{t}+1)^2 - \log t}{(\sqrt{t}+1)^2 - t}$$
$$= \frac{1}{1+2\sqrt{t}}\log\frac{(\sqrt{t}+1)^2}{t},$$

we get

$$\int \frac{1}{1+2\sqrt{t}}\log\frac{(1+\sqrt{t})^2}{t}\Delta t = \log t + c.$$

Exercise 1.14. Let $\mathbb{T} = \mathbb{N}_0^3$. Prove that

$$\int (2t + 3\sqrt[3]{t^2} + 3\sqrt[3]{t} + 2)\Delta t = t^2 + t + c.$$

Theorem 1.22. *Every rd-continuous function $f : \mathbb{T} \to \mathbb{R}$ has an antiderivative. In particular, if $t_0 \in \mathbb{T}$, then F defined by*

$$F(t) = \int_{t_0}^{t} f(\tau) \Delta \tau \quad \text{for} \quad t \in \mathbb{T}$$

is an antiderivative of f.

Proof. Since f is rd-continuous, it is regulated. Let F be a function guaranteed to exist by Theorem 1.21, together with D, satisfying

$$F^{\Delta}(t) = f(t) \quad \text{for} \quad t \in D.$$

We have that F is predifferentiable with D.
Let $t \in \mathbb{T}^{\kappa} \backslash D$. Then t is right-dense. Since f is rd-continuous, it follows that f is continuous at t. Let $\epsilon > 0$ be arbitrarily chosen. Then there exists a neighborhood U of t such that

$$|f(s) - f(t)| \le \epsilon \quad \text{for} \quad \text{all} \quad s \in U.$$

We define

$$h(\tau) := F(\tau) - f(t)(\tau - t_0) \quad \text{for} \quad \tau \in \mathbb{T}.$$

Then h is predifferentiable with D and

$$h^{\Delta}(\tau) = F^{\Delta}(\tau) - f(t) = f(\tau) - f(t) \quad \text{for} \quad \text{all} \quad \tau \in D.$$

Hence,

$$|h^{\Delta}(s)| = |f(s) - f(t)| \le \epsilon \quad \text{for} \quad \text{all} \quad s \in D \cap U.$$

Therefore,

$$\sup_{s \in D \cap U} |h^{\Delta}(s)| \le \epsilon,$$

whereupon

$$\begin{aligned} |F(t) - F(r) - f(t)(t - r)| &= |h(t) + f(t)(t - t_0) - (h(r) \\ &\quad + f(t)(r - t_0)) - f(t)(t - r)| \\ &= |h(t) - h(r)| \\ &\le \left\{ \sup_{s \in D \cap U} |h^{\Delta}(s)| \right\} |t - r| \\ &\le \epsilon |t - r|, \end{aligned}$$

which shows that F is differentiable at t and $F^{\Delta}(t) = f(t)$. This completes the proof.

Theorem 1.23. *If $f \in \mathscr{C}_{rd}(\mathbb{T})$ and $t \in \mathbb{T}^\kappa$, then*

$$\int_t^{\sigma(t)} f(\tau)\Delta\tau = \mu(t)f(t).$$

Proof. Since $f \in \mathscr{C}_{rd}(\mathbb{T})$, there exists an antiderivative F of f, and

$$\int_t^{\sigma(t)} f(\tau)\Delta\tau = F(\sigma(t)) - F(t)$$

$$= \mu(t)F^\Delta(t)$$

$$= \mu(t)f(t),$$

so that the conclusion follows. This completes the proof.

Theorem 1.24. *If $f^\Delta \geq 0$, then f is nondecreasing.*

Proof. Let $f^\Delta \geq 0$ and let $s, t \in \mathbb{T}$ with $a \leq s \leq t \leq b$. Then

$$f(t) = f(s) + \int_s^t f^\Delta(\tau)\Delta\tau \geq f(s),$$

so that the conclusion follows. This completes the proof.

Theorem 1.25. *If $a, b, c \in \mathbb{T}$, $\alpha \in \mathbb{R}$, and $f, g \in \mathscr{C}_{rd}(\mathbb{T})$, then*

(i) $\int_a^b (f(t) + g(t))\Delta t = \int_a^b f(t)\Delta t + \int_a^b g(t)\Delta t,$

(ii) $\int_a^b (\alpha f)(t)\Delta t = \alpha \int_a^b f(t)\Delta t,$

(iii) $\int_a^b f(t)\Delta t = -\int_b^a f(t)\Delta t,$

(iv) $\int_a^b f(t)\Delta t = \int_a^c f(t)\Delta t + \int_c^b f(t)\Delta t,$

(v) $\int_a^b f(\sigma(t))g^\Delta(t)\Delta t = (fg)(b) - (fg)(a) - \int_a^b f^\Delta(t)g(t)\Delta t,$

(vi) $\int_a^b f(t)g^\Delta(t)\Delta t = (fg)(b) - (fg)(a) - \int_a^b f^\Delta(t)g(\sigma(t))\Delta t,$

(vii) $\int_a^a f(t)\Delta t = 0,$

(viii) *if $|f(t)| \leq g(t)$ on $[a, b]$, then*

$$\left|\int_a^b f(t)\Delta t\right| \leq \int_a^b g(t)\Delta t,$$

(ix) if $f(t) \geq 0$ for all $a \leq t < b$, then $\displaystyle\int_a^b f(t)\Delta t \geq 0$.

Proof. Since $f, g \in \mathscr{C}_{rd}(\mathbb{T})$, they possess antiderivatives F and G. We have

$$F^\Delta(t) = f(t) \quad \text{and} \quad G^\Delta(t) = g(t) \quad \text{for} \quad \text{all} \quad t \in \mathbb{T}^\kappa.$$

(i) For all $t \in \mathbb{T}^\kappa$ we have

$$(F+G)^\Delta(t) = F^\Delta(t) + G^\Delta(t).$$

Hence,

$$\int_a^b (f(t) + g(t))\Delta t = (F+G)(b) - (F+G)(a)$$
$$= F(b) - F(a) + G(b) - G(a)$$
$$= \int_a^b f(t)\Delta t + \int_a^b g(t)\Delta t.$$

(ii) Since

$$\alpha F^\Delta(t) = (\alpha F)^\Delta(t) = \alpha f(t) \quad \text{for} \quad \text{all} \quad t \in \mathbb{T}^\kappa,$$

we get

$$\int_a^b \alpha f(t)\Delta t = (\alpha F)(b) - (\alpha F)(a)$$
$$= \alpha(F(b) - F(a))$$
$$= \alpha \int_a^b f(t)\Delta t.$$

(iii) We have

$$\int_a^b f(t)\Delta t = F(b) - F(a)$$
$$= -(F(b) - F(a))$$
$$= -\int_b^a f(t)\Delta t.$$

(iv) We have

$$\int_a^b f(t)\Delta t = F(b) - F(a)$$

$$= F(c) - F(a) + F(b) - F(c)$$

$$= \int_a^c f(t)\Delta t + \int_c^b f(t)\Delta t.$$

(v) For all $t \in \mathbb{T}^\kappa$ we have

$$(fg)^\Delta(t) = f^\Delta(t)g(t) + f(\sigma(t))g^\Delta(t)$$

or

$$f(\sigma(t))g^\Delta(t) = (fg)^\Delta(t) - f^\Delta(t)g(t).$$

Hence and (i), (ii), we get

$$\int_a^b f(\sigma(t))g^\Delta(t)\Delta t = \int_a^b \left((fg)^\Delta(t) - f^\Delta(t)g(t)\right)\Delta t$$

$$= \int_a^b (fg)^\Delta(t)\Delta t - \int_a^b f^\Delta(t)g(t)\Delta t$$

$$= (fg)(b) - (fg)(a) - \int_a^b f^\Delta(t)g(t)\Delta t.$$

(vi) For all $t \in \mathbb{T}^\kappa$ we have

$$(fg)^\Delta(t) = f(t)g^\Delta(t) + f^\Delta(t)g(\sigma(t)),$$

or

$$f(t)g^\Delta(t) = (fg)^\Delta(t) - f^\Delta(t)g(\sigma(t)).$$

From this and (i), (ii), we obtain

$$\int_a^b f(t)g^\Delta(t)\Delta t = \int_a^b (fg)^\Delta(t)\Delta t - \int_a^b f^\Delta(t)g(\sigma(t))\Delta t$$

$$= (fg)(b) - (fg)(a) - \int_a^b f^\Delta(t)g(\sigma(t))\Delta t.$$

(vii)

$$\int_a^a f(t)\Delta t = F(a) - F(a) = 0.$$

(viii)   We note that

$$|F^\Delta(t)| \le G^\Delta(t) \quad \text{on} \quad [a, b].$$

From this and Theorem 1.16, we get

$$|F(b) - F(a)| \le G(b) - G(a),$$

or

$$\left| \int_a^b f(t)\Delta t \right| \le \int_a^b g(t)\Delta t.$$

(ix)   This property follows directly from property (viii). This completes the proof.

Exercise 1.15.   Let $a, b \in \mathbb{T}$ and $f \in \mathscr{C}_{rd}(\mathbb{T})$.

(i)   If $\mathbb{T} = \mathbb{R}$, prove that

$$\int_a^b f(t)\Delta t = \int_a^b f(t)dt,$$

where the integral on the right is the usual Riemann integral from calculus.

(ii)   If $[a, b]$ consists of only isolated points, then

$$\int_a^b f(t)\Delta t = \begin{cases} \displaystyle\sum_{t\in[a,b)} \mu(t)f(t) & \text{if} \quad a < b \\ 0 & \text{if} \quad a = b, \\ -\displaystyle\sum_{t\in[b,a)} \mu(t)f(t) & \text{if} \quad a > b. \end{cases}$$

Definition 1.17 (Improper Integral).   If $a \in \mathbb{T}$, $\sup \mathbb{T} = \infty$, and f is rd-continuous on $[a, \infty)$, then we define the improper integral by

$$\int_a^\infty f(t)\Delta t := \lim_{b \to \infty} \int_a^b f(t)\Delta t,$$

provided this limit exists, and we say that the improper integral converges in this case. If this limit does not exist, then we say that the improper integral diverges.

Example 1.42. Let $\mathbb{T} = q^{\mathbb{N}_0}$, $q > 1$. Then $\sigma(t) = qt$. Since all points are isolated, we have

$$\int_1^\infty \frac{1}{t^2} \Delta t = \sum_{t \in q^{\mathbb{N}_0}} \frac{1}{t^2} \mu(t) = \sum_{t \in q^{\mathbb{N}_0}} \frac{q-1}{t} = q.$$

1.5   The Exponential Function

1.5.1   Hilger's Complex Plane

Definition 1.18. Let $h > 0$.

1. The Hilger complex numbers are defined as follows:

$$\mathbb{C}_h = \{z \in \mathbb{C} : z \neq -\frac{1}{h}\}.$$

2. The Hilger real axis is defined as follows:

$$\mathbb{R}_h = \{z \in \mathbb{C} : z > -\frac{1}{h}\}.$$

3. The Hilger alternative axis is defined as follows:

$$A_h = \{z \in \mathbb{C} : z < -\frac{1}{h}\}.$$

4. The Hilger imaginary circle is defined as follows:

$$\mathbb{I}_h = \left\{z \in \mathbb{C} : \left|z + \frac{1}{h}\right| = \frac{1}{h}\right\}.$$

For $h = 0$ we set

$$\mathbb{C}_0 = \mathbb{C}, \quad \mathbb{R}_0 = \mathbb{R}, \quad A_0 = \emptyset, \quad \mathbb{I}_0 = i\mathbb{R}.$$

Definition 1.19. Let $h > 0$ and $z \in \mathbb{C}_h$. We define the Hilger real part of z by

$$\mathrm{Re}_h(z) = \frac{|zh + 1| - 1}{h}$$

1.5   The Exponential Function

and the Hilger imaginary part by

$$\text{Im}_h(z) = \frac{\text{Arg}(zh + 1)}{h},$$

where $\text{Arg}(z)$ denotes the principal argument of z, i.e.,

$$-\pi < \text{Arg}(z) \leq \pi.$$

We note that

$$-\frac{1}{h} < \text{Re}_h(z) < \infty \quad \text{and} \quad -\frac{\pi}{h} < \text{Im}_h(z) < \frac{\pi}{h}.$$

In particular, $\text{Re}_h(z) \in \mathbb{R}_h$.

Definition 1.20. Let $-\dfrac{\pi}{h} < w \leq \dfrac{\pi}{h}$. We define the Hilger purely imaginary number by

$$\overset{\circ}{i}\, w = \frac{e^{iwh} - 1}{h}.$$

Theorem 1.26. *Let $z \in \mathbb{C}_h$. Then $\overset{\circ}{i}\,\text{Im}_h(z) \in \mathbb{I}_h$.*

Proof. We have

$$\overset{\circ}{i}\,\text{Im}_h(z) = \frac{e^{i\text{Im}_h(z)} - 1}{h}$$

and

$$\begin{aligned}
\left| \overset{\circ}{i}\,\text{Im}_h(z) + \frac{1}{h} \right| &= \left| \frac{e^{i\text{Im}_h(z)} - 1}{h} + \frac{1}{h} \right| \\
&= \frac{\left| e^{i\text{Im}_h(z)} \right|}{h} \\
&= \frac{1}{h}.
\end{aligned}$$

This completes the proof.

Theorem 1.27. *We have*

$$\lim_{h \to 0} [\text{Re}_h(z) + \overset{\circ}{i}\,\text{Im}_h(z)] = \text{Re}(z) + i\text{Im}(z).$$

Proof. We have

$$z = \text{Re}(z) + i\text{Im}(z),$$

$$zh + 1 = (\text{Re}(z) + i\text{Im}(z))h + 1$$

$$= h\text{Re}(z) + 1 + ih\text{Im}(z),$$

$$\text{Arg}(zh + 1) = \arcsin \frac{h\text{Im}(z)}{\sqrt{(h\text{Re}(z) + 1)^2 + h^2 (\text{Im}(z))^2}},$$

$$\text{Im}_h(z) = \frac{\text{Arg}(zh + 1)}{h}$$

$$= \frac{1}{h} \arcsin \frac{h\text{Im}(z)}{\sqrt{(h\text{Re}(z) + 1)^2 + h^2 (\text{Im}(z))^2}},$$

$$|zh + 1| = \sqrt{(h\text{Re}(z) + 1)^2 + h^2 (\text{Im}(z))^2},$$

$$\text{Re}_h(z) = \frac{|zh + 1| - 1}{h}$$

$$= \frac{\sqrt{(h\text{Re}(z) + 1)^2 + h^2 (\text{Im}(z))^2} - 1}{h}.$$

Hence,

$$\lim_{h \to 0} \text{Re}_h(z) = \lim_{h \to 0} \frac{\sqrt{(h\text{Re}(z) + 1)^2 + h^2 (\text{Im}(z))^2} - 1}{h}$$

$$= \lim_{h \to 0} \frac{(h\text{Re}(z) + 1)\text{Re}(z) + h (\text{Im}(z))^2}{\sqrt{(h\text{Re}(z) + 1)^2 + h^2 (\text{Im}(z))^2}}$$

$$= \text{Re}(z),$$

$$\lim_{h \to 0} \text{Im}_h(z) = \lim_{h \to 0} \frac{1}{h} \arcsin \frac{h\text{Im}(z)}{\sqrt{(h\text{Re}(z) + 1)^2 + h^2 (\text{Im}(z))^2}}$$

$$= \lim_{h \to 0} \frac{1}{\sqrt{1 - \frac{h^2 (\text{Im}(z))^2}{(h\text{Re}(z)+1)^2 + h^2 (\text{Im}(z))^2}}} \times$$

$$\times \frac{\text{Im}(z)\sqrt{(h\text{Re}(z) + 1)^2 + h^2 (\text{Im}(z))^2} - h\text{Im}(z) \frac{(h\text{Re}(z)+1)\text{Re}(z)+h(\text{Im}(z))^2}{\sqrt{(h\text{Re}(z)+1)^2 + h^2 (\text{Im}(z))^2}}}{(h\text{Re}(z) + 1)^2 + h^2 (\text{Im}(z))^2}$$

$$= \text{Im}(z),$$

which completes the proof.

Theorem 1.28. *Let* $-\dfrac{\pi}{h} < w \le \dfrac{\pi}{h}$. *Then*

$$|\overset{\circ}{i}\, w|^2 = \frac{4}{h^2}\sin^2\frac{wh}{2}.$$

Proof. We have

$$\overset{\circ}{i}\, w = \frac{e^{iwh}-1}{h} = \frac{\cos(wh)-1+i\sin(wh)}{h}.$$

Hence,

$$
\begin{aligned}
|\overset{\circ}{i}\, w|^2 &= \frac{(\cos(wh)-1)^2}{h^2} + \frac{\sin^2(wh)}{h^2} \\
&= \frac{\cos^2(wh) - 2\cos(wh) + 1 + \sin^2(wh)}{h^2} \\
&= \frac{2(1-\cos(wh))}{h^2} \\
&= \frac{4}{h^2}\sin^2\frac{wh}{2}.
\end{aligned}
$$

This completes the proof.

Definition 1.21. The "circle plus" addition $\oplus$ on $\mathbb{C}_h$ is defined by

$$z \oplus w = z + w + zwh.$$

Theorem 1.29. $(\mathbb{C}_h, \oplus)$ *is an abelian group.*

Proof. Let $z, w \in \mathbb{C}_h$. Then $z, w \in \mathbb{C}$ and $z, w \ne -\dfrac{1}{h}$. Therefore, $z \oplus w \in \mathbb{C}$. Since

$$
\begin{aligned}
h(z \oplus w) + 1 &= h(z + w + zwh) + 1 \\
&= 1 + hz + hw + zwh^2 \\
&= 1 + hz + hw(1 + hz) \\
&= (1 + hw)(1 + hz) \\
&\ne 0,
\end{aligned}
$$

we conclude that $z \oplus w \in \mathbb{C}_h$.

Also,

$$0 \oplus z = z \oplus 0 = z,$$

i.e., 0 is the additive identity for $\oplus$.

For $z \in \mathbb{C}_h$ we have

$$z \oplus \left(-\frac{z}{1+zh}\right) = z - \frac{z}{1+zh} - z\frac{z}{1+zh}h$$

$$= \frac{z^2 h}{1+zh} - \frac{z^2 h}{1+zh}$$

$$= 0,$$

i.e., the additive inverse of z under the addition $\oplus$ is $-\dfrac{z}{1+zh}$. We note that

$-\dfrac{z}{1+zh} \in \mathbb{C}$ and

$$1 - \frac{zh}{1+zh} = \frac{1}{1+zh} \neq 0,$$

i.e., $-\dfrac{z}{1+zh} \neq -\dfrac{1}{h}$. Therefore, $-\dfrac{z}{1+zh} \in \mathbb{C}_h$.

For $z, w, v \in \mathbb{C}_h$, we have

$$(z \oplus w) \oplus v = (z + w + zwh) \oplus v$$

$$= z + w + zwh + v + (z + w + zwh)vh$$

$$= z + w + zwh + v + zvh + wvh + zwvh^2$$

and

$$z \oplus (w \oplus v) = z + (w \oplus v) + z(w \oplus v)h$$

$$= z + w + v + wvh + z(w + v + wvh)h$$

$$= z + w + v + wvh + zwh + zvh + zwvh^2.$$

Consequently,

$$z \oplus (w \oplus v) = (z \oplus w) \oplus v,$$

i.e., in $(\mathbb{C}_h, \oplus)$ the associative law holds.

For $z, w \in \mathbb{C}_h$ we have

$$z \oplus w = z + w + zwh$$

$$= w + z + wzh$$

$$= w \oplus z,$$

which completes the proof.

Example 1.43. Let $z \in \mathbb{C}_h$ and $w \in \mathbb{C}$. We will simplify the expression

$$A = z \oplus \frac{w}{1 + hz}.$$

We have

$$A = z + \frac{w}{1 + hz} + \frac{zw}{a + hz}h$$

$$= z + \frac{(1 + hz)w}{1 + hz}$$

$$= z + w.$$

Theorem 1.30. *For $z \in \mathbb{C}_h$ we have*

$$z = \text{Re}_h(z) \oplus \overset{\circ}{i}\, \text{Im}_h(z).$$

Proof. We have

$$\text{Re}_h(z) \oplus \overset{\circ}{i}\, \text{Im}_h(z) = \frac{|zh + 1| - 1}{h} \oplus \overset{\circ}{i}\, \frac{\text{Arg}(zh + 1)}{h}$$

$$= \frac{|zh + 1| - 1}{h} \oplus \frac{e^{i\text{Arg}(zh+1)} - 1}{h}$$

$$= \frac{|zh + 1| - 1}{h} + \frac{e^{i\text{Arg}(zh+1)} - 1}{h}$$

$$+ \frac{|zh + 1| - 1}{h} \cdot \frac{e^{i\text{Arg}(zh+1)} - 1}{h}h$$

$$= \frac{1}{h}\Big(|zh + 1| - 1 + e^{i\text{Arg}(zh+1)} - 1 + |zh + 1|e^{i\text{Arg}(zh+1)}$$

$$- |zh + 1| - e^{i\text{Arg}(zh+1)} + 1\Big)$$

$$= \frac{1}{h}\Big(|zh + 1|e^{i\text{Arg}(zh+1)} - 1\Big)$$

$$= \frac{1}{h}(zh + 1 - 1)$$

$$= z.$$

This completes the proof.

Definition 1.22. Let $n \in \mathbb{N}$ and $z \in \mathbb{C}_h$. We define "circle dot" multiplication $\odot$ by

$$n \odot z = \underbrace{z \oplus z \oplus z \oplus \ldots \oplus z}_{n}.$$

Theorem 1.31. *Let $n \in \mathbb{N}$ and $z \in \mathbb{C}_h$. Then*

$$n \odot z = \frac{(zh+1)^n - 1}{h}.$$ (1.13)

Proof. 1. Let $n = 2$. Then

$$
\begin{aligned}
2 \odot z &= z \oplus z \\
&= z + z + z^2 h \\
&= 2z + z^2 h \\
&= \frac{1}{h}(z^2 h^2 + 2zh) \\
&= \frac{1}{h}(z^2 h^2 + 2zh + 1 - 1) \\
&= \frac{(zh+1)^2 - 1}{h}.
\end{aligned}
$$

2. Assume that

$$n \odot z = \frac{(zh+1)^n - 1}{h}$$

for some $n \in \mathbb{N}$.
3. We will prove that

$$(n+1) \odot z = \frac{(zh+1)^{n+1} - 1}{h}.$$

Indeed,

$$
\begin{aligned}
(n+1) \odot z &= (n \odot z) \oplus z \\
&= \frac{(zh+1)^n - 1}{h} \oplus z \\
&= \frac{(zh+1)^n - 1}{h} + z + \frac{(zh+1)^n - 1}{h} zh \\
&= \frac{(zh+1)^n - 1 + zh + zh(zh+1)^n - zh}{h} \\
&= \frac{(zh+1)^{n+1} - 1}{h}.
\end{aligned}
$$

Hence, we conclude that (1.13) holds for all $n \in \mathbb{N}$. This completes the proof.

Definition 1.23. Let $z \in \mathbb{C}_h$. The additive inverse of z under the operation $\oplus$ is defined as follows:

$$\ominus z = \frac{-z}{1 + zh}.$$

Theorem 1.32. *Let* $z \in \mathbb{C}_h$. *Then*

$$\ominus(\ominus z) = z.$$

Proof. We have

$$\ominus(\ominus z) = -\frac{\ominus z}{1 + (\ominus z)h}$$

$$= -\frac{\frac{-z}{1+zh}}{1 + \frac{-z}{1+zh}h}$$

$$= \frac{\frac{z}{1+zh}}{\frac{1+zh-zh}{1+zh}}$$

$$= z.$$

This completes the proof.

Definition 1.24. Let $z, w \in \mathbb{C}_h$. We define "circle minus" subtraction as follows:

$$z \ominus w = z \oplus (\ominus w).$$

For $z, w \in \mathbb{C}_h$ we have

$$z \ominus w = z \oplus (\ominus w)$$

$$= z + (\ominus w) + z(\ominus w)h$$

$$= z - \frac{w}{1 + wh} - \frac{zwh}{1 + wh}$$

$$= \frac{z + zwh - w - zwh}{1 + wh}$$

$$= \frac{z - w}{1 + wh},$$

i.e.,

$$z \ominus w = \frac{z - w}{1 + wh}. \tag{1.14}$$

Theorem 1.33. *Let* $z \in \mathbb{C}_h$. *Then* $\bar{z} = \ominus z$ *iff* $z \in \mathbb{I}_h$.

Proof. By (1.14), we have

$$\bar{z} = \ominus z \quad \Longleftrightarrow$$
$$\bar{z} = -\frac{z}{1+zh} \quad \Longleftrightarrow$$
$$\bar{z} + \bar{z}zh = -z \quad \Longleftrightarrow$$
$$2\mathrm{Re}(z) + |z|^2 h = 0.$$

Also,

$$\left| z + \tfrac{1}{h} \right| = \tfrac{1}{h} \quad \Longleftrightarrow$$
$$\left| z + \tfrac{1}{h} \right|^2 = \tfrac{1}{h^2} \quad \Longleftrightarrow$$
$$\left(\mathrm{Re}(z) + \tfrac{1}{h} \right)^2 + (\mathrm{Im}(z))^2 = \tfrac{1}{h^2} \quad \Longleftrightarrow$$
$$(\mathrm{Re}(z))^2 + \tfrac{2}{h}\mathrm{Re}(z) + \tfrac{1}{h^2} + (\mathrm{Im}(z))^2 = \tfrac{1}{h^2} \quad \Longleftrightarrow$$
$$|z|^2 + \tfrac{2}{h}\mathrm{Re}(z) = 0 \quad \Longleftrightarrow$$
$$2\mathrm{Re}(z) + |z|^2 h = 0,$$

which completes the proof.

Theorem 1.34. *Let* $-\dfrac{\pi}{h} < w \le \dfrac{\pi}{h}$. *Then*

$$\ominus(\overset{\circ}{i} w) = \overline{\overset{\circ}{i} w}.$$

Proof. We have

$$\ominus(\overset{\circ}{i} w) = -\frac{\overset{\circ}{i} w}{1 + \overset{\circ}{i} wh}$$

$$= -\frac{\frac{e^{iwh}-1}{h}}{1 + \frac{e^{iwh}-1}{h}h}$$

$$= -\frac{e^{iwh} - 1}{he^{iwh}}$$

$$= \frac{e^{-iwh} - 1}{h}$$

$$= \overline{\overset{\circ}{i} w}.$$

This completes the proof.

Definition 1.25. Let $z \in \mathbb{C}_h$. The generalized square of z is defined as follows:

$$z^{\circleddash} = -z(\ominus z).$$

We have

$$z^{\odot} = -z \frac{-z}{1+zh} = \frac{z^2}{1+zh}.$$

Theorem 1.35. *For $z \in \mathbb{C}_h$ we have*

$$(\ominus z)^{\odot} = z^{\odot}.$$

Proof. We have

$$(\ominus z)^{\odot} = -(\ominus z)(\ominus(\ominus z))$$

$$= \frac{z}{1+zh} z$$

$$= \frac{z^2}{1+zh}$$

$$= z^{\odot}.$$

This completes the proof.

Theorem 1.36. *For $z \in \mathbb{C}_h$ we have*

$$1 + zh = \frac{z^2}{z^{\odot}}.$$

Proof. We have

$$\frac{z^2}{z^{\odot}} = \frac{z^2}{\frac{z^2}{1+zh}}$$

$$= 1 + zh.$$

This completes the proof.

Theorem 1.37. *For $z \in \mathbb{C}_h$ we have*

$$z + (\ominus z) = z^{\odot} h.$$

Proof. We have

$$z^{\odot} h = \frac{z^2}{1+zh} h,$$

$$z + (\ominus z) = z - \frac{z}{1 + zh}$$

$$= \frac{z^2 h}{1 + zh},$$

which completes the proof.

Theorem 1.38. *For $z \in \mathbb{C}_h$ we have*

$$z \oplus z^{\circ} = z + z^2.$$

Proof. We have

$$z \oplus z^{\circ} = z + z^{\circ} + zz^{\circ}h$$

$$= z + \frac{z^2}{1 + zh} + \frac{z^3 h}{1 + zh}$$

$$= z + \frac{z^2(1 + zh)}{1 + zh}$$

$$= z + z^2.$$

This completes the proof.

Theorem 1.39. *Let* $-\dfrac{\pi}{h} < w \le \dfrac{\pi}{h}$. *Then*

$$-(\overset{\circ}{i})^{\circ} = \frac{4}{h^2} \sin^2\left(\frac{wh}{2}\right).$$

Proof. We have

$$-(\overset{\circ}{i})^{\circ} = -(\overset{\circ}{i})(\ominus \overset{\circ}{i} w)$$

$$= (\overset{\circ}{i} w)\overline{\overset{\circ}{i} w}$$

$$= |\overset{\circ}{i} w|^2$$

$$= \frac{4}{h^2} \sin^2\left(\frac{wh}{2}\right).$$

This completes the proof.

Exercise 1.16. Let $z \in \mathbb{C}_h$. Prove that

$$z^{\circ} \in \mathbb{R} \quad \text{iff} \quad z \in \mathbb{R}_h \cup \mathbb{A}_h \cup \mathbb{I}_h.$$

1.5.2 Definition and Properties of the Exponential Function

For $h > 0$, we define the strip

$$\mathbb{Z}_h := \{z \in \mathbb{C} : -\frac{\pi}{h} < \text{Im}(z) \le \frac{\pi}{h}\}.$$

For $h = 0$, we set $\mathbb{Z}_0 := \mathbb{C}$.

Definition 1.26. For $h > 0$, we define the cylindrical transformation $\xi_h : \mathbb{C}_h \to \mathbb{Z}_h$ by

$$\xi_h(z) := \frac{1}{h}\text{Log}(1 + zh),$$

where Log is the principal branch of the logarithm function. For $h = 0$, we define $\xi_0(z) = z$ for all $z \in \mathbb{C}$.

We note that

$$\xi_h^{-1}(z) = \frac{e^{zh} - 1}{h}$$

for $z \in \mathbb{Z}_h$.

Definition 1.27. We say that a function $f : \mathbb{T} \to \mathbb{R}$ is regressive if

$$1 + \mu(t)p(t) \ne 0 \quad \text{for} \quad \text{all} \quad t \in \mathbb{T}^\kappa$$

holds. The set of all regressive and rd-continuous functions $f : \mathbb{T} \to \mathbb{R}$ will be denoted by $\mathcal{R}(\mathbb{T})$ or $\mathcal{R}$.

In $\mathcal{R}$ we define "circle plus" addition as follows:

$$(f \oplus g)(t) = f(t) + g(t) + \mu(t)f(t)g(t).$$

Exercise 1.17. Prove that $(\mathcal{R}, \oplus)$ is an Abelian group.

Definition 1.28. The group $(\mathcal{R}, \oplus)$ will be called the regressive group.

Definition 1.29. For $f \in \mathcal{R}$ we define

$$(\ominus f)(t) = -\frac{f(t)}{1 + \mu(t)f(t)} \quad \text{for} \quad \text{all} \quad t \in \mathbb{T}^\kappa.$$

Exercise 1.18. Let $f \in \mathcal{R}$. Prove that $(\ominus f)(t) \in \mathcal{R}$ for all $t \in \mathbb{T}^\kappa$.

Definition 1.30. We define "circle minus" subtraction $\ominus$ on $\mathscr{R}$ as follows:

$$(f \ominus g)(t) = (f \oplus (\ominus g))(t) \quad \text{for} \quad \text{all} \quad t \in \mathbb{T}^\kappa.$$

For $f, g \in \mathscr{R}$, we have

$$f \ominus g = f \oplus (\ominus g)$$

$$= f \oplus \left(-\frac{g}{1+\mu g}\right)$$

$$= f - \frac{g}{1+\mu g} - \frac{\mu f g}{1+\mu g}$$

$$= \frac{f - g}{1+\mu g}.$$

Theorem 1.40. *Let* $f, g \in \mathscr{R}$. *Then*

1. $f \ominus f = 0,$
2. $\ominus(\ominus f) = f,$
3. $f \ominus g \in \mathscr{R},$
4. $\ominus(f \ominus g) = g \ominus f,$
5. $\ominus(f \oplus g) = (\ominus f) \oplus (\ominus g),$
6. $f \oplus \dfrac{g}{1+\mu f} = f + g.$

Proof. 1.

$$f \ominus f = f \oplus (\ominus f)$$

$$= f \oplus \left(-\frac{f}{1+\mu f}\right)$$

$$= f - \frac{f}{1+\mu f} - \frac{f^2 \mu}{1+\mu f}$$

$$= \frac{f + \mu f^2 - f - \mu f^2}{1+\mu f}$$

$$= 0.$$

2.

$$\ominus(\ominus f) = \ominus\left(-\frac{f}{1+\mu f}\right)$$

$$= \frac{\frac{f}{1+\mu f}}{1 - \frac{\mu f}{1+\mu f}}$$

$$= f.$$

3.

$$1 + \mu f \ominus g = 1 + \frac{\mu f - \mu g}{1 + \mu g}$$

$$= \frac{1 + \mu f}{1 + \mu g} \neq 0.$$

We note that $\dfrac{f - g}{1 + \mu g}$ is rd-continuous. Therefore, $f \ominus g \in \mathscr{R}$.

4.

$$\ominus(f \ominus g) = \ominus \left(\frac{f - g}{1 + \mu g} \right)$$

$$= -\frac{\frac{f-g}{1+\mu g}}{1 + \mu \frac{f-g}{1+\mu g}}$$

$$= -\frac{f - g}{1 + \mu f}$$

$$= \frac{g - f}{1 + \mu f}$$

$$= g \ominus f.$$

5.

$$\ominus(f \oplus g) = \ominus(f + g + \mu f g)$$

$$= -\frac{f + g + \mu f g}{1 + \mu f + \mu g + \mu^2 f g}$$

$$= -\frac{f + g + \mu f g}{(1 + \mu f)(1 + \mu g)},$$

$$\ominus f = -\frac{f}{1 + \mu f},$$

$$\ominus g = -\frac{g}{1 + \mu g},$$

$$(\ominus f) \oplus (\ominus g) = \ominus f + (\ominus g) + \mu(\ominus f)(\ominus g)$$

$$= -\frac{f}{1 + \mu f} - \frac{g}{1 + \mu g} + \frac{\mu f g}{(1 + \mu f)(1 + \mu g)}$$

$$= \frac{-f(1 + \mu g) - g(1 + \mu f) + \mu f g}{(1 + \mu f)(1 + \mu g)}$$

$$= -\frac{f + g + \mu f g}{(1 + \mu f)(1 + \mu g)}.$$

6.

$$f \oplus \frac{g}{1 + \mu f} = f + \frac{g}{1 + \mu f} + \frac{\mu f g}{1 + \mu f}$$
$$= f + g.$$

This completes the proof.

Definition 1.31. If $f \in \mathcal{R}$, then we define the generalized exponential function by

$$e_f(t, s) = e^{\int_s^t \xi_{\mu(\tau)}(f(\tau))\Delta\tau} \quad \text{for} \quad s, t \in \mathbb{T}.$$

In fact, using the definition for the cylindrical transformation, we have

$$e_f(t, s) = e^{\int_s^t \frac{1}{\mu(\tau)}\mathrm{Log}(1+\mu(\tau)f(\tau))\Delta\tau} \quad \text{for} \quad s, t \in \mathbb{T}.$$

Theorem 1.41 (*Semigroup Property*). *If $f \in \mathcal{R}$, then*

$$e_f(t, r)e_f(r, s) = e_f(t, s) \quad \text{for} \quad \text{all} \quad t, r, s \in \mathcal{R}.$$

Proof. We have

$$e_f(t, r)e_f(r, s) = e^{\int_r^t \xi_{\mu(\tau)}(f(\tau))\Delta\tau} e^{\int_s^r \xi_{\mu(\tau)}(f(\tau))\Delta\tau}$$
$$= e^{\int_r^t \xi_{\mu(\tau)}(f(\tau))\Delta\tau + \int_s^r \xi_{\mu(\tau)}(f(\tau))\Delta\tau}$$
$$= e^{\int_s^t \xi_{\mu(\tau)}(f(\tau))\Delta\tau}$$
$$= e_f(t, s).$$

This completes the proof.

Exercise 1.19. Let $f \in \mathcal{R}$. Prove that

$$e_0(t, s) = 1 \quad \text{and} \quad e_f(t, t) = 1.$$

Theorem 1.42. *Let $f \in \mathcal{R}$ and fix $t_0 \in \mathbb{T}$. Then*

$$e_f^{\Delta}(t, t_0) = f(t)e_f(t, t_0).$$

Proof. 1. Let $\sigma(t) > t$. Then

$$e_f^{\Delta}(t, t_0) = \frac{e_f(\sigma(t), t_0) - e_f(t, t_0)}{\mu(t)}$$
$$= \frac{e^{\int_{t_0}^{\sigma(t)} \xi_{\mu(\tau)}(f(\tau))\Delta\tau} - e^{\int_{t_0}^{t} \xi_{\mu(\tau)}(f(\tau))\Delta\tau}}{\mu(t)}$$

$$= \frac{e^{\int_{t_0}^t \xi_{\mu(\tau)}(f(\tau))\Delta\tau + \int_t^{\sigma(t)} \xi_{\mu(\tau)}(f(\tau))\Delta\tau} - e^{\int_{t_0}^t \xi_{\mu(\tau)}(f(\tau))\Delta\tau}}{\mu(t)}$$

$$= \frac{e^{\int_t^{\sigma(t)} \xi_{\mu(\tau)}(f(\tau))\Delta\tau} - 1}{\mu(t)} e^{\int_{t_0}^t \xi_{\mu(\tau)}(f(\tau))\Delta\tau}$$

$$= \frac{e^{\mu(t)\xi_{\mu(t)}(f(t))} - 1}{\mu(t)} e_f(t, t_0)$$

$$= f(t)e_f(t, t_0).$$

2. Let $\sigma(t) = t$. Then

$$|e_f(t, t_0) - e_f(s, t_0) - f(t)e_f(t, t_0)(t - s)|$$
$$= |e_f(t, t_0) - e_f(t, t_0)e_f(s, t) - f(t)e_f(t, t_0)(t - s)|$$
$$= |e_f(t, t_0)||1 - e_f(s, t) - f(t)(t - s)|$$
$$= |e_f(t, t_0)|\left|1 - \int_s^t \xi_{\mu(\tau)}(f(\tau))\Delta\tau - e_f(s, t) + \int_s^t \xi_{\mu(\tau)}(f(\tau))\Delta\tau - f(t)(t - s)\right|$$
$$\leq |e_f(t, t_0)|\left(\left|1 - \int_s^t \xi_{\mu(\tau)}(f(\tau))\Delta\tau - e_f(s, t)\right|\right.$$
$$+ \left.\left|\int_s^t \xi_{\mu(\tau)}(f(\tau))\Delta\tau - f(t)(t - s)\right|\right)$$
$$\leq |e_f(t, t_0)|\left(\left|1 - \int_s^t \xi_{\mu(\tau)}(f(\tau))\Delta\tau - e_f(s, t)\right|\right.$$
$$+ \left.\left|\int_s^t (\xi_{\mu(\tau)}(f(\tau)) - \xi_0(f(t)))\Delta\tau\right|\right). \qquad (1.15)$$

Since $\sigma(t) = t$ and $f \in \mathbb{C}_{rd}$, it follows that

$$\lim_{r \to t} \xi_{\mu(r)}(f(r)) = \xi_0(f(t)).$$

Therefore, there is a neighborhood U_1 of t such that

$$|\xi_{\mu(\tau)}(f(\tau)) - \xi_0(f(t))| < \frac{\epsilon}{3|e_f(t, t_0)|} \quad \text{for all} \quad \tau \in U_1.$$

Let $s \in U_1$. Then

$$|e_f(t, t_0)|\left|\int_s^t (\xi_{\mu(\tau)}(f(\tau)) - \xi_0(f(t)))\Delta\tau\right| \leq \frac{\epsilon}{3}|t - s|. \qquad (1.16)$$

Also, using that

$$\lim_{z \to 0} \frac{1 - z - e^{-z}}{z} = 0,$$

we conclude that there is a neighborhood U_2 of t such that if $s \in U_2$, $s \leq t$, we have

$$\left| \frac{1 - \int_s^t \xi_{\mu(\tau)}(f(\tau)) \Delta\tau - e_f(s,t)}{\int_s^t \xi_{\mu(\tau)}(f(\tau)) \Delta\tau} \right| < \epsilon^*,$$

where

$$\epsilon^* = \min\left\{ 1, \frac{\epsilon}{1 + 3|f(t)||e_f(t,t_0)|} \right\}.$$

Let $s \in U := U_1 \cap U_2$. Then

$$|e_f(t,t_0)| \left| 1 - \int_s^t \xi_{\mu(\tau)}(f(\tau)) \Delta\tau - e_f(s,t) \right|$$

$$= |e_f(t,t_0)| \frac{\left| 1 - \int_s^t \xi_{\mu(\tau)}(f(\tau)) \Delta\tau - e_f(s,t) \right|}{\left| \int_s^t \xi_{\mu(\tau)}(f(\tau)) \Delta\tau \right|} \left| \int_s^t \xi_{\mu(\tau)}(f(\tau)) \Delta\tau \right|$$

$$\leq |e_f(t,t_0)| \epsilon^* \left| \int_s^t \xi_{\mu(\tau)}(f(\tau)) \Delta\tau \right|$$

$$\leq |e_f(t,t_0)| \epsilon^* \left\{ \left| \int_s^t (\xi_{\mu(\tau)}(f(\tau)) - \xi_0(f(t))) \Delta\tau \right| + |f(t)||t - s| \right\}$$

$$\leq |e_f(t,t_0)| \left| \int_s^t (\xi_{\mu(\tau)}(f(\tau)) - \xi_0(f(t))) \Delta\tau \right| + |e_f(t,t_0)| \epsilon^* |f(t)||t - s|$$

$$\leq \tfrac{\epsilon}{3}|t - s| + \tfrac{\epsilon}{3}|t - s|$$

$$= \tfrac{2\epsilon}{3}|t - s|.$$

From the last inequality and from (1.15) and (1.16), we conclude that

$$|e_f(t,t_0) - e_f(s,t_0) - f(t)e_f(t,t_0)(t - s)| \leq \frac{\epsilon}{3}|t - s| + \frac{\epsilon}{3}|t - s| + \frac{\epsilon}{3}|t - s|$$

$$= \epsilon|t - s|,$$

which completes the proof.

Corollary 1.3. *Let $f \in \mathscr{R}$ and fix $t_0 \in \mathbb{T}$. Then $e_f(t,t_0)$ is a solution to the Cauchy problem*

$$y^\Delta(t) = f(t)y(t), \quad y(t_0) = 1. \tag{1.17}$$

Corollary 1.4. *Let $f \in \mathscr{R}$ and fix $t_0 \in \mathbb{T}$. Then $e_f(t,t_0)$ is the unique solution of problem (1.17).*

Proof. Let $y(t)$ be any solution of problem (1.17). Then

$$\left(\frac{y(t)}{e_f(t, t_0)}\right)^{\Delta} = \frac{y^{\Delta}(t)e_f(t, t_0) - y(t)e_f^{\Delta}(t, t_0)}{e_f(\sigma(t), t_0)e_f(t, t_0)}$$

$$= \frac{f(t)y(t)e_f(t, t_0) - y(t)f(t)e_f(t, t_0)}{e_f(\sigma(t), t_0)e_f(t, t_0)}$$

$$= 0.$$

Consequently, $y(t) = ce_f(t, t_0)$, where c is a constant. From this and

$$1 = y(t_0) = ce_f(t_0, t_0) = c,$$

we conclude that $y(t) = e_f(t, t_0)$. This completes the proof.

Theorem 1.43. *Let* $f \in \mathscr{R}$. *Then*

$$e_f(\sigma(t), s) = (1 + \mu(t)f(t))e_f(t, s).$$

Proof. We have

$$e_f(\sigma(t), s) - e_f(t, s) = \mu(t)e_f^{\Delta}(t, s)$$
$$= \mu(t)f(t)e_f(t, s),$$

which completes the proof.

Theorem 1.44. *Let* $f \in \mathscr{R}$. *Then*

$$e_f(t, s) = \frac{1}{e_f(s, t)} = e_{\ominus f}(s, t).$$

Proof. We have

$$e_f(t, s) = e^{\int_s^t \xi_{\mu(\tau)}(f(\tau))\Delta\tau}$$

$$= e^{-\int_t^s \xi_{\mu(\tau)}(f(\tau))\Delta\tau}$$

$$= \frac{1}{e^{\int_t^s \xi_{\mu(\tau)}(f(\tau))\Delta\tau}}$$

$$= \frac{1}{e_f(s, t)}.$$

Now we fix $t_0 \in \mathbb{T}$ and consider the problem

$$y^{\Delta}(t) = \ominus f(t)y(t), \quad y(t_0) = 1.$$

Its solution is $e_{\ominus f}(t, s)$.

Also,

$$\left(\frac{1}{e_f(t,s)}\right)^{\Delta} = -\frac{e_f^{\Delta}(t,s)}{e_f(\sigma(t),s)e_f(t,s)}$$

$$= -\frac{f(t)e_f(t,s)}{(1+\mu(t)f(t))e_f(t,s)e_f(t,s)}$$

$$= -\frac{f(t)}{(1+\mu(t)f(t))e_f(t,s)}$$

$$= (\ominus f)(t)\frac{1}{e_f(t,s)}.$$

Therefore,

$$\frac{1}{e_f(t,s)} = e_{\ominus f}(t,s).$$

This completes the proof.

Theorem 1.45. *Let* $f, g \in \mathcal{R}$. *Then*

$$e_f(t,s)e_g(t,s) = e_{f\oplus g}(t,s).$$

Proof. We have

$$e_f(t,s)e_g(t,s) = e^{\int_s^t \xi_{\mu(\tau)}(f(\tau))\Delta\tau}e^{\int_s^t \xi_{\mu(\tau)}(g(\tau))\Delta\tau}$$

$$= e^{\int_s^t (\xi_{\mu(\tau)}(f(\tau))+\xi_{\mu(\tau)}(g(\tau)))\Delta\tau}$$

$$= e^{\int_s^t \frac{1}{\mu(\tau)}(\text{Log}(1+\mu(\tau)f(\tau))+\text{Log}(1+\mu(\tau)g(\tau)))\Delta\tau}$$

$$= e^{\int_s^t \frac{1}{\mu(\tau)}\text{Log}((1+\mu(\tau)f(\tau))(1+\mu(\tau)g(\tau)))\Delta\tau}$$

$$= e^{\int_s^t \frac{1}{\mu(\tau)}\text{Log}(1+\mu(\tau)(f(\tau)+g(\tau)+\mu(\tau)f(\tau)g(\tau)))\Delta\tau}$$

$$= e^{\int_s^t \xi_{\mu(\tau)}((f\oplus g)(\tau))\Delta\tau}$$

$$= e_{f\oplus g}(t,s).$$

This completes the proof.

Theorem 1.46. *Let* $f, g \in \mathcal{R}$. *Then*

$$\frac{e_f(t,s)}{e_g(t,s)} = e_{f\ominus g}(t,s).$$

Proof. We have

$$\frac{e_f(t,s)}{e_g(t,s)} = e_f(t,s)e_{\ominus g}(t,s)$$

$$= e_{f\oplus(\ominus g)}(t,s)$$

$$= e_{f\ominus g}(t,s).$$

This completes the proof.

Theorem 1.47. *Let $f \in \mathscr{R}$. Then*

$$e_f(t,\sigma(s))e_f(s,r) = \frac{1}{1+\mu(s)f(s)}e_f(t,r).$$

Proof. We have

$$e_f(t,\sigma(s))e_f(s,r) = e^{\int_{\sigma(s)}^{t}\xi_{\mu(\tau)}(f(\tau))\Delta\tau}e^{\int_{r}^{s}\xi_{\mu(\tau)}(f(\tau))\Delta\tau}$$

$$= e^{\int_{\sigma(s)}^{s}\xi_{\mu(\tau)}(f(\tau))\Delta\tau+\int_{s}^{t}\xi_{\mu(\tau)}(f(\tau))\Delta\tau+\int_{r}^{s}\xi_{\mu(\tau)}(f(\tau))\Delta\tau}$$

$$= e^{-\xi_{\mu(s)}(f(s))\mu(s)+\int_{r}^{t}\xi_{\mu(\tau)}(f(\tau))\Delta\tau}$$

$$= e^{-\mathrm{Log}(1+f(s)\mu(s))}e^{\int_{r}^{t}\xi_{\mu(\tau)}(f(\tau))\Delta\tau}$$

$$= \frac{1}{1+\mu(s)f(s)}e_f(t,r).$$

This completes the proof.

Theorem 1.48. *Let $f,g \in \mathscr{R}$. Then*

$$e_{f\ominus g}^{\Delta}(t,t_0) = \frac{(f(t)-g(t))e_f(t,t_0)}{e_g(\sigma(t),t_0)}.$$

Proof. We have

$$e_{f\ominus g}^{\Delta}(t,t_0) = \left(\frac{e_f(t,t_0)}{e_g(t,t_0)}\right)^{\Delta}$$

$$= \frac{e_f^{\Delta}(t,t_0)e_g(t,t_0)-e_f(t,t_0)e_g^{\Delta}(t,t_0)}{e_g(t,t_0)e_g(\sigma(t),t_0)}$$

$$= \frac{f(t)e_f(t,t_0)e_g(t,t_0)-g(t)e_f(t,t_0)e_g(t,t_0)}{e_g(t,t_0)e_g(\sigma(t),t_0)}$$

$$= \frac{(f(t)-g(t))e_f(t,t_0)}{e_g(\sigma(t),t_0)}.$$

This completes the proof.

Theorem 1.49. *Let $f \in \mathscr{R}$ and $a, b, c \in \mathbb{T}$. Then*

$$(e_f(c, t))^{\Delta} = -f(t)(e_f(c, t))^{\sigma} = -f(t)e_f(c, \sigma(t))$$

and

$$\int_a^b f(t)e_f(c, \sigma(t))\Delta t = e_f(c, a) - e_f(c, b).$$

Proof. We have

$$\begin{aligned}
f(t)e_f(c, \sigma(t)) &= f(t)e_{\ominus f}(\sigma(t), c)\\
&= f(t)(1 + \mu(t) \ominus f(t))e_{\ominus f}(t, c)\\
&= f(t)\left(1 - \frac{\mu(t)f(t)}{1 + \mu(t)f(t)}\right)e_{\ominus f}(t, c)\\
&= \frac{f(t)}{1 + \mu(t)f(t)}e_{\ominus f}(t, c)\\
&= -(\ominus f)(t)e_{\ominus f}(t, c)\\
&= -e_{\ominus f}^{\Delta}(t, c)\\
&= -e_f^{\Delta}(c, t).
\end{aligned}$$

Hence,

$$\begin{aligned}
\int_a^b f(t)e_f(c, \sigma(t))\Delta t &= -\int_a^b e_f^{\Delta}(c, t)\Delta t\\
&= -e_f(c, t)\Big|_{t=a}^{t=b}\\
&= e_f(c, a) - e_f(c, b).
\end{aligned}$$

This completes the proof.

Exercise 1.20. Assume $1 + \mu(t)\dfrac{2}{t} \neq 0$, $1 + \mu(t)\dfrac{5}{t} \neq 0$ for all $t \in \mathbb{T} \cap (0, \infty)$. Let also $t_0 \in \mathbb{T} \cap (0, \infty)$. Evaluate the integral

$$I = \int_{t_0}^t \frac{e_{\frac{5}{s}}(s, t_0)}{se_{\frac{2}{s}}^{\sigma}(s, t_0)}\Delta s.$$

Solution. We have

$$\left(\frac{5}{s} - \frac{2}{s}\right) \frac{e_{\frac{5}{s}}(s, t_0)}{e_{\frac{2}{s}}^{\sigma}(s, t_0)} = \frac{3}{s} \frac{e_{\frac{5}{s}}(s, t_0)}{e_{\frac{2}{s}}^{\sigma}(s, t_0)}$$

$$= e_{\frac{5}{s} \ominus \frac{2}{s}}^{\Delta}(s, t_0)$$

$$= e_{\frac{\frac{3}{s}}{1+\mu(s)\frac{2}{s}}}^{\Delta}(s, t_0)$$

$$= e_{\frac{3}{s+2\mu(s)}}^{\Delta}(s, t_0).$$

Hence,

$$I = \frac{1}{3} \int_{t_0}^{t} e_{\frac{3}{s+2\mu(s)}}^{\Delta}(s, t_0) \Delta s$$

$$= \frac{1}{3} e_{\frac{3}{s+2\mu(s)}}(s, t_0) \Big|_{s=t_0}^{s=t}$$

$$= \frac{1}{3} e_{\frac{3}{t+2\mu(t)}}(t, t_0) - \frac{1}{3}.$$

Exercise 1.21. Let $\alpha \in \mathbb{R}$. Let also the exponentials

$$e_{\frac{\alpha^2}{t} - \frac{(\alpha-1)^2}{\sigma(t)}}(t, t_0) \quad \text{and} \quad e_{\frac{\alpha}{t}}(t, t_0)$$

exist for all $t \in \mathbb{T} \cap (0, \infty)$. Prove that

1.

$$\frac{e_{\frac{\alpha^2}{t} - \frac{(\alpha-1)^2}{\sigma(t)}}(t, t_0)}{e_{\frac{\alpha}{t}}(t, t_0)} = e_{\frac{\alpha-1}{\sigma(t)}}(t, t_0),$$

2.

$$\frac{e_{\frac{\alpha-1}{\sigma(t)}}(t, t_0)}{e_{\frac{\alpha}{t}}(t, t_0)} = e_{-\frac{1}{\sigma(t)}}(t, t_0) = \frac{t_0}{t}$$

for all $t, t_0 \in \mathbb{T} \cap (0, \infty)$.

Solution.

1. We have

$$\left(\frac{\alpha^2}{t} - \frac{(\alpha-1)^2}{\sigma(t)}\right) \ominus \frac{\alpha}{t} = \left(\frac{\alpha^2}{t} - \frac{(\alpha-1)^2}{\sigma(t)}\right) \oplus \left(\ominus\frac{\alpha}{t}\right)$$

$$= \frac{\alpha^2}{t} - \frac{(\alpha-1)^2}{\sigma(t)} + \left(\ominus\frac{\alpha}{t}\right) + \left(\frac{\alpha^2}{t} - \frac{(\alpha-1)^2}{\sigma(t)}\right)\left(\ominus\frac{\alpha}{t}\right)\mu(t)$$

$$= \frac{\alpha^2}{t} - \frac{(\alpha-1)^2}{\sigma(t)} - \frac{\frac{\alpha}{t}}{1+\frac{\alpha}{t}\mu(t)}$$

$$+ \left(\frac{\alpha^2}{t} - \frac{(\alpha-1)^2}{\sigma(t)}\right)\left(-\frac{\frac{\alpha}{t}}{1+\frac{\alpha}{t}\mu(t)}\right)\mu(t)$$

$$= \frac{\alpha^2}{t} - \frac{(\alpha-1)^2}{\sigma(t)} - \frac{\alpha}{t+\alpha\mu(t)} - \left(\frac{\alpha^2}{t} - \frac{(\alpha-1)^2}{\sigma(t)}\right)\frac{\alpha}{t+\alpha\mu(t)}\mu(t)$$

$$= \left(\frac{\alpha^2}{t} - \frac{(\alpha-1)^2}{\sigma(t)}\right)\left(1 - \frac{\alpha\mu(t)}{t+\alpha\mu(t)}\right) - \frac{\alpha}{t+\alpha\mu(t)}$$

$$= \left(\frac{\alpha^2}{t} - \frac{(\alpha-1)^2}{\sigma(t)}\right)\frac{t}{t+\alpha\mu(t)} - \frac{\alpha}{t+\alpha\mu(t)}$$

$$= \frac{\alpha^2\sigma(t)-t(\alpha-1)^2}{t\sigma(t)}\frac{t}{t+\alpha\mu(t)} - \frac{\alpha}{t+\alpha\mu(t)}$$

$$= \frac{\alpha^2(t+\mu(t))-t\alpha^2+2\alpha t-t}{\sigma(t)}\frac{1}{t+\alpha\mu(t)} - \frac{\alpha}{t+\alpha\mu(t)}$$

$$= \frac{\alpha^2\mu(t)+2\alpha t-t}{\sigma(t)(t+\alpha\mu(t))} - \frac{\alpha}{t+\alpha\mu(t)}$$

$$= \frac{\alpha^2\mu(t)+2\alpha t-t-\alpha\mu(t)-\alpha t}{\sigma(t)(t+\alpha\mu(t))}$$

$$= \frac{\alpha\mu(t)(\alpha-1)+(\alpha-1)t}{\sigma(t)(t+\alpha\mu(t))}$$

$$= \frac{(\alpha-1)(t+\alpha\mu(t))}{\sigma(t)(t+\alpha\mu(t))}$$

$$= \frac{\alpha-1}{\sigma(t)}.$$

Hence,

$$\frac{e_{\frac{\alpha^2}{t}-\frac{(\alpha-1)^2}{\sigma(t)}}(t,t_0)}{e_{\frac{\alpha}{t}}(t,t_0)} = e_{\left(\frac{\alpha^2}{t}-\frac{(\alpha-1)^2}{\sigma(t)}\right)\ominus\frac{\alpha}{t}}(t,t_0)$$

$$= e_{\frac{\alpha-1}{t}}(t,t_0).$$

2. We have

$$\frac{\alpha-1}{\sigma(t)} \ominus \frac{\alpha}{t} = \frac{\alpha-1}{\sigma(t)} \oplus \left(\ominus\frac{\alpha}{t}\right)$$

$$= \frac{\alpha-1}{\sigma(t)} + \left(\ominus\frac{\alpha}{t}\right) + \frac{\alpha-1}{\sigma(t)}\left(\ominus\frac{\alpha}{t}\right)\mu(t)$$

$$= \frac{\alpha-1}{\sigma(t)} - \frac{\frac{\alpha}{t}}{1+\mu(t)\frac{\alpha}{t}} - \frac{\alpha-1}{\sigma(t)}\frac{\frac{\alpha}{t}}{1+\mu(t)\frac{\alpha}{t}}\mu(t)$$

$$= \frac{\alpha-1}{\sigma(t)} - \frac{\alpha}{t+\alpha\mu(t)} - \frac{\alpha-1}{\sigma(t)}\frac{\alpha}{t+\alpha\mu(t)}\mu(t)$$

$$= \frac{\alpha-1}{\sigma(t)}\left(1 - \frac{\alpha\mu(t)}{t+\alpha\mu(t)}\right) - \frac{\alpha}{t+\alpha\mu(t)}$$

$$= \frac{t(\alpha-1)}{\sigma(t)(t+\alpha\mu(t))} - \frac{\alpha}{t+\alpha\mu(t)}$$

$$= \frac{\alpha t - t - \alpha\mu(t) - \alpha t}{\sigma(t)(t+\alpha\mu(t))}$$

$$= -\frac{t+\alpha\mu(t)}{\sigma(t)(t+\alpha\mu(t))}$$

$$= -\frac{1}{\sigma(t)}.$$

Therefore,

$$\frac{e_{\frac{\alpha-1}{\sigma(t)}}(t,t_0)}{e_{\frac{\alpha}{t}}(t,t_0)} = e_{\frac{\alpha-1}{\sigma(t)}\ominus\frac{\alpha}{t}}(t,t_0)$$

$$= e_{-\frac{1}{\sigma(t)}}(t,t_0)$$

$$= e^{\int_{t_0}^t \xi_{\mu(\tau)}\left(-\frac{1}{\sigma(\tau)}\right)\Delta\tau}$$

$$= e^{\int_{t_0}^t \frac{1}{\mu(\tau)}\mathrm{Log}\left(1-\frac{\mu(\tau)}{\sigma(\tau)}\right)\Delta\tau}$$

$$= e^{\int_{t_0}^t \frac{1}{\mu(\tau)}\mathrm{Log}\frac{\tau}{\sigma(\tau)}\Delta\tau}$$

$$= e^{-\mathrm{Log}\tau\Big|_{\tau=t_0}^{\tau=t}}$$

$$= e^{-\mathrm{Log}\frac{t}{t_0}}$$

$$= \frac{t_0}{t}.$$

1.5.3   Examples for Exponential Functions

Let $\alpha : \mathbb{T} \to \mathbb{R}$ be regulated and suppose that $1 + \alpha(t)\mu(t) \neq 0$ for all $t \in \mathbb{T}$. Let also $t_0, t \in \mathbb{T}$, $t_0 < t$.

1. $\mathbb{T} = h\mathbb{Z}$, $\quad h > 0$.   Every point in $\mathbb{T}$ is isolated and $\mu(t) = h$ for every $t \in \mathbb{T}$. Then

$$e_\alpha(t,t_0) = e^{\int_{t_0}^t \frac{1}{\mu(\tau)}\mathrm{Log}(1+\alpha(\tau)\mu(\tau))\Delta\tau}$$

$$= e^{\sum\limits_{s\in[t_0,t)} \frac{1}{\mu(s)}\mathrm{Log}(1+\alpha(s)\mu(s))\mu(s)}$$

$$= e^{\sum\limits_{s\in[t_0,t)} \mathrm{Log}(1+h\alpha(s))}$$

$$= \prod_{s\in[t_0,t)}(1 + h\alpha(s)).$$

If α is a constant, then

$$e_\alpha(t, t_0) = \prod_{s\in[t_0,t)} (1 + h\alpha)$$

$$= (1 + h\alpha)^{t-t_0}.$$

2. $\mathbb{T} = q^{\mathbb{N}_0}$, $q > 1$. Every point of $\mathbb{T}$ is isolated and $\mu(t) = (q - 1)t$ for all $t \in \mathbb{T}$. Then

$$e_\alpha(t, t_0) = e^{\int_{t_0}^t \frac{1}{\mu(\tau)}\text{Log}(1+\alpha(\tau)\mu(\tau))\Delta\tau}$$

$$= e^{\sum_{s\in[t_0,t)} \frac{1}{\mu(s)}\text{Log}(1+\alpha(s)\mu(s))\mu(s)}$$

$$= e^{\sum_{s\in[t_0,t)} \text{Log}(1+\alpha(s)\mu(s))}$$

$$= e^{\sum_{s\in[t_0,t)} \text{Log}(1+(q-1)s\alpha(s))}$$

$$= \prod_{s\in[t_0,t)} (1 + (q - 1)s\alpha(s)).$$

Exercise 1.22. Let $\mathbb{T} = q^{\mathbb{N}_0}$, $0 < q < 1$. Prove that

$$e_\alpha(t, t_0) = \prod_{s\in[t_0,t)} \left(1 + \frac{1-q}{q}\alpha(s)s\right).$$

3. $\mathbb{T} = \mathbb{N}_0^k$, $k \in \mathbb{N}$. Every point of $\mathbb{T}$ is isolated and

$$\mu(t) = \left(\sqrt[k]{t} + 1\right)^{k-t}.$$

Then

$$e_\alpha(t, t_0) = e^{\int_{t_0}^t \frac{1}{\mu(\tau)}\text{Log}(1+\alpha(\tau)\mu(\tau))\Delta\tau}$$

$$= e^{\sum_{s\in[t_0,t)} \frac{1}{\mu(s)}\text{Log}(1+\alpha(s)\mu(s))\mu(s)}$$

$$= e^{\sum_{s\in[t_0,t)} \text{Log}(1+\alpha(s)\mu(s))}$$

$$= \prod_{s\in[t_0,t)} (1 + \alpha(s)\mu(s))$$

$$= \prod_{s\in[t_0,t)} \left(1 + \left(\left(\sqrt[k]{s} + 1\right)^k - s\right)\alpha(s)\right).$$

1.6 Hyperbolic and Trigonometric Functions

Definition 1.32 (Hyperbolic Functions). If $f \in \mathscr{C}_{rd}$ and $-\mu f^2 \in \mathscr{R}$, then we define the hyperbolic functions $\cosh_f$ and $\sinh_f$ by

$$\cosh_f = \frac{e_f + e_{-f}}{2} \quad \text{and} \quad \sinh_f = \frac{e_f - e_{-f}}{2}.$$

Theorem 1.50. *Let $f \in \mathscr{C}_{rd}$. If $-\mu f^2 \in \mathscr{R}$, then we have*

1. $\cosh_f^{\Delta} = f \sinh_f$,
2. $\sinh_f^{\Delta} = f \cosh_f$,
3. $\cosh_f^2 - \sinh_f^2 = e_{-\mu f^2}$.

Proof. 1. We have

$$\cosh_f^{\Delta} = \left(\frac{e_f + e_{-f}}{2}\right)^{\Delta}$$
$$= \frac{f e_f - f e_{-f}}{2}$$
$$= f \sinh_f .$$

2. We have

$$\sinh_f^{\Delta} = \left(\frac{e_f - e_{-f}}{2}\right)^{\Delta}$$
$$= \frac{f e_f + f e_{-f}}{2}$$
$$= f \cosh_f .$$

3. We have

$$\cosh_f^2 - \sinh_f^2 = \left(\frac{e_f + e_{-f}}{2}\right)^2 - \left(\frac{e_f - e_{-f}}{2}\right)^2$$
$$= \frac{e_f^2 + 2 e_f e_{-f} + e_{-f}^2}{4} - \frac{e_f^2 - 2 e_f e_{-f} + e_{-f}^2}{4}$$
$$= e_f e_{-f}$$
$$= e_{f \oplus (-f)}$$
$$= e_{-\mu f^2} .$$

This completes the proof.

Definition 1.33 (Trigonometric Functions). If $f \in \mathscr{C}_{rd}$ and $\mu f^2 \in \mathscr{R}$, then we define the trigonometric functions $\cos_f$ and $\sin_f$ by

$$\cos_f = \frac{e_{if} + e_{-if}}{2} \quad \text{and} \quad \sin_f = \frac{e_{if} - e_{-if}}{2i}.$$

Theorem 1.51. *Let $f \in \mathscr{C}_{rd}$ and $-\mu f^2 \in \mathscr{R}$. Prove that*

1. $\cos_f^\Delta = -f \sin_f$.
2. $\sin_f^\Delta = f \cos_f$.
3. $\cos_f^2 + \sin_f^2 = e_{\mu f^2}$.

Proof. 1. We have

$$\cos_f^\Delta = \left(\frac{e_{if} + e_{-if}}{2} \right)^\Delta$$
$$= \frac{ife_{if} - ife_{-if}}{2}$$
$$= -f \frac{e_{if} - e_{-if}}{2i}$$
$$= -f \sin_f^\Delta.$$

2. We have

$$\sin_f^\Delta = \left(\frac{e_{if} - e_{-if}}{2i} \right)^\Delta$$
$$= \frac{ife_{if} + ife_{-if}}{2i}$$
$$= f \cos_f^\Delta.$$

3. We have

$$\cos_f^2 + \sin_f^2 = \left(\frac{e_{if} + e_{-if}}{2} \right)^2 + \left(\frac{e_{if} - e_{-if}}{2i} \right)^2$$
$$= \frac{e_{if}^2 + 2e_{if}e_{-if} + e_{-if}^2}{4} - \frac{e_{if}^2 - 2e_{if}e_{-if} + e_{-if}^2}{4}$$
$$= e_{if}e_{-if}$$
$$= e_{if \oplus (-if)}$$
$$= e_{\mu f^2}.$$

This completes the proof.

Exercise 1.23. Let $f \in \mathscr{C}_{rd}$ and $\mu f^2 \in \mathscr{R}$. Prove Euler's formula

$$e_{if} = \cos_f + i \sin_f .$$

1.7   Dynamic Equations

Theorem 1.52. *Let* $p : \mathbb{T} \to \mathbb{R}$ *be rd-continuous and suppose that* $1 + \mu(t)p(t)$ $\neq 0$ *for all* $t \in \mathbb{T}$. *Let also* $t_0 \in \mathbb{T}$ *and* $\phi_0 \in \mathbb{C}$. *Then the unique solution to the initial value problem*

$$\phi^{\Delta} = p(t)\phi, \quad \phi(t_0) = \phi_0, \tag{1.18}$$

is given by

$$\phi(t) = \phi_0 e_p(t, t_0) \quad \text{on} \quad \mathbb{T}. \tag{1.19}$$

Proof. We have

$$\begin{aligned}
\phi^{\Delta}(t) &= \left(\phi_0 e_p(t, t_0)\right)^{\Delta} \\
&= \phi_0 e_p^{\Delta}(t, t_0) \\
&= \phi_0 p(t) e_p(t, t_0) \\
&= p(t)\phi, \quad t \in \mathbb{T}
\end{aligned}$$

and

$$\phi(t_0) = e_p(t_0, t_0) = 1.$$

Consequently, $\phi(t)$, defined by (1.19), satisfies (1.18).
Now we will prove that problem (1.18) has a unique solution. Let $\phi_1(t)$ and $\phi_2(t)$ be two solutions of (1.18). Then

$$\psi(t) = \phi_1(t) - \phi_2(t), \quad t \in \mathbb{T},$$

satisfies the problem

$$\psi^{\Delta} = p(t)\psi, \quad \psi(t_0) = 0.$$

Therefore, $\psi(t) \equiv 0$ on $\mathbb{T}$. This completes the proof.

Theorem 1.53 (*Variation of Constants*). *Let* $f, p : \mathbb{T} \to \mathbb{R}$ *be rd-continuous functions and suppose that* $1 + \mu(t)p(t) \neq 0$ *for all* $t \in \mathbb{T}$, $t_0 \in \mathbb{T}$, *and* $\phi_0 \in \mathbb{C}$. *Then the unique solution of the initial value problem*

$$\phi^{\Delta}(t) = -p(t)\phi(\sigma(t)) + f(t), \quad \phi(t_0) = \phi_0, \tag{1.20}$$

is given by

$$\phi(t) = e_{\ominus p}(t, t_0)\phi_0 + \int_{t_0}^{t} e_{\ominus p}(t, \tau)f(\tau)\Delta\tau. \tag{1.21}$$

Proof. We begin by showing that $\phi(t)$ given by (1.21) is a solution to (1.20). Indeed,

$$\phi^{\Delta}(t) = \left(e_{\ominus p}(t, t_0)\phi_0 + \int_{t_0}^{t} e_{\ominus p}(t, \tau)f(\tau)\Delta\tau\right)^{\Delta}$$

$$= \left(e_{\ominus p}(t, t_0)\phi_0\right)^{\Delta} + \left(\int_{t_0}^{t} e_{\ominus p}(t, \tau)f(\tau)\Delta\tau\right)^{\Delta}$$

$$= \phi_0 e_{\ominus p}^{\Delta}(t, t_0) + \int_{t_0}^{t} e_{\ominus p}^{\Delta}(t, \tau)f(\tau)\Delta\tau + e_{\ominus p}(\sigma(t), t)f(t)$$

$$= \phi_0 \ominus p(t) e_{\ominus p}(t, t_0) + \ominus p(t) \int_{t_0}^{t} e_{\ominus p}(t, \tau)f(\tau)\Delta\tau + e_{\ominus p}(\sigma(t), t)f(t)$$

$$= \phi_0 \ominus p(t) e_{\ominus p}(t, t_0) + \ominus p(t) \int_{t_0}^{t} e_{\ominus p}(t, \tau)f(\tau)\Delta\tau$$

$$\quad + (1 + \mu(t) \ominus p(t)) e_{\ominus p}(t, t)f(t)$$

$$= \phi_0 \ominus p(t) e_{\ominus p}(t, t_0) + \ominus p(t) \int_{t_0}^{t} e_{\ominus p}(t, \tau)f(\tau)\Delta\tau$$

$$\quad + \frac{1}{1 + \mu(t)p(t)}f(t)$$

$$= -\frac{p(t)}{1 + \mu(t)p(t)}\phi_0 e_{\ominus p}(t, t_0) - \frac{p(t)}{1 + \mu(t)p(t)} \int_{t_0}^{t} e_{\ominus p}(t, \tau)\Delta\tau$$

$$\quad + \frac{1}{1 + \mu(t)p(t)}f(t).$$

Multiplying both sides by $1 + \mu(t)p(t)$ gives

$$(1 + \mu(t)p(t))\phi^{\Delta}(t) = -\phi_0 p(t) e_{\ominus p}(t, t_0) - p(t) \int_{t_0}^{t} e_{\ominus p}(t, \tau)f(\tau)\Delta\tau + f(t)$$

$$= -p(t)\left(\phi_0 e_{\ominus p}(t, t_0) + \int_{t_0}^{t} e_{\ominus p}(t, \tau)f(\tau)\Delta\tau\right) + f(t)$$

$$= -p(t)\phi(t) + f(t).$$

Hence,

$$\phi^{\Delta}(t) = -\mu(t)p(t)\phi^{\Delta}(t) - p(t)\phi(t) + f(t)$$

$$= -p(t)\left(\mu(t)\phi^{\Delta}(t) + \phi(t)\right) + f(t)$$

$$= -p(t)\phi(\sigma(t)) + f(t).$$

Also,

$$\phi(t_0) = e_{\ominus p}(t_0, t_0)\phi_0 + \int_{t_0}^{t_0} e_{\ominus p}(t_0, \tau)f(\tau)\Delta\tau = \phi_0.$$

Consequently, $\phi(t)$ satisfies (1.20).

Now we proceed to prove the uniqueness of the solution. Suppose that $\phi(t)$ is a solution of (1.20). Then

$$f(t) = \phi^{\Delta}(t) + p(t)\phi(\sigma(t))$$

and

$$
\begin{aligned}
e_p(t, t_0)f(t) &= e_p(t, t_0)\left(\phi^{\Delta}(t) + p(t)\phi(\sigma(t))\right)\\
&= e_p(t, t_0)\phi^{\Delta}(t) + p(t)e_p(t, t_0)\phi(\sigma(t))\\
&= e_p(t, t_0)\phi^{\Delta}(t) + e_p^{\Delta}(t, t_0)\phi(\sigma(t))\\
&= \left(e_p(t, t_0)\phi(t)\right)^{\Delta},
\end{aligned}
$$

whereupon

$$
\begin{aligned}
\int_{t_0}^{t} e_p(\tau, t_0)f(\tau)\Delta\tau &= \int_{t_0}^{t}\left(e_p(\tau, t_0)\phi(\tau)\right)^{\Delta}\Delta\tau\\
&= e_p(\tau, t_0)\phi(\tau)\Big|_{\tau=t_0}^{\tau=t}\\
&= e_p(t, t_0)\phi(t) - \phi(t_0),
\end{aligned}
$$

i.e.,

$$e_p(t, t_0)\phi(t) = \phi_0 + \int_{t_0}^{t} e_p(\tau, t_0)f(\tau)\Delta\tau.$$

Solving for $\phi(t)$, we get

$$
\begin{aligned}
\phi(t) &= \phi_0 e_{\ominus p}(t, t_0) + \int_{t_0}^{t} e_{\ominus p}(t, t_0)e_p(\tau, t_0)f(\tau)\Delta\tau\\
&= \phi_0 e_{\ominus p}(t, t_0) + \int_{t_0}^{t} e_{\ominus p}(t, t_0)e_{\ominus p}(t_0, \tau)f(\tau)\Delta\tau\\
&= \phi_0 e_{\ominus p}(t, t_0) + \int_{t_0}^{t} e_{\ominus p}(t, \tau)f(\tau)\Delta\tau.
\end{aligned}
$$

Consequently, if $\phi_1(t)$ and $\phi_2(t)$ are solutions to the problem (1.20), we find that $\psi(t) = \phi_1(t) - \phi_2(t)$ satisfies the problem

$$\psi^\Delta(t) = -p(t)\psi(\sigma(t)), \quad \psi(t_0) = 0,$$

whereupon $\psi(t) \equiv 0$ on $\mathbb{T}$. This completes the proof.

Corollary 1.5 (*Variation of Constants*). *Let $f, p : \mathbb{T} \to \mathbb{R}$ be rd-continuous functions and $1 + \mu(t)p(t) \neq 0$ for all $t \in \mathbb{T}$, $t_0 \in \mathbb{T}$, and $\phi_0 \in \mathbb{R}$. Then the unique solution of the initial value problem*

$$\phi^\Delta(t) = p(t)\phi(t) + f(t), \quad \phi(t_0) = \phi_0, \tag{1.22}$$

is given by

$$\phi(t) = e_p(t, t_0)\phi_0 + \int_{t_0}^t e_p(t, \sigma(\tau))f(\tau)\Delta\tau.$$

Proof. Let ϕ be a solution to the problem (1.22). We have

$$\phi(t) = \phi(\sigma(t)) - \mu(t)\phi^\Delta(t).$$

Then (1.22) takes the form

$$\phi^\Delta(t) = p(t)\left(\phi(\sigma(t)) - \mu(t)\phi^\Delta(t)\right) + f(t)$$
$$= p(t)\phi(\sigma(t)) - p(t)\mu(t)\phi^\Delta(t) + f(t),$$

or

$$(1 + p(t)\mu(t))\phi^\Delta(t) = p(t)\phi(\sigma(t)) + f(t).$$

Hence,

$$\phi^\Delta(t) = \frac{p(t)}{1 + \mu(t)p(t)}\phi(\sigma(t)) + \frac{f(t)}{1 + \mu(t)p(t)}$$
$$= -\frac{-p(t)}{1 + \mu(t)p(t)}\phi(\sigma(t)) + \frac{f(t)}{1 + \mu(t)p(t)}$$
$$= -\ominus p(t)\phi(\sigma(t)) + \frac{f(t)}{1 + p(t)\mu(t)}.$$

Therefore, $\phi(t)$ satisfies the problem

$$\phi^\Delta(t) = -\ominus p(t)\phi(\sigma(t)) + \frac{f(t)}{1 + \mu(t)p(t)}, \quad \phi(t_0) = \phi_0. \tag{1.23}$$

From this, Theorem 1.53, and $\ominus(\ominus p)(t) = p(t)$, we get that the problem (1.23) has a unique solution given by

$$\phi(t) = \phi_0 e_p(t, t_0) + \int_{t_0}^t e_p(t, \tau) \frac{f(\tau)}{1 + \mu(\tau)p(\tau)} \Delta\tau$$

$$= \phi_0 e_p(t, t_0) + \int_{t_0}^t \frac{f(\tau)}{e_p(\tau, t)(1 + \mu(\tau)p(\tau))} \Delta\tau$$

$$= \phi_0 e_p(t, t_0) + \int_{t_0}^t \frac{f(\tau)}{e_p(\sigma(\tau), t)} \Delta\tau$$

$$= \phi_0 e_p(t, t_0) + \int_{t_0}^t e_p(t, \sigma(\tau))f(\tau)\Delta\tau,$$

which completes the proof.

Example 1.44. Consider the equation

$$\phi^\Delta = 2\phi + 3^t, \quad \phi(0) = 0, \quad \mathbb{T} = \mathbb{Z}.$$

Here

$$p(t) = 2, \quad f(t) = 3^t, \quad \phi_0 = 0,$$
$$\sigma(t) = t + 1, \quad \mu(t) = 1, \quad t \in \mathbb{T}.$$

Then, using Corollary 1.5, we obtain

$$\phi(t) = \int_0^t e_2(t, \sigma(\tau))3^\tau \Delta\tau$$

$$= \int_0^t e_2(t, \tau + 1)3^\tau \Delta\tau$$

$$= \int_0^t 3^{t-\tau-1}3^\tau \Delta\tau$$

$$= \int_0^t 3^{t-1} \Delta\tau$$

$$= 3^{t-1} \int_0^t \Delta\tau$$

$$= t3^{t-1}.$$

Example 1.45. Consider the equation

$$\phi^\Delta = 4\phi + t, \quad \phi(0) = 1, \quad \mathbb{T} = 2\mathbb{Z}.$$

We will find its solution for $t > 1$, $t \in \mathbb{T}$.

Here

$$\sigma(t) = t + 2, \quad \mu(t) = 2, \quad p(t) = 4,$$
$$f(t) = t, \quad \phi_0 = 1, \quad t \in \mathbb{T}.$$

Then, using Corollary 1.5, we obtain

$$\phi(t) = e_4(t, 0) + \int_0^t e_4(t, \sigma(\tau))\tau \Delta \tau$$

$$= 9^{\frac{t}{2}} + \int_0^t e_4(t, \tau + 2)\tau \Delta \tau$$

$$= 3^t + \int_0^t 9^{\frac{t-\tau-2}{2}} \tau \Delta \tau$$

$$= 3^t + 9^{\frac{t}{2}-1} \int_0^t 9^{-\frac{\tau}{2}} \tau \Delta \tau$$

$$= 3^t + 3^{t-2} \sum_{s \in [0, t-1]} \mu(s)s 3^{-s}$$

$$= 3^t + 2 \cdot 3^{t-2} \sum_{s \in [0, t-1]} s 3^{-s}.$$

Example 1.46. Consider the equation

$$\phi^{\Delta} = p(t)\phi + e_p(t, t_0), \quad \phi(t_0) = 0,$$

where $p : \mathbb{T} \to \mathbb{R}$ is rd-continuous and $1 + \mu(t)p(t) \neq 0$ for all $t \in \mathbb{T}$.
Using Corollary 1.5, we obtain

$$\phi(t) = \int_{t_0}^t e_p(t, \sigma(\tau))e_p(\tau, t_0)\Delta \tau$$

$$= \int_{t_0}^t \frac{1}{e_p(\sigma(\tau), t)} e_p(\tau, t_0)\Delta \tau$$

$$= e_p(t, t_0) \int_{t_0}^t \frac{1}{(1 + p(\tau)\mu(\tau))e_p(\tau, t_0)} e_p(\tau, t_0)\Delta \tau$$

$$= e_p(t, t_0) \int_{t_0}^t \frac{1}{1 + p(\tau)\mu(\tau)} \Delta \tau.$$

Exercise 1.24. Let $p : \mathbb{T} \to \mathbb{R}$ be rd-continuous and $1 + \mu(t)p(t) \neq 0$ for all $t \in \mathbb{T}$.
Let also $\phi(t)$ be a solution to the equation

$$\phi^{\Delta} - p(t)\phi = 0,$$

and let $\psi(t)$ be a solution to the equation

$$\psi^\Delta + p(t)\psi^\sigma = 0.$$

Prove that

$$\phi(t)\psi(t) = c, \quad t \in \mathbb{T},$$

where c is a constant.

Now we define the operator $L : \mathbb{T}^{\kappa^n} \to \mathbb{R}$ as follows:

$$Ly = y^{\Delta^n} + \sum_{i=1}^{n} a_i y^{\Delta^{n-i}},$$

where $a_i \in \mathbb{R}$, $i \in \{1, \dots, n\}$, are given constants.

Definition 1.34. One defines the Cauchy function $y : \mathbb{T} \times \mathbb{T}^{\kappa^n} \to \mathbb{R}$ for the linear dynamic equation

$$Ly = 0 \tag{1.24}$$

to be for each fixed $s \in \mathbb{T}^{\kappa^n}$ the solution of the initial value problem for (1.24) with initial conditions

$$y^{\Delta^i}(\sigma(s), s) = 0, \quad i \in \{0, \dots, n-2\}, \quad y^{\Delta^{n-1}}(\sigma(s), s) = 1.$$

Theorem 1.54. $y(t, s) = h_{n-1}(t, \sigma(s))$ *is the Cauchy function of* $y^{\Delta^n} = 0$, *where*

$$h_0(t, s) = 1, \quad h_n(t, s) = \int_s^t h_{n-1}(\tau, s)\Delta\tau, \quad n \in \mathbb{N}.$$

Proof. We have

$$y^{\Delta_t}(t, s) = h_{n-2}(t, \sigma(s)),$$
$$y^{\Delta_t^2}(t, s) = h_{n-3}(t, \sigma(s)),$$
$$\vdots$$
$$y^{\Delta_t^{n-1}}(t, s) = h_0(t, \sigma(s))$$
$$= 1,$$
$$y^{\Delta_t^n}(t, s) = 0,$$

and

$$y(\sigma(s), s) = h_{n-1}(\sigma(s), \sigma(s))$$
$$= 0,$$

$$y^{\Delta_t}(\sigma(s), s) = h_{n-2}(\sigma(s), \sigma(s))$$
$$= 0,$$

$$\vdots$$

$$y^{\Delta_t^{n-2}}(\sigma(s), s) = h_1(\sigma(s), \sigma(s))$$
$$= 0,$$

$$y^{\Delta_t^{n-1}}(\sigma(s), s) = h_0(\sigma(s), \sigma(s))$$
$$= 1.$$

This completes the proof.

Theorem 1.55. *Let $f \in \mathscr{C}_{rd}$. Then the solution of the initial value problem*

$$Ly = f(t), \quad y^{\Delta^i}(t_0) = 0, \quad i \in \{0, \ldots, n-1\},$$

is given by

$$y(t) = \int_{t_0}^{t} y(t, s) f(s) \Delta s, \quad t \in \mathbb{T},$$

where $y(t, s)$ is the Cauchy function for (1.24).

Proof. We have

$$y(t_0) = 0,$$
$$y^{\Delta}(t) = \int_{t_0}^{t} y^{\Delta}(t, s) f(s) \Delta s + y(\sigma(t), t) f(t)$$
$$= \int_{t_0}^{t} y^{\Delta_t}(t, s) f(s) \Delta s,$$
$$y^{\Delta}(t_0) = 0,$$
$$y^{\Delta^2}(t) = \int_{t_0}^{t} y^{\Delta_t^2}(t, s) f(s) \Delta s + y^{\Delta_t}(\sigma(t), t) f(t)$$
$$= \int_{t_0}^{t} y^{\Delta_t^2}(t, s) f(s) \Delta s,$$
$$y^{\Delta^2}(t_0) = 0,$$

$$\vdots$$

$$y^{\Delta^{n-1}}(t) = \int_{t_0}^t y^{\Delta_t^{n-1}}(t,s)f(s)\Delta s + y^{\Delta_t^{n-2}}(\sigma(t),t)f(t)$$

$$= \int_{t_0}^t y^{\Delta_t^{n-1}}(t,s)f(s)\Delta s,$$

$$y^{\Delta^{n-1}}(t_0) = 1,$$

$$y^{\Delta^n}(t) = \int_{t_0}^t y^{\Delta_t^n}(t,s)f(s)\Delta s + y^{\Delta_t^{n-1}}(\sigma(t),t)f(t)$$

$$= \int_{t_0}^t y^{\Delta_t^n}(t,s)f(s)\Delta s + f(t), \quad t \in \mathbb{T}^{\kappa^n}.$$

Hence,

$$y^{\Delta^n}(t) + \sum_{i=1}^{n-1} a_i y^{\Delta_t^{n-1}}(t) = \int_{t_0}^t y^{\Delta_t^n}(t,s)f(s)\Delta s + f(t)$$

$$+ \sum_{i=1}^n a_i \int_{t_0}^t y^{\Delta_t^{n-i}}(t,s)f(s)\Delta s$$

$$= \int_{t_0}^t \left(y^{\Delta_t^n}(t,s) + \sum_{i=1}^n a_i y^{\Delta_t^{n-i}}(t,s) \right) f(s)\Delta s + f(t)$$

$$= \int_{t_0}^t Ly(t,s)f(s)\Delta s + f(t)$$

$$= f(t), \quad t \in \mathbb{T}^{\kappa^n}.$$

This completes the proof.

1.8 Power Series on Time Scales

We introduce the generalized monomials $h_k : \mathbb{T} \times \mathbb{T} \to \mathbb{R}$, $k \in \mathbb{N}_0$, defined recurrently by

$$h_0(t,s) = 1,$$

$$h_k(t,s) = \int_s^t h_{k-1}(\tau,s)\Delta\tau, \quad k \in \mathbb{N}, \quad t,s \in \mathbb{T}.$$

Then

$$h_1(t, s) = \int_s^t \Delta\tau$$

$$= t - s,$$

$$h_k^{\Delta_t}(t, s) = h_{k-1}(t, s), \quad k \in \mathbb{N}, \quad t, s \in \mathbb{T}.$$

Example 1.47. Let $\mathbb{T} = \mathbb{R}$. Then

$$h_k(t, s) = \frac{(t - s)^k}{k!}, \quad k \in \mathbb{N}, \quad t, s \in \mathbb{T}.$$

Example 1.48. Let $\mathbb{T} = \mathbb{Z}$. We define

$$t^{(0)} = 1,$$

$$t^{(k)} = \prod_{j=0}^{k-1} (t - j), \quad k \in \mathbb{N}, \quad t \in \mathbb{T}.$$

Then

$$h_0(t, s) = (t - s)^{(0)},$$

$$h_1(t, s) = \int_s^t h_0(\tau, s)\Delta\tau$$

$$= \int_s^t \Delta\tau$$

$$= t - s$$

$$= \frac{(t - s)^{(1)}}{1!}.$$

Assume that

$$h_k(t, s) = \frac{(t - s)^{(k)}}{k!}, \quad t, s \in \mathbb{T}, \tag{1.25}$$

for some $k \in \mathbb{N}$. We will prove that

$$h_{k+1}(t, s) = \frac{(t - s)^{(k+1)}}{(k + 1)!}.$$

Indeed, we have

$$
\left(\frac{(t-s)^{(k+1)}}{(k+1)!}\right)^{\Delta_t} = \frac{1}{(k+1)!}\left((\sigma(t)-s)^{(k+1)} - (t-s)^{(k+1)}\right)
$$

$$
= \frac{1}{(k+1)!}\Big((\sigma(t)-s)(\sigma(t)-s-1)\cdots(\sigma(t)-s-k)
$$

$$
-(t-s)(t-s-1)\cdots(t-s-k)\Big)
$$

$$
= \frac{1}{(k+1)!}\Big((t+1-s)(t+1-s-1)\cdots(t+1-s-k)
$$

$$
-(t-s)(t-s-1)\cdots(t-s-k)\Big)
$$

$$
= \frac{1}{(k+1)!}\Big((t+1-s)(t-s)\cdots(t-s-k+1)
$$

$$
-(t-s)(t-s-1)\cdots(t-s-k)\Big)
$$

$$
= \frac{1}{(k+1)!}(t-s)\cdots(t-s-k+1)(t+1-s-t+s+k)
$$

$$
= \frac{1}{(k+1)!}(t-s)\cdots(t-s-k+1)(k+1)
$$

$$
= \frac{1}{k!}(t-s)(t-s-1)\cdots(t-s-k+1)
$$

$$
= \frac{(t-s)^{(k)}}{k!}
$$

$$
= h_k(t,s), \quad t,s \in \mathbb{T}.
$$

Therefore, (1.25) holds for all $k \in \mathbb{N}$.

Theorem 1.56. *We have*

$$
h_{k+m+1}(t,t_0) = \int_{t_0}^{t} h_k(t,\sigma(s))h_m(s,t_0)\Delta s, \quad t,t_0 \in \mathbb{T},
$$

$k,m \in \mathbb{N}_0$.

Proof. Let

$$
g(t) = \int_{t_0}^{t} h_k(t,\sigma(s))h_m(s,t_0)\Delta s.
$$

Then, using the chain rule, we get

$$g^{\Delta}(t) = h_k(\sigma(t), \sigma(t))h_m(\sigma(t), t_0)$$

$$+ \int_{t_0}^{t} h_{k-1}(t, \sigma(s)), h_m(s, t_0)\Delta s$$

$$= \int_{t_0}^{t} h_{k-1}(t, \sigma(s))h_m(s, t_0)\Delta s,$$

$$g^{\Delta^2}(t) = h_{k-1}(\sigma(t), \sigma(t))h_m(t, t_0)$$

$$+ \int_{t_0}^{t} h_{k-2}(t, \sigma(s))h_m(s, t_0)\Delta s$$

$$= \int_{t_0}^{t} h_{k-2}(t, \sigma(s))h_m(s)\Delta s,$$

$$\vdots$$

$$g^{\Delta^k}(t) = \int_{t_0}^{t} h_m(s, t_0)\Delta s$$

$$= h_{m+1}(t, t_0),$$

$$g^{\Delta^{k+1}}(t) = h_m(t, t_0),$$

$$\vdots$$

$$g^{\Delta^{k+m}}(t) = h_0(t, t_0).$$

This completes the proof.

We define

$$g_0(t, s) = 1, \quad g_{k+1}(t, s) = \int_{s}^{t} g_k(\sigma(\tau), s)\Delta\tau, \quad k \in \mathbb{N}_0, \quad t, s \in \mathbb{T}.$$

Lemma 1.1. *Let $n \in \mathbb{N}$. If f is n-times differentiable and p_k, $0 \leq k \leq n - 1$, are differentiable at some $t \in \mathbb{T}$ with*

$$p_{k+1}^{\Delta}(t) = p_k^{\sigma}(t) \quad \text{for} \quad \text{all} \quad 0 \leq k \leq n - 2, \quad n \geq 2,$$

then we have

$$\left(\sum_{k=0}^{n-1}(-1)^k f^{\Delta^k}(t)p_k(t)\right)^{\Delta} = (-1)^{n-1}f^{\Delta^n}(t)p_{n-1}^{\sigma}(t) + f(t)p_0^{\Delta}(t).$$

Proof. We have

$$\left(\sum_{k=0}^{n-1}(-1)^k f^{\Delta^k}(t)p_k(t)\right)^{\Delta} = \sum_{k=0}^{n-1}(-1)^k \left(f^{\Delta^k}(t)p_k(t)\right)^{\Delta}$$

$$= \sum_{k=0}^{n-1}(-1)^k \left(f^{\Delta^{k+1}}(t)p_k^{\sigma}(t) + f^{\Delta^k}(t)p_k^{\Delta}(t)\right)$$

$$= \sum_{k=0}^{n-1}(-1)^k f^{\Delta^{k+1}}(t)p_k^{\sigma}(t) + \sum_{k=0}^{n-1}(-1)^k f^{\Delta^k}(t)p_k^{\Delta}(t)$$

$$= \sum_{k=0}^{n-2}(-1)^k f^{\Delta^{k+1}}(t)p_k^{\sigma}(t) + (-1)^{n-1}f^{\Delta^n}(t)p_{n-1}^{\sigma}(t)$$

$$+ \sum_{k=1}^{n-1}(-1)^k f^{\Delta^k}(t)p_k^{\Delta}(t) + f^{\Delta^0}(t)p_0^{\Delta}(t)$$

$$= \sum_{k=0}^{n-2}(-1)^k f^{\Delta^{k+1}}(t)p_{k+1}^{\Delta}(t) + (-1)^{n-1}f^{\Delta^n}(t)p_{n-1}^{\sigma}(t)$$

$$+ \sum_{k=0}^{n-2}(-1)^{k+1}f^{\Delta^{k+1}}(t)p_{k+1}^{\Delta}(t) + f(t)p_0^{\Delta}(t)$$

$$= (-1)^{n-1}f^{\Delta^n}(t)p_{n-1}^{\sigma}(t) + f(t)p_0^{\Delta}(t).$$

This completes the proof.

Lemma 1.2. *The functions $g_n(t, s)$ satisfy for all $t \in \mathbb{T}$ the relationship*

$$g_n(\rho^k(t), t) = 0 \quad \text{for} \quad \text{all} \quad n \in \mathbb{N} \quad \text{and} \quad \text{all} \quad 0 \leq k \leq n - 1.$$

Proof. Let $n \in \mathbb{N}$ be arbitrarily chosen. Then

$$g_n(\rho^0(t), t) = g_n(t, t)$$

$$= \int_t^t g_{n-1}(\sigma(\tau), t)\Delta\tau$$

$$= 0.$$

Assume that

$$g_{n-1}(\rho^k(t), t) = 0 \quad \text{and} \quad g_n(\rho^k(t), t) = 0$$

for some $0 \leq k < n - 1$.

We will prove that

$$g_n(\rho^{k+1}(t), t) = 0.$$

Case 1. $\rho^k(t)$ is left-dense. Then

$$\rho^{k+1}(t) = \rho(\rho^k(t)) = \rho^k(t).$$

Consequently, using the induction hypothesis, we have

$$g_n(\rho^{k+1}(t), t) = g_n(\rho^k(t), t) = 0.$$

Case 2. $\rho^k(t)$ is left-scattered. Then

$$\rho(\rho^k(t)) < \rho^k(t),$$

and there is no $s \in \mathbb{T}$ such that $\rho^{k+1}(t) < s < \rho^k(t)$. Hence,

$$\sigma(\rho^{k+1}(t)) = \rho^k(t).$$

Therefore,

$$g_n(\sigma(\rho^{k+1}(t)), t) = g_n(\rho^{k+1}(t), t) + \mu(\rho^{k+1}(t))g_n^\Delta(\rho^{k+1}(t), t),$$

or

$$g_n(\rho^k(t), t) = g_n(\rho^{k+1}(t), t) + \mu(\rho^{k+1}(t))g_n^\Delta(\rho^{k+1}(t), t),$$

whereupon

$$\begin{aligned}
g_n(\rho^{k+1}(t), t) &= g_n(\rho^k(t), t) - \mu(\rho^{k+1}(t))g_n^\Delta(\rho^{k+1}(t), t) \\
&= g_n(\rho^k(t), t) - \mu(\rho^{k+1}(t))g_{n-1}(\sigma(\rho^{k+1}(t)), t) \\
&= g_n(\rho^k(t), t) - \mu(\rho^{k+1}(t))g_{n-1}(\rho^k(t), t) \\
&= 0.
\end{aligned}$$

This completes the proof.

Lemma 1.3. *Let $n \in \mathbb{N}$, and suppose that f is $(n-1)$-times differentiable at $\rho^{n-1}(t)$. Then*

$$f(t) = \sum_{k=0}^{n-1}(-1)^k f^{\Delta^k}(\rho^{n-1}(t))g_k(\rho^{n-1}(t), t).$$

Proof. 1. Let $n = 1$. Then

$$\sum_{k=0}^{0}(-1)^k f^{\Delta^k}(\rho^0(t))g_k(\rho^0(t), t) = (-1)^0 f^{\Delta^0}(t)g_0(t, t)$$

$$= f(t).$$

2. Assume that

$$f(t) = \sum_{k=0}^{m-1}(-1)^k f^{\Delta^k}(\rho^{m-1}(t))g_k(\rho^{m-1}(t), t)$$

for some $m \in \mathbb{N}$.
3. We will prove that

$$f(t) = \sum_{k=0}^{m}(-1)^k f^{\Delta^k}(\rho^m(t))g_k(\rho^m(t), t).$$

Case 1. $\rho^{m-1}(t)$ is left-dense. Then

$$\rho^m(t) = \rho(\rho^{m-1}(t)) = \rho^{m-1}(t).$$

By this and the induction hypothesis, we obtain

$$\sum_{k=0}^{m}(-1)^k f^{\Delta^k}(\rho^m(t))g_k(\rho^m(t), t)$$

$$= \sum_{k=0}^{m-1}(-1)^k f^{\Delta^k}(\rho^m(t))g_k(\rho^m(t), t) + (-1)^m f^{\Delta^m}(\rho^m(t))g_m(\rho^m(t), t)$$

$$= \sum_{k=0}^{m-1}(-1)^k f^{\Delta^k}(\rho^{m-1}(t))g_k(\rho^{m-1}(t), t)$$

$$+ (-1)^m f^{\Delta^m}(\rho^{m-1}(t))g_m(\rho^{m-1}(t), t)$$

now we apply Lemma 1.2 $(g_m(\rho^{m-1}(t), t) = 0)$

$$= \sum_{k=0}^{m-1}(-1)^k f^{\Delta^k}(\rho^{m-1}(t))g_k(\rho^{m-1}(t), t)$$

now we apply the induction assumption
$$= f(t).$$

Case 2. $\rho^{m-1}(t)$ is left-scattered. Then

$$\rho^m(t) = \rho(\rho^{m-1}(t)) < \rho^{m-1}(t),$$

and there is no $s \in \mathbb{T}$ such that

$$\rho^m(t) < s < \rho^{m-1}(t).$$

Also,

$$\sigma(\rho^m(t)) = \rho^{m-1}(t).$$

Hence,

$$g_k(\sigma(\rho^m(t)), t) = g_k(\rho^{m-1}(t), t).$$

Therefore,

$$\begin{aligned}
g_k(\rho^{m-1}(t), t) &= g_k(\sigma(\rho^m(t)), t) \\
&= g_k(\rho^m(t), t) + \mu(\rho^m(t))g_k^{\Delta}(\rho^m(t), t) \\
&= g_k(\rho^m(t), t) + \mu(\rho^m(t))g_{k-1}(\sigma(\rho^m(t)), t) \\
&= g_k(\rho^m(t), t) + \mu(\rho^m(t))g_{k-1}(\rho^{m-1}(t), t),
\end{aligned}$$

whereupon

$$g_k(\rho^m(t), t) = g_k(\rho^{m-1}(t), t) - \mu(\rho^m(t))g_{k-1}(\rho^{m-1}(t), t).$$

Consequently,

$$\sum_{k=0}^{m}(-1)^k f^{\Delta^k}(\rho^m(t))g_k(\rho^m(t), t)$$

$$= f(\rho^m(t)) + \sum_{k=1}^{m}(-1)^k f^{\Delta^k}(\rho^m(t))g_k(\rho^m(t), t)$$

$$= f(\rho^m(t)) + \sum_{k=1}^{m}(-1)^k f^{\Delta^k}(\rho^m(t))g_k(\rho^{m-1}(t), t)$$

$$+ \sum_{k=1}^{m}(-1)^{k-1} f^{\Delta^k}(\rho^m(t))\mu(\rho^m(t))g_{k-1}(\rho^{m-1}(t), t)$$

$$= f(\rho^m(t)) + \sum_{k=1}^{m-1}(-1)^k f^{\Delta^k}(\rho^m(t))g_k(\rho^{m-1}(t), t)$$

$$+(-1)^m f^{\Delta^m}(\rho^m(t))g_m(\rho^{m-1}(t), t)$$

$$+ \sum_{k=0}^{m-1}(-1)^k f^{\Delta^{k+1}}(\rho^m(t))\mu(\rho^m(t))g_k(\rho^{m-1}(t), t)$$

$$= \sum_{k=0}^{m-1} (-1)^k f^{\Delta^k}(\rho^m(t)) g_k(\rho^{m-1}(t), t)$$

$$+ \sum_{k=0}^{m-1} (-1)^k \mu(\rho^m(t)) f^{\Delta^{k+1}}(\rho^m(t)) g_k(\rho^{m-1}(t), t)$$

$$= \sum_{k=0}^{m-1} (-1)^k \left(f^{\Delta^k}(\rho^m(t)) + \mu(\rho^m(t))(f^{\Delta^k})^{\Delta}(\rho^m(t)) \right) g_k(\rho^{m-1}(t), t)$$

$$= \sum_{k=0}^{m-1} (-1)^k f^{\Delta^k}(\sigma(\rho^m(t))) g_k(\rho^{m-1}(t), t)$$

$$= \sum_{k=0}^{m-1} (-1)^k f^{\Delta^k}(\rho^{m-1}(t)) g_k(\rho^{m-1}(t), t)$$

$$= f(t).$$

This completes the proof.

Theorem 1.57 (*Taylor's Formula*).   *Let* $n \in \mathbb{N}$. *Suppose* f *is* n-*times differentiable on* $\mathbb{T}^{\kappa^n}$. *Let* $\alpha \in \mathbb{T}^{\kappa^{n-1}}$, $t \in \mathbb{T}$. *Then*

$$f(t) = \sum_{k=0}^{n-1} (-1)^k g_k(\alpha, t) f^{\Delta^k}(\alpha)$$

$$+ \int_{\alpha}^{\rho^{n-1}(t)} (-1)^{n-1} g_{n-1}(\sigma(\tau), t) f^{\Delta^n}(\tau) \Delta \tau.$$

Proof. We note that on applying Lemma 1.1 for $p_k = g_k$, we have

$$\left(\sum_{k=0}^{n-1} (-1)^k g_k(\tau, t) f^{\Delta^k}(\tau) \right)_{\tau}^{\Delta}$$
$$= (-1)^{n-1} f^{\Delta^n}(\tau) g_{n-1}(\sigma(\tau), t) + f(\tau) g_0^{\Delta}(\tau, t)$$
$$= (-1)^{n-1} f^{\Delta^n}(\tau) g_{n-1}(\sigma(\tau), t) \quad \text{for all} \quad \tau \in \mathbb{T}^{\kappa^n}.$$

We integrate the last relation from α to $\rho^{n-1}(t)$, and we get

$$\int_{\alpha}^{\rho^{n-1}(t)} \left(\sum_{k=0}^{n-1} (-1)^k g_k(\tau, t) f^{\Delta^k}(\tau) \right)_{\tau}^{\Delta}$$

$$\Delta \tau = \int_{\alpha}^{\rho^{n-1}(t)} (-1)^{n-1} f^{\Delta^n}(\tau) g_{n-1}(\sigma(\tau), t) \Delta \tau,$$

or

$$\sum_{k=0}^{n-1}(-1)^k g_k(\rho^{n-1}(t), t) f^{\Delta^k}(\rho^{n-1}(t)) - \sum_{k=0}^{n-1}(-1)^k g_k(\alpha, t) f^{\Delta^k}(\alpha)$$
$$= \int_\alpha^{\rho^{n-1}(t)} (-1)^{n-1} f^{\Delta^n}(\tau) g_{n-1}(\sigma(\tau), t) \Delta\tau.$$

Hence, applying Lemma 1.3, we obtain

$$f(t) - \sum_{k=0}^{n-1}(-1)^k g_k(\alpha, t) f^{\Delta^k}(\alpha) = \int_\alpha^{\rho^{n-1}(t)} (-1)^{n-1} f^{\Delta^n}(\tau) g_{n-1}(\sigma(\tau), t) \Delta\tau.$$

This completes the proof.

Theorem 1.58. *The functions g_n and h_n satisfy the relationship*

$$h_n(t, s) = (-1)^n g_n(s, t)$$

for all $t \in \mathbb{T}$ and all $s \in \mathbb{T}^{\kappa^n}$.

Proof. Let $t \in \mathbb{T}$ and $s \in \mathbb{T}^{\kappa^n}$ be arbitrarily chosen. We apply Theorem 1.57 for $\alpha = s$ and $f(\tau) = h_n(\tau, s)$. We observe that

$$f^{\Delta^k}(\tau) = h_{n-k}(\tau, s), \quad 0 \le k \le n.$$

Hence,

$$f^{\Delta^k}(s) = h_{n-k}(s, s) = 0, \quad 0 \le k \le n - 1,$$
$$f^{\Delta^n}(s) = h_0(s, s) = 1, \quad f^{\Delta^{n+1}}(\tau) = 0.$$

From this, using Taylor's formula, we get

$$f(t) = h_n(t, s)$$
$$= \sum_{k=0}^{n}(-1)^k g_k(\alpha, t) f^{\Delta^k}(\alpha) + \int_\alpha^{\rho^n(t)} (-1)^n g_n(\sigma(\tau), t) f^{\Delta^{n+1}}(\tau) \Delta\tau$$
$$= \sum_{k=0}^{n}(-1)^k g_k(s, t) f^{\Delta^k}(s) + \int_s^{\rho^n(t)} (-1)^n g_n(\sigma(\tau), t) f^{\Delta^{n+1}}(\tau) \Delta\tau$$
$$= \sum_{k=0}^{n-1}(-1)^k g_k(s, t) f^{\Delta^k}(s) + (-1)^n g_n(s, t) f^{\Delta^n}(s)$$
$$= (-1)^n g_n(s, t) f^{\Delta^n}(s)$$
$$= (-1)^n g_n(s, t),$$

i.e.,

$$h_n(t, s) = (-1)^n g_n(s, t).$$

This completes the proof.

From Theorem 1.57 and Theorem 1.58 we deduce the following theorem.

Theorem 1.59 (*Taylor's Formula*). *Let* $n \in \mathbb{N}$. *Suppose* f *is* n-*times differentiable on* $\mathbb{T}^{\kappa^n}$. *Let also* $\alpha \in \mathbb{T}^{\kappa^{n-1}}$, $t \in \mathbb{T}$. *Then*

$$f(t) = \sum_{k=0}^{n-1} h_k(t, \alpha) f^{\Delta^k}(\alpha) + \int_\alpha^{\rho^{n-1}(t)} h_{n-1}(t, \sigma(\tau)) f^{\Delta^n}(\tau) \Delta\tau.$$

Now we will formulate and prove another variant of Taylor's formula.

Theorem 1.60 (*Taylor's Formula*). *Let* $n \in \mathbb{N}$. *Suppose that the function* f *is* $n+1$-*times differentiable on* $\mathbb{T}^{\kappa^{n+1}}$. *Let* $\alpha \in \mathbb{T}^{\kappa^{n+1}}$, $t \in \mathbb{T}$, *and* $t > \alpha$. *Then*

$$f(t) = \sum_{k=0}^{n} h_k(t, \alpha) f^{\Delta^k}(\alpha) + \int_\alpha^t h_n(t, \sigma(\tau)) f^{\Delta^{n+1}}(\tau) \Delta\tau. \qquad (1.26)$$

Proof. Let

$$g(t) = f^{\Delta^{n+1}}(t).$$

Then f solves the problem

$$x^{\Delta^{n+1}} = g(t), \quad x^{\Delta^k}(\alpha) = f^{\Delta^k}(\alpha), \quad k \in \{0, \dots, n\}.$$

By Theorem 1.54, we have that

$$y(t, s) = h_n(t, \sigma(s))$$

is the Cauchy function for $y^{\Delta^{n+1}} = 0$. By this and Theorem 1.55, it follows that

$$
\begin{aligned}
f(t) &= u(t) + \int_\alpha^t y(t, \sigma(\tau)) g(\tau) \Delta\tau \\
&= u(t) + \int_\alpha^t h_n(t, \sigma(s)) g(s) \Delta s,
\end{aligned}
\qquad (1.27)
$$

where u solves the initial value problem

$$u^{\Delta^{n+1}} = 0, \quad u^{\Delta^m}(\alpha) = f^{\Delta^m}(\alpha), \quad m \in \{0, \dots, n\}.$$

We set

$$w(t) = \sum_{k=0}^{n} h_k(t, \alpha) f^{\Delta^k}(\alpha). \tag{1.28}$$

We have

$$w^{\Delta^m}(t) = \sum_{k=0}^{n} h_{k-m}(t, \alpha) f^{\Delta^k}(\alpha), \quad m \in \{0, \ldots, n\},$$

and hence

$$w^{\Delta^m}(\alpha) = \sum_{k=0}^{n} h_{k-m}(\alpha, \alpha) f^{\Delta^k}(\alpha)$$

$$= f^{\Delta^m}(\alpha), \quad m \in \{0, \ldots, n\},$$

i.e., w solves (1.27). Consequently, $w = u$. From this and (1.28), we obtain (1.26). This completes the proof.

Theorem 1.61. *For all $k \in \mathbb{N}_0$ we have*

$$0 \le h_k(t, s) \le \frac{(t - s)^k}{k!}, \quad t \ge s. \tag{1.29}$$

Proof. Let

$$g(t) = (t - s)^{k+1}, \quad t, s \in \mathbb{T}, \quad k \in \mathbb{N}.$$

Then

$$g^{\Delta}(t) = \lim_{y \to t} \frac{g(\sigma(t)) - g(y)}{\sigma(t) - y}$$

$$= \lim_{y \to t} \frac{(\sigma(t) - s)^{k+1} - (y - s)^{k+1}}{\sigma(t) - y}$$

$$= \lim_{y \to t} \frac{(\sigma(t) - y) \sum_{v=0}^{k} (\sigma(t) - s)^v (y - s)^{k-v}}{\sigma(t) - y}$$

$$= \lim_{y \to t} \sum_{v=0}^{k} (\sigma(t) - s)^v (y - s)^{k-v}$$

$$= \sum_{v=0}^{k} (\sigma(t) - s)^v (t - s)^{k-v}, \quad t, s \in \mathbb{T}, \quad k \in \mathbb{N}.$$

Note that the inequalities (1.29) are true for $k = 0$. Assume that the inequalities (1.29) are true for some $k \in \mathbb{N}$. We will prove the inequalities (1.29) for $k + 1$. We have

$$0 \leq h_{k+1}(t, s)$$

$$= \int_s^t h_k(\tau.s)\Delta\tau$$

$$\leq \frac{1}{k!} \int_s^t (\tau - s)^k \Delta\tau$$

$$= \frac{1}{(k+1)!} \int_s^t \sum_{\nu=0}^k (\tau - s)^k \Delta\tau$$

$$= \frac{1}{(k+1)!} \int_s^t \sum_{\nu=0}^k (\tau - s)^\nu (\tau - s)^{k-\nu} \Delta\tau$$

$$\leq \frac{1}{(k+1)!} \int_s^t \sum_{\nu=0}^k (\sigma(\tau) - s)^\nu (\tau - s)^{k-\nu} \Delta\tau$$

$$= \frac{1}{(k+1)!} \int_s^t g^\Delta(\tau) \Delta\tau$$

$$= \frac{1}{(k+1)!} g(\tau) \Big|_{\tau=s}^{\tau=t}$$

$$= \frac{1}{(k+1)!} (\tau - s)^{k+1} \Big|_{\tau=s}^{\tau=t}$$

$$= \frac{(t - s)^{k+1}}{(k+1)!}, \quad t, s \in \mathbb{T}, \quad t \geq s.$$

By the principle of mathematical induction, it follows that (1.29) is true for all $k \in \mathbb{N}$. This completes the proof.

Let

$$R_n(t, \alpha) = \int_\alpha^{\rho^{n-1}(t)} h_{n-1}(t, \sigma(\tau)) f^{\Delta^n}(\tau)\Delta\tau.$$

Theorem 1.62. *Let $t \in \mathbb{T}$, $t \geq \alpha$, and*

$$M_n(t) = \sup \left\{ \left| f^{\Delta^n}(\tau) \right| : \tau \in [\alpha, t] \right\}.$$

Then

$$|R_n(t, \alpha)| \leq M_n(t) \frac{(t - \alpha)^n}{(n - 1)!}.$$

Proof. Let $\tau \in [\alpha, t)$. Then $\alpha \leq \sigma(\tau) \leq t$, and applying (1.29), we get

$$0 \leq h_{n-1}(t, \sigma(\tau))$$

$$\leq \frac{(t - \sigma(\tau))^{n-1}}{(n-1)!}$$

$$\leq \frac{(t - \tau)^{n-1}}{(n-1)!}$$

$$\leq \frac{(t - \alpha)^{n-1}}{(n-1)!}.$$

Hence,

$$|R_n(t, \alpha)| = \left| \int_{\alpha}^{\rho^{n-1}(t)} h_{n-1}(t, \sigma(\tau)) f^{\Delta^n}(\tau) \Delta \tau \right|$$

$$\leq \int_{\alpha}^{t} h_{n-1}(t, \sigma(\tau)) \left| f^{\Delta^n}(\tau) \right| \Delta \tau$$

$$\leq M_n(t) \int_{\alpha}^{t} \frac{(t - \alpha)^{n-1}}{(n-1)!} \Delta \tau$$

$$= M_n(t) \frac{(t - \alpha)^n}{(n-1)!}.$$

This completes the proof.

If a function $f : \mathbb{T} \rightarrow \mathbb{R}$ is infinitely Δ-differentiable at a point $\alpha \in \mathbb{T}^{\infty} = \bigcap_{n=1}^{\infty} \mathbb{T}^{\kappa^n}$, then we can formally write

$$\sum_{k=0}^{\infty} h_k(t, \alpha) f^{\Delta^k}(\alpha) = f(\alpha) + h_1(t, \alpha) f^{\Delta}(\alpha) + h_2(t, \alpha) f^{\Delta^2}(\alpha) + \cdots . \qquad (1.30)$$

Definition 1.35. The series (1.30) is called the Taylor series for the function f at the point α.

For given values of α and t, the Taylor series can be convergent or divergent. The Taylor series (1.30) is convergent if and only if the remainder in Taylor's formula

$$f(t) = \sum_{k=0}^{n-1} h_k(t, \alpha) f^{\Delta^k}(\alpha) + R_n(t, \alpha)$$

tends to zero as $n \rightarrow \infty$, that is, $\lim_{n \to \infty} R_n(t, \alpha) = 0$. It may turn out that the series (1.30) is convergent for some values of t but its sum is not equal to $f(t)$.

Theorem 1.63. *For all $z \in \mathbb{C}$ and $\alpha, R \in \mathbb{T}, R > \alpha$, the initial value problem*

$$y^{\Delta} = zy, \quad y(\alpha) = 1, \quad t \in [\alpha, R], \tag{1.31}$$

has a unique solution y that is represented in the form

$$y(t) = \sum_{k=0}^{\infty} z^k h_k(t, \alpha), \quad t \in [\alpha, R]$$

and satisfies the inequality

$$|y(t)| \le e^{|z|(t-\alpha)}, \quad t \in [\alpha, R].$$

Proof. The problem (1.31) is equivalent to finding a continuous solution of the integral equation

$$y(t) = 1 + \int_{\alpha}^{t} y(\tau)\Delta\tau, \quad t \in [\alpha, R]. \tag{1.32}$$

We will solve equation (1.32) using the method of successive approximations. Let

$$y_0(t) = 1, \quad y_k(t) = z \int_{\alpha}^{t} y_{k-1}(\tau)\Delta\tau, \quad t \in [\alpha, R], \quad k \in \mathbb{N}.$$

Note that

$$y_0(t) = h_0(t, \alpha),$$

$$y_1(t) = z \int_{\alpha}^{t} h_0(\tau, \alpha)\Delta\tau$$

$$= z \int_{\alpha}^{t} \Delta\tau$$

$$= z(t - \alpha)$$

$$= z h_1(t, \alpha),$$

$$y_2(t) = z \int_{\alpha}^{t} y_1(\tau)\Delta\tau$$

$$= z^2 \int_{\alpha}^{t} h_1(\tau, \alpha)\Delta\tau$$

$$= z^2 h_2(t, \alpha), \quad t \in [\alpha, R].$$

Assume that

$$y_k(t) = z^k h_k(t, \alpha), \quad t \in [\alpha, R], \tag{1.33}$$

for some $k \in \mathbb{N}$. We will prove that

$$y_{k+1}(t) = z^{k+1} h_{k+1}(t, \alpha), \quad t \in [\alpha, R].$$

Indeed, we have

$$\begin{aligned}
y_{k+1}(t) &= z \int_{\alpha}^{t} y_k(\tau) \Delta \tau \\
&= z^{k+1} \int_{\alpha}^{t} h_k(\tau, \alpha) \Delta \tau \\
&= z^{k+1} h_{k+1}(t, \alpha), \quad t \in [\alpha, R].
\end{aligned}$$

Therefore, (1.33) holds for all $k \in \mathbb{N}$. Using Theorem 1.61, we get

$$\begin{aligned}
|y_k(t)| &= \left| z^k h_k(t, \alpha) \right| \\
&= |z|^k |h_k(t, \alpha)| \\
&\le |z|^k \frac{(t - \alpha)^k}{k!}, \quad k \in \mathbb{N}_0
\end{aligned}$$

and

$$\begin{aligned}
\sum_{k=0}^{\infty} |y_k(t)| &\le \sum_{k=0}^{\infty} |z|^k \frac{(t - \alpha)^k}{k!} \\
&= e^{|z|(t-\alpha)}, \quad t \in [\alpha, R].
\end{aligned}$$

Therefore, the series $\sum_{k=0}^{\infty} y_k(t)$ converges uniformly with respect to $t \in [\alpha, R]$. Consequently, its sum is a continuous solution of the problem (1.32). Suppose that (1.32) has two solutions y and x. Let $u = y - x$. Then

$$u(t) = z \int_{\alpha}^{t} u(\tau) \Delta \tau, \quad t \in [\alpha, R].$$

We set

$$M = \sup\{|u(t)| : t \in [\alpha, R]\}.$$

Hence,

$$|u(t)| = \left| z \int_\alpha^t u(\tau) \Delta \tau \right|$$

$$\leq |z| \int_\alpha^t |u(\tau)| \Delta \tau$$

$$\leq M|z|(t - \alpha)$$

$$= M|z| h_1(t, \alpha), \quad t \in [\alpha, R].$$

Hence,

$$|u(t)| \leq |z|^2 M \int_\alpha^t h_1(\tau, \alpha) \Delta \tau$$

$$= M|z|^2 h_2(t, \alpha), \quad t \in [\alpha, R].$$

Repeating this procedure, we obtain

$$|u(t)| \leq M|z|^k h_k(t, \alpha),$$

and applying Theorem 1.61, we get

$$|u(t)| \leq M|z|^k \frac{(t - \alpha)^k}{k!}, \quad t \in [\alpha, R],$$

for all $k \in \mathbb{N}$. Passing to the limit as $k \to \infty$, we get $u(t) = 0$ for all $t \in [\alpha, R]$. This completes the proof.

Let y be the solution of the problem (1.31). Then

$$y^{\Delta^2} = z y^\Delta$$

$$= z(zy)$$

$$= z^2 y.$$

Assume that

$$y^{\Delta^k} = z^k y \tag{1.34}$$

for some $k \in \mathbb{N}$. We will prove that

$$y^{\Delta^{k+1}} = z^{k+1} y.$$

Indeed, we have

$$y^{\Delta^{k+1}} = \left(y^{\Delta^k}\right)^{\Delta}$$
$$= \left(z^k y\right)^{\Delta}$$
$$= z^{k+1} y.$$

Therefore, (1.34) holds for all $k \in \mathbb{N}$. Note that $e_z(t, \alpha)$, $t \in [\alpha, R]$, coincides with the unique solution of the problem (1.31). Therefore,

$$e_z(t, \alpha) = \sum_{k=0}^{\infty} z^k h_k(t, \alpha), \quad t \in [\alpha, R].$$

Hence,

$$\cosh_z(t, \alpha) = \frac{e_z(t, \alpha) + e_{-z}(t, \alpha)}{2}$$

$$= \frac{1}{2} \left(\sum_{k=0}^{\infty} z^k h_k(t, \alpha) + \sum_{k=0}^{\infty} (-z)^k h_k(t, \alpha) \right)$$

$$= \frac{1}{2} \sum_{k=0}^{\infty} \left(z^k + (-z)^k \right) h_k(t, \alpha)$$

$$= \sum_{k=0}^{\infty} z^{2k} h_{2k}(t, \alpha),$$

$$\sinh_z(t, \alpha) = \frac{e_z(t, \alpha) - e_{-z}(t, \alpha)}{2}$$

$$= \frac{1}{2} \left(\sum_{k=0}^{\infty} z^k h_k(t, \alpha) - \sum_{k=0}^{\infty} (-z)^k h_k(t, \alpha) \right)$$

$$= \frac{1}{2} \sum_{k=0}^{\infty} \left(z^k - (-z)^k \right) h_k(t, \alpha)$$

$$= \sum_{k=0}^{\infty} z^{2k+1} h_{2k+1}(t, \alpha),$$

$$\cos_z(t, \alpha) = \frac{e_{iz}(t, \alpha) + e_{-iz}(t, \alpha)}{2}$$

$$= \sum_{k=0}^{\infty} (iz)^{2k} h_{2k}(t, \alpha)$$

$$= \sum_{k=0}^{\infty} (-1)^k z^{2k} h_{2k}(t, \alpha),$$

$$\sin_z(t, \alpha) = \frac{e_{iz}(t, \alpha) - e_{-iz}(t, \alpha)}{2i}$$

$$= \frac{1}{i} \sum_{k=0}^{\infty} (iz)^{2k+1} h_{2k+1}(t, \alpha)$$

$$= \sum_{k=0}^{\infty} (-1)^k z^{2k+1} h_{2k+1}(t, \alpha), \quad t \in [\alpha, R].$$

Definition 1.36. Assume that $\sup \mathbb{T} = \infty$ and let $t_0 \in \mathbb{T}$ be fixed. A series of the form

$$\sum_{k=0}^{\infty} a_k h_k(t, t_0) = a_0 + a_1 h_1(t, t_0) + a_2 h_2(t, t_0) + \cdots, \tag{1.35}$$

where $a_k \in \mathbb{C}$, $k \in \mathbb{N}_0$, $t \in \mathbb{T}$, is called a power series on the time scale $\mathbb{T}$. The numbers a_k, $k \in \mathbb{N}_0$, are referred to as its coefficients.

We denote by $\mathscr{P}$ the set of all functions $f : [t_0, \infty) \to \mathbb{C}$ of the form

$$f(t) = \sum_{k=0}^{\infty} a_k h_k(t, t_0), \quad t \in [t_0, L], \tag{1.36}$$

where the coefficients a_k, $k \in \mathbb{N}_0$, satisfy

$$|a_k| \leq M R^k, \quad k \in \mathbb{N}_0, \tag{1.37}$$

for some constants $M > 0$ and $R > 0$ depending only on the series (1.36). Here $L \in \mathbb{T}, L > t_0$. Note that under the given condition (1.37), the series (1.36) converges uniformly on $[t_0, L]$. Note that $\mathscr{P}$ is a linear space. Also, every function $f \in \mathscr{P}$ can be uniquely represented in the form of a power series (1.36). Indeed, if we Δ-differentiate the series (1.36) term by term n times, we get

$$f^{\Delta^n}(t) = a_n + a_{n+1} h_1(t, t_0) + \cdots, \quad t \in [t_0, L],$$

whereupon

$$f^{\Delta^n}(t_0) = a_n, \quad n \in \mathbb{N}_0. \tag{1.38}$$

Thus, the coefficients of the power series (1.36) are defined uniquely by the formula (1.38).

1.9 Advanced Practical Problems

Problem 1.1. Classify each point $t \in \mathbb{T} = \{\sqrt[8]{15n} : n \in \mathbb{N}_0\}$ as left-dense, left-scattered, right-dense, or right-scattered.

Answer. Each of the points $t = \sqrt[8]{15n}$, $n \in \mathbb{N}$, is isolated; $t = 0$ is right-scattered.

Problem 1.2. Let $\mathbb{T} = \left\{ \sqrt[3]{n+2} : n \in \mathbb{N}_0 \right\}$. Find $\mu(t), t \in \mathbb{T}$.

Answer. $\mu\left(\sqrt[3]{n+2}\right) = \sqrt[3]{n+3} - \sqrt[3]{n+2}$.

Problem 1.3. Let $\mathbb{T} = \left\{ t = \sqrt[3]{n+1} : n \in \mathbb{N} \right\}$, $f(t) = 1 + t^3$, $t \in \mathbb{T}$. Find $f(\sigma(t))$, $t \in \mathbb{T}$.

Answer. $2 + t^3$.

Problem 1.4. Let $\mathbb{T} = \left\{ \dfrac{7}{8n + 11} : n \in \mathbb{N} \right\} \cup \{0\}$. Find $\mathbb{T}^\kappa$.

Answer. $\left\{ \dfrac{7}{8n + 11} : n \in \mathbb{N}, n \geq 2 \right\} \cup \{0\}$.

Problem 1.5. Let $f(t) = 1 + t^4$, $t \in \mathbb{T}$. Prove that

$$f^{\Delta}(t) = (\sigma(t))^3 + t(\sigma(t))^2 + t^2\sigma(t) + t^3, \quad t \in \mathbb{T}^\kappa.$$

Problem 1.6. Let $\mathbb{T} = 3^{\mathbb{N}_0}$, $f(t) = \frac{t-1}{t+2}$, $t \in \mathbb{T}$. Find $f^{\Delta}(t), t \in \mathbb{T}$.

Answer.

$$f^{\Delta}(t) = \frac{3}{(t+2)(3t+2)}, \quad t \in \mathbb{T}.$$

Problem 1.7. Let $\mathbb{T} = \{2n + 5 : n \in \mathbb{N}_0\}$, $f(t) = t + 2$, $g(t) = t - 2$. Find a constant $c \in [7, \sigma(7)]$ such that

$$(f \circ g)^{\Delta}(7) = f'(g(c))g^{\Delta}(7).$$

Answer. Every $c \in [7, 9]$.

Problem 1.8. Let $\mathbb{T} = \left\{3^{n+1} : n \in \mathbb{N}_0\right\}$, $v(t) = t^5$, $w(t) = t^2 + t$. Prove

$$(w \circ v)^{\Delta}(t) = (w^{\tilde{\Delta}} \circ v(t))v^{\Delta}(t), \quad t \in \mathbb{T}^{\kappa}.$$

Problem 1.9. Let $\mathbb{T} = \mathbb{R}$ and

$$f(t) = \begin{cases} 4 & \text{for} \quad t = 3, \\ \frac{1}{t-3} & \text{for} \quad t \in \mathbb{R}\backslash\{3\}. \end{cases}$$

Determine whether f is regulated.

Answer. It is not.

Problem 1.10. Let $\mathbb{T} = \mathbb{R}$ and

$$f(t) = \begin{cases} 0 & \text{if} \quad t = 1, \\ \frac{11}{2t-2} & \text{if} \quad \mathbb{R}\backslash\{1\}. \end{cases}$$

Check whether $f : \mathbb{T} \to \mathbb{R}$ is predifferentiable, and if it is, find the region of differentiation.

Answer. It is not.

Problem 1.11. Let $\mathbb{T} = 3^{\mathbb{N}_0}$. Prove that

$$\int \left(t^2 + t\right) \Delta t = \frac{1}{13}t^3 + \frac{1}{4}t^2 + c, \quad t \in \mathbb{T}.$$

Problem 1.12. Let $\mathbb{T} = 2^{\mathbb{N}_0}$. Find

1. $\int \left(t^2 - 3t + 1\right) \Delta t$,
2. $\int \left(t^3 - 2t^2 + t + 3\right) \Delta t$,
3. $\int (t+1)(t+5) \Delta t$.

Problem 1.13. Let $\mathbb{T} = 3^{\mathbb{N}_0}$. Find

1. $\int_1^9 \left(3t^2 + 1\right) \Delta t$,
2. $\int_1^9 \left(2t^3 + t^2 + 5t + 1\right) \Delta t$,
3. $\int_1^3 t(t-1)(t-5) \Delta t$.

Problem 1.14. Let $\mathbb{T} = 3^{\mathbb{N}_0}$. Find the solution of the Cauchy problem

$$\phi^{\Delta}(t) = \phi(t), \quad t \in \mathbb{T}, \quad \phi(1) = 1.$$

Answer. $e_1(t, 1), t \in \mathbb{T}$.

Problem 1.15. Let $p : \mathbb{T} \to \mathbb{R}$ be rd-continuous and $1 + \mu(t)p(t) \neq 0$ for all $t \in \mathbb{T}$. Let also $\phi(t)$ be a nontrivial solution to the equation

$$\phi^\Delta - p(t)\phi = 0.$$

Prove that $\dfrac{1}{\phi(t)}$ is a solution to the equation

$$\psi^\Delta + p(t)\psi^\sigma = 0.$$

Chapter 2
The Laplace Transform on Time Scales

2.1 Definition and Properties

Let $\mathbb{T}_i$, $i \in \{1, \ldots, n\}$, be time scales with forward jump operators, backward jump operators, delta differentiation operators, and nabla differentiation operators σ_i, ρ_i, Δ_i, and ∇_i, respectively. Assume that $0 \in \mathbb{T}_i$, $i \in \{1, \ldots, n\}$. If $1 + \mu_i(t)z_i \neq 0$ for all $t \in \mathbb{T}_i^\kappa$, then

$$\ominus z_i = -\frac{z_i}{1 + \mu_i(t_i)\, z_i}$$

and

$$1 + \mu_i(t_i)\,(\ominus z_i) = 1 - \frac{\mu_i(t)z_i}{1 + \mu_i(t_i)\, z_i}$$

$$= \frac{1}{1 + \mu_i(t_i)\, z_i}$$

for all $t_i \in \mathbb{T}_i^\kappa$, and hence $e_{\ominus z_i}(t_i, 0)$ is well defined on $\mathbb{T}_i^\kappa$, $i \in \{1, \ldots, n\}$. Let $\Lambda^n = \mathbb{T}_1 \times \ldots \times \mathbb{T}_n$.

Definition 2.1. A function $f : \Lambda^n \to \mathbb{R}$ will be called regulated with respect to t_i if the function $f : \mathbb{T}_i \to \mathbb{R}$ is regulated.

Definition 2.2. Let $i \in \{1, \ldots, n\}$ be fixed. Suppose that $f : \Lambda^n \to \mathbb{R}$ is regulated with respect to t_i and $\sup \mathbb{T}_i = \infty$. Then the Laplace transform of f with respect to t_i is defined by

© Springer International Publishing AG, part of Springer Nature 2018
S. G. Georgiev, *Fractional Dynamic Calculus and Fractional Dynamic Equations on Time Scales*, https://doi.org/10.1007/978-3-319-73954-0_2

$$\mathscr{L}_i(f)(t_1,\ldots,t_{i-1},z_i,t_{i+1},\ldots,t_n) = \int_0^\infty e^{\sigma_i}_{\ominus z_i}(t_i,0)f(t)\Delta_i t_i, \qquad (2.1)$$

$t = (t_1,\ldots,t_n)$, for $z_i \in \mathbb{C}$ for which $1 + \mu_i(t_i)z_i \neq 0$ for all $t_i \in \mathbb{T}_i^\kappa$ and the improper integral (2.1) exists.

Remark 2.1. Let $i \in \{1,\ldots,n\}$ be fixed and suppose that $f : \Lambda^n \to \mathbb{R}$ is regulated with respect to t_i. By $\mathscr{D}_i\{f\}$ we will denote the set of all complex numbers z_i for which $1 + \mu_i(t_i)z_i \neq 0$ for all $t_i \in \mathbb{T}_i^\kappa$ and the improper integral (2.1) exists.

Theorem 2.1 (Linearity). *Assume that $f, g : \Lambda^n \to \mathbb{R}$ are regulated with respect to t_i and $\sup \mathbb{T}_i = \infty$ for some $i \in \{1,\ldots,n\}$. Then for all constants α and β, we have*

$$\mathscr{L}_i(\alpha f + \beta g)(t_1,\ldots,t_{i-1},z_i,t_{i+1},\ldots,t_n) = \alpha\mathscr{L}_i(f)(t_1,\ldots,t_{i-1},z_i,t_{i+1},\ldots,t_n)$$
$$+\beta\mathscr{L}_i(g)(t_1,\ldots,t_{i-1},z_i,t_{i+1},\ldots,t_n)$$

for $z_i \in \mathscr{D}_i\{f\} \bigcap \mathscr{D}_i\{g\}$, $t_j \in \mathbb{T}_j, j \in \{1,\ldots,n\}, j \neq i$.

Proof. Let $z_i \in \mathscr{D}_i\{f\} \bigcap \mathscr{D}_i\{g\}$, $t_j \in \mathbb{T}_j, j \in \{1,\ldots,n\}, j \neq i$. Then we have

$$\mathscr{L}_i(\alpha f + \beta g)(t_1,\ldots,t_{i-1},z_i,t_{i+1},\ldots,t_n) = \int_0^\infty e^{\sigma_i}_{\ominus z_i}(t_i,0)(\alpha f + \beta g)(t)\Delta_i t_i$$

$$= \int_0^\infty e^{\sigma_i}_{\ominus z_i}(t_i,0)(\alpha f + \beta g)(t)\Delta_i t_i$$

$$= \int_0^\infty e^{\sigma_i}_{\ominus z_i}(t_i,0)\alpha f(t)\Delta_i t_i + \int_0^\infty e^{\sigma_i}_{\ominus z_i}(t_i,0)\beta g(t)\Delta_i t_i$$

$$= \alpha\int_0^\infty e^{\sigma_i}_{\ominus z_i}(t_i,0)f(t)\Delta_i t_i + \beta\int_0^\infty e^{\sigma_i}_{\ominus z_i}(t_i,0)g(t)\Delta_i t_i$$

$$= \alpha\mathscr{L}_i(f)(t_1,\ldots,t_{i-1},z_i,t_{i+1},\ldots,t_n) + \beta\mathscr{L}_i(g)(t_1,\ldots,t_{i-1},z_i,t_{i+1},\ldots,t_n).$$

This completes the proof.

Lemma 2.1. *Let $i \in \{1,\ldots,n\}$ and $z_i \in \mathbb{C}$. Then*

$$e^{\sigma_i}_{\ominus z_i}(t_i,0) = \frac{e_{\ominus z_i}(t_i,0)}{1 + \mu_i(t_i)z_i}$$

$$= -\frac{(\ominus z_i)(t_i)}{z_i}e_{\ominus z_i}(t_i,0), \quad t_i \in \mathbb{T}_i.$$

Proof. For $z_i \in \mathbb{C}$ and $t_i \in \mathbb{T}_i$, we have

$$e^{\sigma_i}_{\ominus z_i}(t_i,0) = e_{\ominus z_i}(t_i,0) + \mu_i(t_i)e^{\Delta_i}_{\ominus z_i}(t_i,0)$$

$$= e_{\ominus z_i}(t_i,0) + (\ominus z_i)(t_i)\mu_i(t_i)e_{\ominus z_i}(t_i,0)$$

$$= (1 + (\ominus z_i)(t_i)\mu_i(t_i))e_{\ominus z_i}(t_i,0)$$

$$= \left(1 - \frac{\mu_i\,(t_i)\,z_i}{1 + \mu_i\,(t_i)\,z_i}\right) e_{\ominus z_i}\,(t_i,0)$$

$$= \frac{1}{1 + \mu_i\,(t_i)\,z_i}\,e_{\ominus z_i}\,(t_i,0)$$

$$= -\frac{1}{z_i}\left(-\frac{z_i}{1 + \mu_i\,(t_i)\,z_i}\right) e_{\ominus z_i}\,(t_i,0)$$

$$= -\frac{(\ominus z_i)\,(t_i)}{z_i}\,e_{\ominus z_i}\,(t_i,0)\,.$$

This completes the proof.

Theorem 2.2. *Assume that* $f : \Lambda^n \to \mathbb{C}$ *is such that* $f_{t_i}^{\Delta_i^l}$, $l \in \{0, \dots, k\}$, *is regulated with respect to* t_i *and* $\sup \mathbb{T}_i = \infty$ *for some* $i \in \{1, \dots, n\}$ *and for some* $k \in \mathbb{N}$. *Then*

$$\mathscr{L}_i\left(f_{t_i}^{\Delta_i^k}\right)(t_1, \dots, t_{i-1}, z_i, t_{i+1}, \dots, t_n)$$

$$= z_i^k\,\mathscr{L}_i\,(f)\,(t_1, \dots, t_{i-1}, z_i, t_{i+1}, \dots, t_n)$$

$$- \sum_{l=0}^{k-1} z_i^l f_{t_i}^{\Delta_i^{k-1-l}}\,(t_1, \dots, t_{i-1}, 0, t_{i+1}, \dots, t_n) \tag{2.2}$$

for all $z_i \in \mathscr{D}_i\{f\} \cap \mathscr{D}_i\left\{f_{t_i}^{\Delta_i}\right\} \cap \dots \cap \mathscr{D}\left\{f_{t_i}^{\Delta_i^k}\right\}$ *satisfying*

$$\lim_{t_i \to \infty}\left(f_{t_i}^{\Delta_i^l}(t)e_{\ominus z_i}\,(t_i,0)\right) = 0, \quad l = 0, \dots, k-1. \tag{2.3}$$

Proof. We will use induction.

1. Let $k = 1$. Integrating by parts and using Lemma 2.1 and the condition (2.3), we get

$$\mathscr{L}_i\left(f_{t_i}^{\Delta_i}\right)(t_1, \dots, t_{i-1}, z_i, t_{i+1}, \dots, t_n) = \int_0^\infty f_{t_i}^{\Delta_i}(t)e_{\ominus z_i}^{\sigma_i}\,(t_i,0)\,\Delta_i t_i$$

$$= f(t)e_{\ominus z_i}^{\sigma_i}\,(t_i,0)\,\Big|_{t_i=0}^{t_i\to\infty} - \int_0^\infty f(t)\,(\ominus z_i)\,(t_i)\,e_{\ominus z_i}\,(t_i,0)\,\Delta_i t_i$$

$$= -f\,(t_1, \dots, t_{i-1}, 0, t_{i+1}, \dots, t_n) + z_i \int_0^\infty f(t)e_{\ominus z_i}^{\sigma_i}\,(t_i,0)\,\Delta_i t_i$$

$$= -f\,(t_1, \dots, t_{i-1}, 0, t_{i+1}, \dots, t_n)$$

$$+ z_i\mathscr{L}_i\,(f)\,(t_1, \dots, t_{i-1}, z_i, t_{i+1}, \dots, t_n)\,.$$

2. Assume that (2.2) holds for some $k \in \mathbb{N}$.
3. We will prove

$$\mathscr{L}_i\left(f_{t_i}^{\Delta_i^{k+1}}\right)(t_1, \ldots, t_{i-1}, z_i, t_{i+1}, \ldots, t_n)$$
$$= z_i^{k+1}\mathscr{L}_i(f)(t_1, \ldots, t_{i-1}, z_i, t_{i+1}, \ldots, t_n)$$
$$- \sum_{l=0}^{k} z_i^l f_{t_i}^{\Delta_i^{k-l}}(t_1, \ldots, t_{i-1}, 0, t_{i+1}, \ldots, t_n).$$

Integrating by parts and using Lemma 2.1 and (2.3), we get

$$\mathscr{L}_i\left(f_{t_i}^{\Delta_i^{k+1}}\right)(t_1, \ldots, t_{i-1}, z_i, t_{i+1}, \ldots, t_n) = \int_0^\infty f_{t_i}^{\Delta_i^{k+1}}(t)e_{\ominus z_i}^{\sigma_1}(t_i, 0)\,\Delta_i t_i$$

$$= f_{t_i}^{\Delta_i^k}(t)e_{\ominus z_i}(t_i, 0)\Big|_{t_i=0}^{|t_i \to \infty} - \int_0^\infty f_{t_i}^{\Delta_i^k}(t)\,(\ominus z_i)\,e_{\ominus z_i}(t_i, 0)\,\Delta_i t_i$$

$$= z_i \int_0^\infty f_{t_i}^{\Delta_i^k}(t)e_{\ominus z_i}^{\sigma_i}(t_i, 0)\,\Delta_i t_i - f_{t_i}^{\Delta_i^k}(t_1, \ldots, t_{i-1}, 0, t_{i+1}, \ldots, t_n)$$

$$= z_i^{k+1}\mathscr{L}_i(f)(t_1, \ldots, t_{i-1}, z_i, t_{i+1}, \ldots, t_n)$$

$$- \sum_{l=0}^{k-1} z_i^{l+1} f_{t_i}^{\Delta_i^{k-1-l}}(t_1, \ldots, t_{i-1}, 0, t_{i+1}, \ldots, t_n)$$

$$- f_{t_i}^{\Delta_i^k}(t_1, \ldots, t_{i-1}, 0, t_{i+1}, \ldots, t_n)$$

$$= z_i^{k+1}\mathscr{L}_i(f)(t_1, \ldots, t_{i-1}, z_i, t_{i+1}, \ldots, t_n)$$

$$- \sum_{l=0}^{k} z_i^l f_{t_i}^{\Delta_i^{k-l}}(t_1, \ldots, t_{i-1}, 0, t_{i+1}, \ldots, t_n).$$

This completes the proof.

Example 2.1. Let $n = 1$ and $z \in \mathbb{C}$ be such that

$$1 + \mu(t)z \neq 0 \quad \text{and} \quad \lim_{t \to \infty} e_{\ominus z}(t, 0) = 0.$$

Then

$$\mathscr{L}(1)(z) = \int_0^\infty e_{\ominus z}^\sigma(t, 0)\Delta t$$

$$= \int_0^\infty (1 + \mu(t)(\ominus z))\,e_{\ominus z}(t, 0)\Delta t$$

$$= \int_0^\infty \left(1 - \frac{z\mu(t)}{1 + z\mu(t)}\right)e_{\ominus z}(t, 0)\Delta t$$

$$= -\frac{1}{z} \int_0^\infty \left(-\frac{z}{1 + z\mu(t)} \right) e_{\ominus z}(t, 0) \Delta t$$

$$= -\frac{1}{z} \int_0^\infty (\ominus z) \, e_{\ominus z}(t, 0) \, \Delta t$$

$$= -\frac{1}{z} \int_0^\infty e_{\ominus z}^\Delta(t, 0) \Delta t$$

$$= -\frac{1}{z} e_{\ominus z}(t, 0) \Big|_{t=0}^{t \to \infty}$$

$$= \frac{1}{z}.$$

Example 2.2. Let $\mathbb{T}_1 = \mathbb{T}_2 = \mathbb{Z}$ and $z_1 \in \mathbb{C}$ be such that

$$z_1 \neq -1, 0, \quad \lim_{t_1 \to \infty} \frac{1}{(1 + z_1)^{t_1}} = \lim_{t_1 \to \infty} \frac{t_1}{(1 + z_1)^{t_1}} = 0.$$

Let also

$$f(t_1, t_2) = t_1^2 + t_2^2, \quad (t_1, t_2) \in \mathbb{T}_1 \times \mathbb{T}_2.$$

We will find $\mathscr{L}_1(f)(z_1, t_2)$. We have

$$\sigma_1(t_1) = t_1 + 1, \quad \mu_1(t_1) = 1, \quad t_1 \in \mathbb{T}_1.$$

We set

$$f_1(t_1) = t_1, \quad t_1 \in \mathbb{T}_1.$$

By (2.2), we have

$$\mathscr{L}_1 \left(f_1^{\Delta_1} \right)(z_1, t_2) = z_1 \mathscr{L}_1(f_1)(z_1, t_2) - f_1(0, t_2),$$

or

$$\mathscr{L}_1(1)(z_1, t_2) = z_1 \mathscr{L}(f_1)(z_1, t_2),$$

or

$$\frac{1}{z_1} = z_1 \mathscr{L}_1(f_1)(z_1, t_2),$$

or

$$\mathscr{L}_1(f_1)(z_1, t_2) = \frac{1}{z_1^2}.$$

Note that

$$f_{t_1}^{\Delta_1}(t_1, t_2) = \sigma_1(t_1) + t_1$$
$$= t_1 + 1 + t_1$$
$$= 2t_1 + 1.$$

Therefore,

$$\mathscr{L}_1\left(f_{t_1}^{\Delta_1}\right)(z_1, t_2) = 2\mathscr{L}_1(f_1)(z_1, t_2) + \mathscr{L}_1(1)(z_1, t_2)$$

$$= 2\frac{1}{z_1^2} + \frac{1}{z_1}$$

$$= \frac{z_1 + 2}{z_1^2}.$$

Now we use (2.2), and we get

$$\mathscr{L}_1\left(f_{t_1}^{\Delta_1}\right)(z_1, t_2) = z_1\mathscr{L}_1(f)(z_1, t_2) - f(0, t_2),$$

or

$$\frac{z_1 + 2}{z_1^2} = z_1\mathscr{L}_1(f)(z_1, t_2) - t_2^2,$$

or

$$\frac{z_1 + 2}{z_1^2} + t_2^2 = z_1\mathscr{L}_1(f)(z_1, t_2),$$

or

$$\frac{t_2^2 z_1^2 + z_1 + 2}{z_1^2} = z_1\mathscr{L}_1(f)(z_1, t_2),$$

or

$$\mathscr{L}_1(f)(z_1, t_2) = \frac{t_2^2 z_1^2 + z_1 + 2}{z_1^3}.$$

Example 2.3. Let $\mathbb{T}_1 = \mathbb{T}_2 = \mathbb{N}_0^2$ and

$$f(t_1, t_2) = t_1\frac{2t_2 + 2\sqrt{t_2} + 3}{(t_2 + 1)^2\left(t_2 + 2\sqrt{t_2} + 2\right)^2}.$$

We will find $\mathcal{L}_2(f)(t_1, z_2)$ for all $z_2 \in \mathbb{C}$ such that

$$1 + z_2\left(1 + 2\sqrt{t_2}\right) \neq 0$$

for all $t_2 \in \mathbb{T}_2$, and

$$\lim_{t_2 \to \infty} \left(\frac{1}{(t_2+1)^2} \prod_{\tau_2 \in [0,t_2)} \frac{1}{\left(1 + z_2\left(1 + 2\sqrt{\tau_2}\right)\right)^{1+2\sqrt{\tau_2}}} \right) = 0.$$

Let

$$g(t_2) = \frac{1}{(t_2+1)^2}, \quad h(t_1, t_2) = \frac{t_1}{(t_2+1)^2}.$$

We have

$$\sigma_2(t_2) = t_2 + 2\sqrt{t_2} + 1, \quad \mu_2(t_2) = 2\sqrt{t_2} + 1, \quad t_2 \in \mathbb{T}_2.$$

Note that

$$\begin{aligned}
e^{\sigma_2}_{\ominus z_2}(t_2, 0) &= e_{-\frac{z_2}{1+z_2\mu_2(t_2)}}(\sigma_2(t_2), 0) \\
&= e^{-\int_0^{\sigma_2(t_2)} \mathrm{Log}(1+z_2\mu_2(\tau_2))\Delta_2\tau_2} \\
&= e^{-\sum_{\tau_2 \in [0,t_2]} \mu_2(\tau_2)\mathrm{Log}(1+z_2\mu_2(\tau_2))} \\
&= \prod_{\tau_2 \in [0,t_2]} \frac{1}{(1 + z_2\mu_2(\tau_2))^{\mu_2(\tau_2)}} \\
&= \prod_{\tau_2 \in [0,t_2]} \frac{1}{\left(1 + z_2\left(1 + 2\sqrt{\tau_2}\right)\right)^{1+2\sqrt{\tau_2}}}.
\end{aligned}$$

Then

$$\begin{aligned}
\mathcal{L}_2(g)(t_1, z_2) &= \int_0^\infty \frac{1}{(1+t_2)^2} e^{\sigma_2}_{\ominus z_2}(t_2, 0)\, \Delta_2 t_2 \\
&= \sum_{t_2 \in [0,\infty)} \frac{2\sqrt{t_2} + 1}{(t_2+1)^2} \prod_{\tau_2 \in [0,t_2]} \frac{1}{\left(1 + z_2\left(1 + 2\sqrt{\tau_2}\right)\right)^{1+2\sqrt{\tau_2}}}.
\end{aligned}$$

Next,

$$h_{t_2}^{\Delta_2}(t_1, t_2) = -t_1 \frac{\sigma_2(t_2) + t_2 + 2}{(t_2 + 1)^2 (\sigma_2(t_2) + 1)^2}$$

$$= -t_1 \frac{2t_2 + 2\sqrt{t_2} + 3}{(t_2 + 1)^2 (t_2 + 2\sqrt{t_2} + 2)}$$

$$= -f(t_1, t_2).$$

From this and (2.2), we obtain

$$\mathscr{L}_2\left(h_{t_2}^{\Delta_2}\right)(t_1, z_2) = z_2 \mathscr{L}_2(h)(t_1, z_2) - h(t_1, 0),$$

or

$$-\mathscr{L}_2(f)(t_1, z_2) = z_2 t_1 \mathscr{L}_2(g)(t_1, z_2) - t_1,$$

or

$$\mathscr{L}_2(f)(t_1, z_2) = -z_2 t_1 \sum_{t_2 \in [0,\infty)} \frac{2\sqrt{t_2} + 1}{(t_2 + 1)^2} \prod_{\tau_2 \in [0,t_2]} \frac{1}{\left(1 + z_2\left(1 + 2\sqrt{\tau_2}\right)\right)^{1 + 2\sqrt{\tau_2}}} + t_1.$$

Example 2.4. Let $\mathbb{T}_1 = \mathbb{T}_2 = 3^{\mathbb{N}_0} \bigcup \{0\}$,

$$f(t_1, t_2) = \begin{cases} \dfrac{4t_2}{(t_1 + 1)(3t_1 + 1)(9t_1 + 1)}, & t_1 \neq 0, \quad (t_1, t_2) \in \mathbb{T}_1 \times \mathbb{T}_2, \\ \dfrac{3}{8}t_2, & t_1 = 0, \quad t_2 \in \mathbb{T}_2. \end{cases}$$

We will find $\mathscr{L}_1(f)(z_1, t_2)$ for all $z_1 \in \mathbb{C}$ such that

$$z_1 \neq -1, \quad 1 + 2z_1 t_1 \neq 0,$$

for all $t_1 \in \mathbb{T}_1$, and

$$\lim_{t_1 \to \infty} \prod_{\tau_1 \in [1,t_1]} \frac{1}{(1 + 2z_1\tau_1)^{2\tau_1}} = 0.$$

Let

$$g(t_1) = \frac{1}{1 + t_1}, \quad h(t_1, t_2) = \frac{t_2}{1 + t_1}, \quad (t_1, t_2) \in \mathbb{T}_1 \times \mathbb{T}_2.$$

We have

$$\sigma_1(t_1) = 3t_1, \quad \mu_1(t_1) = 2t_1, \quad \sigma_1(0) = 1, \quad \mu_1(0) = 1, \quad t_1 \in 3^{\mathbb{N}_0}.$$

Next,

$$
\begin{aligned}
e^{\sigma_1}_{\ominus z_1}(t_1, 0) &= e^{\int_0^{\sigma_1(t_1)} \mathrm{Log}(1+(\ominus z_1)\mu_1(\tau_1))\Delta_1\tau_1} \\
&= e^{\int_0^{\sigma_1(t_1)} \mathrm{Log}\left(1-\frac{z_1\mu_1(\tau_1)}{1+z_1\mu_1(\tau_1)}\right)\Delta_1\tau_1} \\
&= e^{\int_0^{\sigma_1(t_1)} \mathrm{Log}\frac{1}{1+z_1\mu_1(\tau_1)}\Delta_1\tau_1} \\
&= e^{\sum\limits_{\tau_1 \in [0,t_1]} \mu_1(\tau_1)\mathrm{Log}\frac{1}{1+z_1\mu_1(\tau_1)}} \\
&= \prod_{\tau_1 \in [0,t_1]} \frac{1}{(1+z_1\mu_1(\tau_1))^{\mu_1(\tau_1)}} \\
&= \begin{cases} \frac{1}{1+z_1} \prod_{\tau_1 \in [1,t_1]} \frac{1}{(1+2z_1\tau_1)^{2\tau_1}}, & t_1 \in 3^{\mathbb{N}_0}, \\ \frac{1}{z_1+1}, & t_1 = 0. \end{cases}
\end{aligned}
$$

Hence,

$$
\begin{aligned}
\mathscr{L}_1(g)(z_1) &= \int_0^\infty \frac{1}{t_1+1} e^{\sigma_1}_{\ominus z_1}(t_1, 0)\, \Delta_1 t_1 \\
&= \sum_{t_1 \in [0,\infty)} \frac{\mu_1(t_1)}{t_1+1} e^{\sigma_1}_{\ominus z_1}(t_1, 0) \\
&= \frac{1}{1+z_1}\left(1 + \sum_{t_1 \in [1,\infty)} \frac{2t_1}{t_1+1} \prod_{\tau_1 \in [1,t_1]} \frac{1}{(1+2z_1\tau_1)^{2\tau_1}}\right).
\end{aligned}
$$

Note that

$$
\begin{aligned}
h_{t_1}^{\Delta_1}(t_1, t_2) &= -t_2 \frac{1}{(1+t_1)(1+\sigma_1(t_1))} \\
&= \begin{cases} -\frac{t_2}{2}, & t_1 = 0, \\ -\frac{t_2}{(1+t_1)(1+3t_1)}, & t_1 \neq 0, \end{cases}
\end{aligned}
$$

for $t_1 = 0$

$$
\begin{aligned}
h_{t_1}^{\Delta_1^2}(0, t_2) &= \frac{h_{t_1}^{\Delta_1}(\sigma_1(0), t_2) - h_{t_1}^{\Delta_1}(0, t_2)}{\sigma_1(0) - 0} \\
&= h_{t_1}^{\Delta_1}(1, t_2) + \frac{t_2}{2}
\end{aligned}
$$

$$= -\frac{t_2}{8} + \frac{t_2}{2}$$

$$= \frac{3}{8}t_2$$

$$= f(0, t_2),$$

for $t_1 \neq 0$

$$h_{t_1}^{\Delta_1^2}(t_1, t_2) = \frac{h_{t_1}^{\Delta_1}(\sigma_1(t_1), t_2) - h_{t_1}^{\Delta_1}(t_1, t_2)}{\sigma_1(t_1) - t_1}$$

$$= \frac{-\frac{t_2}{(3t_1+1)(9t_1+1)} + \frac{t_2}{(t_1+1)(3t_1+1)}}{2t_1}$$

$$= \frac{4t_2}{(t_1+1)(3t_1+1)(9t_1+1)}$$

$$= f(t_1, t_2).$$

From this and (2.2), we get

$$\mathscr{L}_1\left(h_{t_1}^{\Delta_1^2}\right)(z_1, t_2) = \mathscr{L}_1(f)(z_1, t_2)$$

$$= z_1^2 \mathscr{L}_1(h)(z_1, t_2) - h_{t_1}^{\Delta_1}(0, t_2) - z_1 h(0, t_2)$$

$$= \frac{z_1^2 t_2}{1 + z_1}\left(1 + \sum_{t_1 \in [1,\infty)} \frac{2t_1}{1 + t_1} \prod_{\tau_1 \in [1,t_1]} \frac{1}{(1 + 2z_1\tau_1)^{2\tau_1}}\right)$$

$$- \frac{t_2}{2} - t_2 z_1.$$

Exercise 2.1. Let $\mathbb{T}_1 = \mathbb{T}_2 = 2^{\mathbb{N}_0} \bigcup\{0\}$ and

$$f(t_1, t_2) = \frac{t_2}{2t_1^3 + 1}, \quad (t_1, t_2) \in \mathbb{T}_1 \times \mathbb{T}_2.$$

Find $\mathscr{L}_1(f)(z_1, t_2)$ for all $z_1 \in \mathbb{C}$ such that $1 + t_2 z_1 \neq 0$ for all $t_2 \in \mathbb{T}_2$.

Answer.

$$\mathscr{L}_1(f)(z_1, t_2) = \frac{t_2}{1 + z_1}\left(1 + \sum_{t_1 \in [1,\infty)} \frac{t_1}{2t_1^3 + 1} \prod_{\tau_1 \in [1,t_1]} \frac{1}{(1 + \tau_1 z_1)^{\tau_1}}\right).$$

Theorem 2.3. *Let $m \in \mathbb{N}$, $i \in \{1, \ldots, n\}$, be fixed and let $f : \Lambda^n \to \mathbb{R}$ be regulated with respect to t_i and $\sup \mathbb{T}_i = \infty$. Suppose that for some $j \in \{1, \ldots, n\}$, $j \neq i$, $f_{t_j}^{\Delta_j^l}$ exist and are continuous at $t \in \Lambda^n$, $t_j \in \mathbb{T}_j^\kappa$, for all $l \in \{0, \ldots, m\}$. Then*

$$\mathscr{L}_i\left(f_{t_j}^{\Delta_j^m}\right)(t_1, \ldots, t_{i-1}, z_i, t_{i+1}, \ldots, t_n)$$

$$= \mathscr{L}_{it_j}^{\Delta_j^m}(f)(t_1, \ldots, t_{i-1}, z_i, t_{i+1}, \ldots, t_n)$$

for $z_i \in \mathscr{D}_i\{f\}$ and for all $t_r \in \mathbb{T}_r$, $r \in \{1, \ldots, n\}$, $r \neq i, j$, $t_j \in \mathbb{T}_j^\kappa$.

Proof. We have

$$\mathscr{L}_i\left(f_{t_j}^{\Delta_j^m}\right)(t_1, \ldots, t_{i-1}, z_i, t_{i+1}, \ldots, t_n)$$

$$= \int_0^\infty f_{t_j}^{\Delta_j^m}(t) e_{\ominus z_i}^{\sigma_i}(t_i, 0)\, \Delta_i t_i$$

$$= \left(\int_0^\infty f(t) e_{\ominus z_i}^{\sigma_i}(t_i, 0)\, \Delta_i t_i\right)_{t_j}^{\Delta_j},$$

for $z_i \in \mathscr{D}_i\{f\}$ and for all $t_r \in \mathbb{T}_r$, $r \in \{1, \ldots, n\}$, $r \neq i, j$, $t_j \in \mathbb{T}_j^\kappa$. In the last equality we have used the Lebesgue dominated convergence theorem. This completes the proof. ∎

Below we will give some important properties of the Laplace transform. For convenience, we consider the case $n = 1$. These properties will be used in the next sections. Suppose that $\mathbb{T}_0$ is a time scale with forward jump operator and delta differentiation operator σ and Δ, respectively, and $0 \in \mathbb{T}_0$, and $\sup \mathbb{T}_0 = \infty$.

Theorem 2.4. *Assume that $f : \mathbb{T}_0 \longmapsto \mathbb{C}$ is regulated. If*

$$F(x) = \int_0^x f(y)\, \Delta y$$

for $x \in \mathbb{T}_0$, then

$$\mathscr{L}(F)(z) = \frac{1}{z}\mathscr{L}(f)(z)$$

for all $z \in \mathscr{D}\{f\}\backslash\{0\}$ satisfying

$$\lim_{x \to \infty}\left(e_{\ominus z}(x, 0)\int_0^x f(y)\, \Delta y\right) = 0.$$

Proof. Using integration by parts, we get

$$
\begin{aligned}
\mathcal{L}(F)(z) &= \int_0^\infty F(y)e_{\ominus z}(\sigma(y), 0)\Delta y \\
&= \int_0^\infty F(y)(1 + \ominus z\mu(y))e_{\ominus z}(y, 0)\Delta y \\
&= \int_0^\infty F(y)\frac{1}{1 + z\mu(y)}e_{\ominus z}(y, 0)\Delta y \\
&= -\frac{1}{z}\int_0^\infty F(y)\frac{-z}{1 + z\mu(y)}e_{\ominus z}(y, 0)\Delta y \\
&= -\frac{1}{z}\int_0^\infty F(y) \ominus z(y)e_{\ominus z}(y, 0)\Delta y \\
&= -\frac{1}{z}\int_0^\infty F(y)e_{\ominus z}^\Delta(y, 0)\Delta y \\
&= -\frac{1}{z}\left(\lim_{y\to\infty} F(y)e_{\ominus z}(y, 0) - F(0)e_{\ominus z}(0, 0)\right) \\
&\quad +\frac{1}{z}\int_0^\infty F^\Delta(y)e_{\ominus z}(\sigma(y), 0)\Delta y \\
&= \frac{1}{z}\int_0^\infty f(y)e_{\ominus z}(\sigma(y), 0)\Delta y \\
&= \frac{1}{z}\mathcal{L}(f)(z),
\end{aligned}
$$

which completes the proof.

Theorem 2.5. *For all $n \in \mathbb{N}_0$ we have*

$$
\mathcal{L}(h_n(x, 0))(z) = \frac{1}{z^{n+1}}, \qquad x \in \mathbb{T}_0, \tag{2.4}
$$

for all $z \in \mathbb{C}\backslash\{0\}$ such that $1 + z\mu(x) \neq 0$, $x \in \mathbb{T}_0$, and

$$
\lim_{x\to\infty} (h_n(x, 0)e_{\ominus z}(x, 0)) = 0. \tag{2.5}
$$

Proof. We note that (2.5) implies

$$
\lim_{x\to\infty} (h_l(x, 0)e_{\ominus z}(x, 0)) = 0 \quad \text{for all} \quad 0 \leq l \leq n.
$$

To prove our assertion we will use mathematical induction.

1. $n = 0$. We have that $h_0(x, 0) = 1$ and

$$
\mathcal{L}(1)(z) = \frac{1}{z}.
$$

2.1 Definition and Properties

2. Assume that (2.4) holds for some $n \in \mathbb{N}_0$.
3. We will prove that

$$\mathscr{L}(h_{n+1}(x, 0))(z) = \frac{1}{z^{n+2}}$$

for all $z \in \mathbb{C} \backslash \{0\}$ such that $1 + z\mu(x) \neq 0$, $x \in \mathbb{T}_0$, and

$$\lim_{x \to \infty} (h_{n+1}(x, 0)e_{\ominus z}(x, 0)) = 0.$$

Indeed,

$$\mathscr{L}(h_{n+1}(x, 0))(z) = \int_0^\infty h_{n+1}(y, 0)e_{\ominus z}(\sigma(y), 0)\Delta y$$

$$= \int_0^\infty \left(\int_0^y h_n(t, 0)\Delta t \right) e_{\ominus z}(\sigma(y), 0)\Delta y.$$

From this, using Theorems 2.4 and (2.4), we obtain

$$\mathscr{L}(h_{n+1}(x, 0))(z) = \frac{1}{z}\mathscr{L}(h_n(x, 0))(z)$$

$$= \frac{1}{z} \frac{1}{z^{n+1}}$$

$$= \frac{1}{z^{n+2}},$$

which completes the proof.

Theorem 2.6. *Let $\alpha \in \mathbb{C}$ and $1 + \alpha\mu(x) \neq 0$ for $x \in \mathbb{T}_0$. Then*

$$\mathscr{L}(e_\alpha(x, 0))(z) = \frac{1}{z - \alpha}, \quad x \in \mathbb{T}_0,$$

provided

$$\lim_{x \to \infty} e_{\alpha \ominus z}(x, 0) = 0.$$

Proof. We have

$$\mathscr{L}(e_\alpha(x, 0))(z) = \int_0^\infty e_\alpha(y, 0)e_{\ominus z}(\sigma(y), 0)\Delta y$$

$$= \int_0^\infty e_\alpha(y, 0) \left(1 + (\ominus z)(y)\mu(y)\right) e_{\ominus z}(y, 0)\Delta y$$

$$= \int_0^\infty \frac{1}{1 + z\mu(y)} e_\alpha(y, 0)e_{\ominus z}(y, 0)\Delta y$$

$$= \int_0^\infty \frac{1}{1 + z\mu(y)} e_{\alpha \ominus z}(y, 0) \Delta y$$

$$= \frac{1}{\alpha - z} \int_0^\infty \frac{\alpha - z}{1 + z\mu(y)} e_{\alpha \ominus z}(y, 0) \Delta y,$$

i.e.,

$$\mathscr{L}(e_\alpha(x, 0))(z) = \frac{1}{\alpha - z} \int_0^\infty \frac{\alpha - z}{1 + z\mu(y)} e_{\alpha \ominus z}(y, 0) \Delta y. \qquad (2.6)$$

We note that

$$\alpha \ominus z = \alpha \oplus (\ominus z)$$

$$= \alpha + (\ominus z) + \alpha(\ominus z)\mu(y)$$

$$= \alpha - \frac{z}{1 + z\mu(y)} - \frac{\alpha z \mu(y)}{1 + z\mu(y)}$$

$$= \frac{\alpha - z}{1 + z\mu(y)}.$$

From this and (2.6), we obtain

$$\mathscr{L}(e_\alpha(x, 0))(z) = \frac{1}{\alpha - z} \int_0^\infty \frac{\alpha - z}{1 + z\mu(y)} e_{\alpha \ominus z}(y, 0) \Delta y$$

$$= \frac{1}{\alpha - z} \int_0^\infty \alpha \ominus z(y) e_{\alpha \ominus z}(y, 0) \Delta y$$

$$= \frac{1}{\alpha - z} \int_0^\infty e_{\alpha \ominus z}^\Delta(y, 0) \Delta y$$

$$= \frac{1}{\alpha - z} \left(\lim_{y \to \infty} e_{\alpha \ominus z}(y, 0) - e_{\alpha \ominus z}(0, 0) \right)$$

$$= \frac{1}{z - \alpha}.$$

This completes the proof.

Corollary 2.1. *We have*

1. $\mathscr{L}(\cos_\alpha(x, 0))(z) = \dfrac{z}{z^2 + \alpha^2}$,
2. $\mathscr{L}(\sin_\alpha(x, 0))(z) = \dfrac{\alpha}{z^2 + \alpha^2}$,

provided that

$$\lim_{x \to \infty} e_{i\alpha \ominus z}(x, 0) = \lim_{x \to \infty} e_{-i\alpha \ominus z}(x, 0) = 0.$$

Proof. 1. From the definition of $\cos_\alpha(x, 0)$ we have

$$\cos_\alpha(x, 0) = \frac{e_{i\alpha}(x, 0) + e_{-i\alpha}(x, 0)}{2}.$$

Hence,

$$\mathcal{L}(\cos_\alpha(x, 0))(z) = \mathcal{L}\left(\frac{e_{i\alpha}(x, 0) + e_{-i\alpha}(x, 0)}{2}\right)(z)$$

$$= \frac{1}{2}\mathcal{L}(e_{i\alpha}(x, 0) + e_{-i\alpha}(x, 0))(z)$$

$$= \frac{1}{2}\left(\mathcal{L}(e_{i\alpha}(x, 0))(z) + \mathcal{L}(e_{-i\alpha}(x, 0))(z)\right)$$

$$= \frac{1}{2}\left(\frac{1}{z - i\alpha} + \frac{1}{z + i\alpha}\right)$$

$$= \frac{1}{2}\frac{z - i\alpha + z + i\alpha}{(z - i\alpha)(z + i\alpha)}$$

$$= \frac{z}{z^2 + \alpha^2}.$$

2. From the definition of $\sin_\alpha(x, 0)$ we have

$$\sin_\alpha(x, 0) = \frac{e_{i\alpha}(x, 0) - e_{-i\alpha}(x, 0)}{2i}.$$

Hence,

$$\mathcal{L}(\sin_\alpha(x, 0))(z) = \mathcal{L}\left(\frac{e_{i\alpha}(x, 0) - e_{-i\alpha}(x, 0)}{2i}\right)(z)$$

$$= \frac{1}{2i}\mathcal{L}(e_{i\alpha}(x, 0) - e_{-i\alpha}(x, 0))(z)$$

$$= \frac{1}{2i}\left(\mathcal{L}(e_{i\alpha}(x, 0))(z) - \mathcal{L}(e_{-i\alpha}(x, 0))(z)\right)$$

$$= \frac{1}{2i}\left(\frac{1}{z - i\alpha} - \frac{1}{z + i\alpha}\right)$$

$$= \frac{1}{2i}\frac{z + i\alpha - z + i\alpha}{(z - i\alpha)(z + i\alpha)}$$

$$= \frac{\alpha}{z^2 + \alpha^2}.$$

This completes the proof.

Definition 2.3. Let $f : \mathbb{N}_0 \longmapsto \mathbb{R}$ and let $z \in \mathbb{R}$. Then the $\mathscr{Z}$-transform is defined by

$$\mathscr{Z}(f)(z) = \sum_{t=0}^{\infty} \frac{f(t)}{(z+1)^{t+1}},$$

provided the series converges.

Theorem 2.7. *Let $\mathbb{T}_0 = \mathbb{N}_0$. Then*

$$\mathscr{L}(f)(z) = \mathscr{Z}(f)(z)$$

for every $f : \mathbb{T}_0 \longmapsto \mathbb{R}$ and every $z \in \mathscr{D}\{f\}$.

Proof. For $y \in \mathbb{T}_0 \cap [0, \infty)$ we have

$$
\begin{aligned}
e_{\ominus z}(\sigma(y), 0) &= (1 + (\ominus z)(y)\mu(y))e_{\ominus z}(y, 0) \\
&= \left(1 - \frac{z\mu(y)}{1 + z\mu(y)}\right) e_{\ominus z}(y, 0) \\
&= \frac{1}{1 + z\mu(y)} e_{\ominus z}(y, 0) \\
&= \frac{1}{1 + z} e_{\ominus z}(y, 0) \\
&= \frac{1}{1 + z} e^{\int_0^y \frac{1}{\mu(\tau)} \mathrm{Log}(1 + (\ominus z)(\tau)\mu(\tau))\Delta\tau} \\
&= \frac{1}{1 + z} e^{\int_0^y \mathrm{Log}\frac{1}{1+z}\Delta\tau} \\
&= \frac{1}{1 + z}\left(\frac{1}{1 + z}\right)^y \\
&= \frac{1}{(1 + z)^{y+1}}.
\end{aligned}
$$

Hence,

$$
\begin{aligned}
\mathscr{L}(f)(z) &= \int_0^{\infty} \frac{f(y)}{(1 + z)^{y+1}} \Delta y \\
&= \sum_{t=0}^{\infty} \frac{f(t)}{(1 + z)^{t+1}} \\
&= \mathscr{Z}(f)(z).
\end{aligned}
$$

This completes the proof.

Exercise 2.2. Let $\alpha > 0$. Prove that

$$\mathscr{L}\left(\alpha^t\right)(z) = \frac{1}{z+1-\alpha}$$

for every $z \in \mathscr{D}\{\alpha^t\}$ such that $|z| > \alpha$.

Exercise 2.3. Let $f : \mathbb{N}_0 \longmapsto \mathbb{R}$. Prove that

1. $\mathscr{L}(f^\sigma)(z) = (z+1)\mathscr{L}(f)(z) - f(0)$,
2. $\mathscr{L}(f^{\sigma\sigma}) = (z+1)^2\mathscr{L}(f)(z) - (z+1)f(0) - f(1)$,
3. $\mathscr{L}\left(f^{\sigma^l}\right)(z) = (z+1)^l\mathscr{L}(f)(z) - \sum_{k=0}^{l-1}(z+1)^{l-1-k}f(k), \quad l \in \mathbb{N}$,

for every $z \in \mathscr{D}\{f\}$.

The usual convolution of two functions f and g on the real interval $[0, \infty)$ is defined by

$$(f \star g)(x) = \int_0^x f(x-y)g(y)dy \quad \text{for} \quad x \geq 0.$$

However, this definition does not work for general time scales, because $x, y \in \mathbb{T}_0$ does not imply that $x - y \in \mathbb{T}_0$.

Definition 2.4. Assume that f is one of the functions $e_\alpha(x, 0)$, $\sinh_\alpha(x, 0)$, $\cosh_\alpha(x, 0)$, $\cos_\alpha(x, 0)$, $\sin_\alpha(x, 0)$, or $h_k(x, 0)$, $k \in \mathbb{N}_0$. If g is a regulated function on $\mathbb{T}_0$, then we define the convolution of f with g by

$$(f \star g)(x) = \int_0^x f(x, \sigma(y))g(y)\Delta y \quad \text{for} \quad x \in \mathbb{T}_0.$$

Theorem 2.8 (*Convolution Theorem*). *Assume that $\alpha \in \mathbb{R}$ and f is one of the functions $e_\alpha(x, 0)$, $\sinh_\alpha(x, 0)$, $\cosh_\alpha(x, 0)$, $\cos_\alpha(x, 0)$, $\sin_\alpha(x, 0)$, or $h_k(x, 0)$, $k \in \mathbb{N}_0$. If g is a regulated function on $\mathbb{T}_0$ such that*

$$\lim_{x \to \infty} e_{\ominus z}(x, 0)(f \star g)(x) = 0,$$

then

$$\mathscr{L}(f \star g)(z) = \mathscr{L}(f)(z)\mathscr{L}(g)(z). \tag{2.7}$$

Proof. 1. $f(x, 0) = e_\alpha(x, 0)$. Consider the initial value problem for the dynamic equation

$$l^\Delta - \alpha l = g(x), \quad l(0) = 0. \tag{2.8}$$

The solution of this problem is given by

$$l(x) = \int_0^x e_\alpha(x, \sigma(y))g(y)\Delta y,$$

which can be rewritten in the form

$$l(x) = (e_\alpha(x, \sigma(y)) \star g)(x).$$

Now we apply the Laplace transform to both sides of equation (2.8) to get

$$\mathscr{L}\left(l^\Delta - \alpha l\right)(z) = \mathscr{L}(g)(z),$$

or

$$\mathscr{L}(l^\Delta)(z) - \alpha\mathscr{L}(l)(z) = \mathscr{L}(g)(z),$$

or

$$z\mathscr{L}(l)(z) - \alpha\mathscr{L}(l)(z) = \mathscr{L}(g)(z),$$

whereupon

$$\mathscr{L}(l)(z) = \frac{1}{z - \alpha}\mathscr{L}(g)(z).$$

Since

$$\mathscr{L}(e_\alpha(x, 0))(z) = \frac{1}{z - \alpha},$$

we conclude that

$$\mathscr{L}(l)(z) = \mathscr{L}(f)(z)\mathscr{L}(g)(z),$$

or we get (2.7).
2. Let $f(x, 0) = \cosh_\alpha(x, 0)$. We have

$$\cosh_\alpha(x, 0) = \frac{e_\alpha(x, 0) + e_{-\alpha}(x, 0)}{2}.$$

Then

$$(f \star g)(x) = \int_0^x \cosh_\alpha(x, \sigma(y))g(y)\Delta y$$

$$= \frac{1}{2}\int_0^x (e_\alpha(x, \sigma(y)) + e_{-\alpha}(x, \sigma(y)))\, g(y)\Delta y$$

$$= \frac{1}{2} \int_0^x e_\alpha(x, \sigma(y)) g(y) \Delta y + \frac{1}{2i} \int_0^x e_{-\alpha}(x, \sigma(y)) g(y) \Delta y$$

$$= \frac{1}{2}(e_\alpha(x, \sigma(y)) \star g)(x) + \frac{1}{2i}(e_{-\alpha}(x, \sigma(y)) \star g)(x).$$

From this, we obtain

$$\mathscr{L}(f \star g)(z) = \mathscr{L}\left(\frac{1}{2}(e_\alpha(x, \sigma(y)) \star g)(x) + \frac{1}{2}(e_{-\alpha}(x, \sigma(y)) \star g)(x)\right)(z)$$

$$= \frac{1}{2}\mathscr{L}(e_\alpha(x, \sigma(y)) \star g)(z) + \frac{1}{2}\mathscr{L}(e_{-\alpha}(x, \sigma(y)) \star g)(z)$$

$$= \frac{1}{2}\mathscr{L}(e_\alpha(x, \sigma(y)))(z)\mathscr{L}(g)(z) + \frac{1}{2}\mathscr{L}(e_{-\alpha}(x, \sigma(y)))(z)\mathscr{L}(g)(z)$$

$$= \frac{1}{2}\frac{1}{z - \alpha}\mathscr{L}(g)(z) + \frac{1}{2}\frac{1}{z + \alpha}\mathscr{L}(g)(z)$$

$$= \frac{1}{2}\left(\frac{1}{z - \alpha} + \frac{1}{z + \alpha}\right)\mathscr{L}(g)(z)$$

$$= \frac{z}{z^2 - \alpha^2}\mathscr{L}(g)(z)$$

$$= \mathscr{L}(\cosh_\alpha(x, 0))(z)\mathscr{L}(g)(z)$$

$$= \mathscr{L}(f)(z)\mathscr{L}(g)(z).$$

3. Let $f(x, 0) = \sinh_\alpha(x, 0)$. Then

$$\sinh_\alpha(x, 0) = \frac{e_\alpha(x, 0) - e_{-\alpha}(x, 0)}{2}$$

and

$$(f \star g)(x) = \int_0^x \sinh_\alpha(x, \sigma(y)) g(y) \Delta y$$

$$= \frac{1}{2} \int_0^x (e_\alpha(x, \sigma(y)) - e_{-\alpha}(x, \sigma(y))) g(y) \Delta y$$

$$= \frac{1}{2} \int_0^x e_\alpha(x, \sigma(y)) g(y) \Delta y - \frac{1}{2} \int_0^x e_{-\alpha}(x, \sigma(y)) g(y) \Delta y$$

$$= \frac{1}{2}(e_\alpha(x, \sigma(y)) \star g)(x) - \frac{1}{2}(e_{-\alpha}(x, \sigma(y)) \star g)(x).$$

Hence,

$$\mathscr{L}(f \star g)(z) = \mathscr{L}\left(\frac{1}{2}(e_\alpha(x, \sigma(y)) \star g)(x) - \frac{1}{2}(e_{-\alpha}(x, \sigma(y)) \star g)(x)\right)(z)$$

$$= \frac{1}{2}\mathscr{L}(e_\alpha(x, \sigma(y)) \star g)(z) - \frac{1}{2}\mathscr{L}(e_{-\alpha}(x, \sigma(y)) \star g)(z)$$

$$= \frac{1}{2}\mathscr{L}(e_\alpha(x, \sigma(y)))(z)\mathscr{L}(g)(z) - \frac{1}{2}\mathscr{L}(e_{-\alpha}(x, \sigma(y)))(z)\mathscr{L}(g)(z)$$

$$= \frac{1}{2}\frac{1}{z-\alpha}\mathscr{L}(g)(z) - \frac{1}{2}\frac{1}{z+\alpha}\mathscr{L}(g)(z)$$

$$= \frac{1}{2}\left(\frac{1}{z-\alpha} - \frac{1}{z+\alpha}\right)\mathscr{L}(g)(z)$$

$$= \frac{\alpha}{z^2-\alpha^2}\mathscr{L}(g)(z)$$

$$= \mathscr{L}(\sinh_\alpha(x, 0))(z)\mathscr{L}(g)(z)$$

$$= \mathscr{L}(f)(z)\mathscr{L}(g)(z).$$

4. Let $f(x, 0) = \cos_\alpha(x, 0)$. We have

$$\cos_\alpha(x, 0) = \frac{e_{i\alpha}(x, 0) + e_{-i\alpha}(x, 0)}{2}.$$

Then

$$(f \star g)(x) = \int_0^x \cos_\alpha(x, \sigma(y))g(y)\Delta y$$

$$= \frac{1}{2}\int_0^x (e_{i\alpha}(x, \sigma(y)) + e_{-i\alpha}(x, \sigma(y)))\, g(y)\Delta y$$

$$= \frac{1}{2}\int_0^x e_{i\alpha}(x, \sigma(y))g(y)\Delta y + \frac{1}{2}\int_0^x e_{-i\alpha}(x, \sigma(y))g(y)\Delta y$$

$$= \frac{1}{2}(e_{i\alpha}(x, \sigma(y)) \star g)(x) + \frac{1}{2}(e_{-i\alpha}(x, \sigma(y)) \star g)(x).$$

From this, we obtain

$$\mathscr{L}(f \star g)(z) = \mathscr{L}\left(\frac{1}{2}(e_{i\alpha}(x, \sigma(y)) \star g)(x) + \frac{1}{2}(e_{-i\alpha}(x, \sigma(y)) \star g)(x)\right)(z)$$

$$= \frac{1}{2}\mathscr{L}(e_{i\alpha}(x, \sigma(y)) \star g)(z) + \frac{1}{2}\mathscr{L}(e_{-i\alpha}(x, \sigma(y)) \star g)(z)$$

$$= \frac{1}{2}\mathscr{L}(e_{i\alpha}(x, \sigma(y)))(z)\mathscr{L}(g)(z) + \frac{1}{2}\mathscr{L}(e_{-i\alpha}(x, \sigma(y)))(z)\mathscr{L}(g)(z)$$

$$= \frac{1}{2}\frac{1}{z-i\alpha}\mathscr{L}(g)(z) + \frac{1}{2}\frac{1}{z+i\alpha}\mathscr{L}(g)(z)$$

$$= \frac{1}{2}\left(\frac{1}{z-i\alpha} + \frac{1}{z+i\alpha}\right)\mathscr{L}(g)(z)$$

$$= \frac{z}{z^2+\alpha^2}\mathscr{L}(g)(z)$$

$$= \mathscr{L}(\cos_\alpha(x,0))(z)\mathscr{L}(g)(z)$$

$$= \mathscr{L}(f)(z)\mathscr{L}(g)(z).$$

5. Let $f(x,0) = \sin_\alpha(x,0)$. Then

$$\sin_\alpha(x,0) = \frac{e_{i\alpha}(x,0) - e_{-i\alpha}(x,0)}{2i}$$

and

$$(f \star g)(x) = \int_0^x \sin_\alpha(x, \sigma(y))g(y)\Delta y$$

$$= \frac{1}{2i}\int_0^x \left(e_{i\alpha}(x, \sigma(y)) - e_{-i\alpha}(x, \sigma(y))\right)g(y)\Delta y$$

$$= \frac{1}{2i}\int_0^x e_{i\alpha}(x, \sigma(y))g(y)\Delta y - \frac{1}{2i}\int_0^x e_{-i\alpha}(x, \sigma(y))g(y)\Delta y$$

$$= \frac{1}{2i}(e_{i\alpha}(x, \sigma(y)) \star g)(x) - \frac{1}{2i}(e_{-i\alpha}(x, \sigma(y)) \star g)(x).$$

Hence,

$$\mathscr{L}(f \star g)(z) = \mathscr{L}\left(\frac{1}{2i}(e_{i\alpha}(x, \sigma(y)) \star g)(x) - \frac{1}{2i}(e_{-i\alpha}(x, \sigma(y)) \star g)(x)\right)(z)$$

$$= \frac{1}{2i}\mathscr{L}(e_{i\alpha}(x, \sigma(y)) \star g)(z) - \frac{1}{2i}\mathscr{L}(e_{-i\alpha}(x, \sigma(y)) \star g)(z)$$

$$= \frac{1}{2i}\mathscr{L}(e_{i\alpha}(x, \sigma(y)))(z)\mathscr{L}(g)(z) - \frac{1}{2i}\mathscr{L}(e_{-i\alpha}(x, \sigma(y)))(z)\mathscr{L}(g)(z)$$

$$= \frac{1}{2i}\frac{1}{z-i\alpha}\mathscr{L}(g)(z) - \frac{1}{2i}\frac{1}{z+i\alpha}\mathscr{L}(g)(z)$$

$$= \frac{1}{2i}\left(\frac{1}{z-i\alpha} - \frac{1}{z+i\alpha}\right)\mathscr{L}(g)(z)$$

$$= \frac{\alpha}{z^2+\alpha^2}\mathscr{L}(g)(z)$$

$$= \mathscr{L}(\sin_\alpha(x,0))(z)\mathscr{L}(g)(z)$$

$$= \mathscr{L}(f)(z)\mathscr{L}(g)(z).$$

6. $f(x, 0) = h_k(x, 0)$, $k \in \mathbb{N}_0$. Consider the initial value problem for the dynamic equation

$$l^{\Delta^{k+1}}(x) = g(x), \quad l^{\Delta^j}(0) = 0, \quad j = 0, 1, \dots, k. \tag{2.9}$$

The solution of equation (2.9) is given by

$$l(x) = \int_0^x h_k(x, \sigma(y))g(y)\Delta y,$$

or

$$l(x) = (h_k(x, \sigma(y)) \star g)(x).$$

Taking the Laplace transform of both sides of (2.9) gives

$$\mathscr{L}\left(l^{\Delta^{k+1}}\right)(z) = \mathscr{L}(g)(z),$$

whereupon

$$z^{k+1}\mathscr{L}(l)(z) = \mathscr{L}(g)(z),$$

or

$$\mathscr{L}(l)(z) = \frac{1}{z^{k+1}}\mathscr{L}(g)(z)$$

$$= \mathscr{L}(h_k(x, 0))(z)\mathscr{L}(g)(z)$$

$$= \mathscr{L}(f)(z)\mathscr{L}(g)(z).$$

This completes the proof.

Theorem 2.9. *Assume that f and g are each one of the functions $e_\alpha(x, 0)$, $\cosh_\alpha(x, 0)$, $\sinh_\alpha(x, 0)$, $\cos_\alpha(x, 0)$, $\sin_\alpha(x, 0)$, $h_k(x, 0)$, not both $h_k(x, 0)$. Then*

$$f \star g = g \star f.$$

Proof. 1. Let $f(x, 0) = e_\alpha(x, 0)$ and $g(x, 0) = e_\beta(x, 0)$. Let also

$$p(x) = e_\alpha(x, 0) \star e_\beta(x, 0) \quad and \quad q(x) = e_\beta(x, 0) \star e_\alpha(x, 0).$$

Note that

$$p(0) = q(0) = 0.$$

Also, $p(x)$ and $q(x)$ are solutions to the initial value problems

$$p^\Delta - \alpha p = e_\beta(x, 0), \quad p(0) = 0,$$

and

$$q^\Delta - \beta q = e_\alpha(x, 0), \quad q(0) = 0,$$

respectively.
Then

$$p^\Delta(0) = \alpha p(0) + e_\beta(0, 0) = 1,$$
$$q^\Delta(0) = \beta q(0) + e_\alpha(0, 0) = 1.$$

We claim that $p(x)$ and $q(x)$ are solutions to the initial value problem

$$m^{\Delta^2} - (\alpha + \beta)m^\Delta + \alpha\beta m = 0, \quad m(0) = 0, \quad m^\Delta(0) = 1. \tag{2.10}$$

We have

$$p^\Delta = \alpha p + e_\beta(x, 0),$$

from which, after differentiating, we obtain

$$p^{\Delta^2} = \alpha p^\Delta + \beta e_\beta(x, 0).$$

Hence,

$$
\begin{aligned}
p^{\Delta^2} - (\alpha + \beta)p^\Delta + \alpha\beta p &= \alpha p^\Delta + \beta e_\beta(x, 0) - (\alpha + \beta)p^\Delta + \alpha\beta p \\
&= \beta e_\beta(x, 0) - \beta p^\Delta + \alpha\beta p \\
&= \beta e_\beta(x, 0) - \beta(\alpha p + e_\beta(x, 0)) + \alpha\beta p \\
&= \beta e_\beta(x, 0) - \alpha\beta p - \beta e_\beta(x, 0) + \alpha\beta p \\
&= 0.
\end{aligned}
$$

Also,

$$q^\Delta = \beta q + e_\alpha(x, 0),$$

whereupon, on differentiating, we obtain

$$q^{\Delta^2} = \beta q^\Delta + \alpha e_\alpha(x, 0).$$

Hence,

$$
\begin{aligned}
q^{\Delta^2} - (\alpha + \beta)q^\Delta + \alpha\beta q &= \beta q^\Delta + \alpha e_\alpha(x, 0) - (\alpha + \beta)q^\Delta + \alpha\beta q \\
&= \alpha e_\alpha(x, 0) - \alpha q^\Delta + \alpha\beta q \\
&= \alpha e_\alpha(x, 0) - \alpha(\beta q + e_\alpha(x, 0)) + \alpha\beta q \\
&= \alpha e_\alpha(x, 0) - \alpha\beta q - \alpha e_\alpha(x, 0) + \alpha\beta q \\
&= 0.
\end{aligned}
$$

Since problem (2.10) has a unique solution, we conclude that

$$p(x) = q(x).$$

2. Next we consider

$$e_\alpha(x, 0) \star \cosh_\beta(x, 0) = e_\alpha(x, 0) \star \frac{e_\beta(x, 0) + e_{-\beta}(x, 0)}{2}$$

$$= \frac{1}{2} \left(e_\alpha(x, 0) \star (e_\beta(x, 0) + e_{-\beta}(x, 0)) \right)$$

$$= \frac{1}{2} \left(e_\alpha(x, 0) \star e_\beta(x, 0) + e_\alpha(x, 0) \star e_{-\beta}(x, 0) \right)$$

$$= \frac{1}{2} \left(e_\beta(x, 0) \star e_\alpha(x, 0) + e_{-\beta}(x, 0) \star e_\alpha(x, 0) \right)$$

$$= \frac{1}{2} \left(e_\beta(x, 0) + e_{-\beta}(x, 0) \right) \star e_\alpha(x, 0)$$

$$= \cosh_\beta(x, 0) \star e_\alpha(x, 0).$$

3. Let

$$z(x) = e_\alpha(x, 0) \star h_k(x, 0) \quad \text{and} \quad q(x) = h_k(x, 0) \star e_\alpha(x, 0).$$

We have that $z(x)$ is the solution to the initial value problem

$$z^\Delta(x) - \alpha z(x) = h_k(x, 0), \quad z(0) = 0.$$

Differentiating this equation i times gives

$$z^{\Delta^{i+1}}(x) - \alpha z^\Delta(x) = h_{k-i}(x, 0), \quad i = 1, 2, \ldots, k.$$

Also,

$$z^{\Delta^i}(0) = 0, \quad 0 \le i \le k, \quad z^{\Delta^{k+1}}(0) = 1.$$

Thus we get that $z(x)$ is the solution to the initial value problem

$$z^{\Delta^{k+2}}(x) - \alpha z^{\Delta^{k+1}}(x) = 0,$$
$$z^{\Delta^i}(0) = 0, \quad 0 \le i \le k, \quad z^{\Delta^{k+1}}(0) = 1.$$

Since $h_k(0, 0) = 0, \quad k > 0$, we obtain that

$$q^{\Delta^i}(0) = 0, \quad 0 \le i \le k.$$

Also, q is the solution of the initial value problem

$$q^{\Delta^{k+1}}(x) = e_\alpha(x, 0), \quad q^{\Delta^i}(0) = 0, \quad 0 \le i \le k.$$

Hence, differentiating the last equation, we obtain

$$q^{\Delta^{k+2}}(x) = e_\alpha^\Delta(x, 0)$$
$$= \alpha e_\alpha(x, 0)$$
$$= \alpha q^{\Delta^{k+1}}(x)$$

and

$$q^{\Delta^{k+1}}(0) = e_\alpha(0, 0) = 1.$$

Consequently, $q(x)$ is the solution of the initial value problem

$$q^{\Delta^{k+2}}(x) - \alpha q^{\Delta^{k+1}}(x) = 0, \quad q^{\Delta^i}(0) = 0, \quad q^{\Delta^{k+1}}(0) = 1, \quad 0 \le i \le k.$$

Therefore, $z(x)$ and $q(x)$ are solutions to the same initial value problem. Hence, they must be equal.

This completes the proof.

Exercise 2.4. Prove that

1. $e_\alpha(x, 0) \star \sinh_\beta(x, 0) = \sinh_\beta(x, 0) \star e_\alpha(x, 0)$,
2. $e_\alpha(x, 0) \star \cos_\beta(x, 0) = \cos_\beta(x, 0) \star e_\alpha(x, 0)$,
3. $e_\alpha(x, 0) \star \sin_\beta(x, 0) = \sin_\beta(x, 0) \star e_\alpha(x, 0)$,
4. $\cosh_\alpha(x, 0) \star \cosh_\beta(x, 0) = \cosh_\beta(x, 0) \star \cosh_\alpha(x, 0)$,
5. $\cosh_\alpha(x, 0) \star \sinh_\beta(x, 0) = \sinh_\beta(x, 0) \star \cosh_\alpha(x, 0)$,
6. $\cosh_\alpha(x, 0) \star \cos_\beta(x, 0) = \cos_\beta(x, 0) \star \cosh_\alpha(x, 0)$,
7. $\cosh_\alpha(x, 0) \star \sin_\beta(x, 0) = \sin_\beta(x, 0) \star \cosh_\alpha(x, 0)$,
8. $\cosh_\alpha(x, 0) \star h_k(x, 0) = h_k(x, 0) \star \cosh_\alpha(x, 0)$,
9. $\sinh_\alpha(x, 0) \star \sinh_\beta(x, 0) = \sinh_\beta(x, 0) \star \sinh_\alpha(x, 0)$,
10. $\sinh_\alpha(x, 0) \star \cos_\beta(x, 0) = \cos_\beta(x, 0) \star \sinh_\alpha(x, 0)$,
11. $\sinh_\alpha(x, 0) \star \sin_\beta(x, 0) = \sin_\beta(x, 0) \star \sinh_\alpha(x, 0)$,
12. $\sinh_\alpha(x, 0) \star h_k(x, 0) = h_k(x, 0) \star \sinh_\alpha(x, 0)$,
13. $\cos_\alpha(x, 0) \star \cos_\beta(x, 0) = \cos_\beta(x, 0) \star \cos_\alpha(x, 0)$,
14. $\cos_\alpha(x, 0) \star \sin_\beta(x, 0) = \sin_\beta(x, 0) \star \cos_\alpha(x, 0)$,
15. $\cos_\alpha(x, 0) \star h_k(x, 0) = h_k(x, 0) \star \cos_\alpha(x, 0)$,
16. $\sin_\alpha(x, 0) \star \sin_\beta(x, 0) = \sin_\beta(x, 0) \star \sin_\alpha(x, 0)$,
17. $\sin_\alpha(x, 0) \star h_k(x, 0) = h_k(x, 0) \star \sin_\alpha(x, 0)$.

Theorem 2.10. *Let* $\alpha, \beta \in \mathbb{R}$ *and* $1 + \alpha\mu(x) \ne 0$, $1 + \beta\mu(x) \ne 0$ *for all* $x \in \mathbb{T}$. *Then*

$$\mathscr{L}\left(e_\alpha(x, 0) \sin_{\frac{\beta}{1+\alpha\mu}}(x, 0)\right)(z) = \frac{\beta}{(z - \alpha)^2 + \beta^2},$$

provided

$$\lim_{x\to\infty} e_\alpha(x,0)\sin_{\frac{\beta}{1+\mu\alpha}}(x,0) = 0 \quad \text{and} \quad \lim_{x\to\infty} e_\alpha(x,0)\left(\sin_{\frac{\beta}{1+\mu\alpha}}(x,0)\right)^\Delta = 0.$$

Proof. Let

$$p(x) = e_\alpha(x,0)\sin_{\frac{\beta}{1+\mu\alpha}}(x,0).$$

Then

$$p^\Delta(x) = e_\alpha^\Delta(x,0)\sin_{\frac{\beta}{1+\mu\alpha}}(x,0) + e_\alpha(\sigma(x),0)\sin_{\frac{\beta}{1+\mu\alpha}}^\Delta(x,0)$$
$$= \alpha e_\alpha(x,0)\sin_{\frac{\beta}{1+\mu\alpha}}(x,0) + (1+\alpha\mu(x))e_\alpha(x,0)\frac{\beta}{1+\alpha\mu(x)}\cos_{\frac{\beta}{1+\mu\alpha}}(x,0)$$
$$= \alpha e_\alpha(x,0)\sin_{\frac{\beta}{1+\mu\alpha}}(x,0) + \beta e_\alpha(x,0)\cos_{\frac{\beta}{1+\mu\alpha}}(x,0).$$

We differentiate the last equation, and we get

$$p^{\Delta^2}(x) = \alpha e_\alpha^\Delta(x,0)\sin_{\frac{\beta}{1+\mu\alpha}}(x,0) + \alpha e_\alpha(\sigma(x),0)\sin_{\frac{\beta}{1+\mu\alpha}}^\Delta(x,0)$$
$$+ \beta e_\alpha^\Delta(x,0)\cos_{\frac{\beta}{1+\mu\alpha}}(x,0) + \beta e_\alpha(\sigma(x),0)\cos_{\frac{\beta}{1+\mu\alpha}}^\Delta(x,0)$$
$$= \alpha^2 e_\alpha(x,0)\sin_{\frac{\beta}{1+\mu\alpha}}(x,0) + \alpha(1+\alpha\mu(x))e_\alpha(x,0)\frac{\beta}{1+\alpha\mu(x)}\cos_{\frac{\beta}{1+\mu\alpha}}(x,0)$$
$$+ \alpha\beta e_\alpha(x,0)\cos_{\frac{\beta}{1+\mu\alpha}}(x,0) - \beta(1+\alpha\mu(x))e_\alpha(x,0)\frac{\beta}{1+\alpha\mu(x)}\sin_{\frac{\beta}{1+\mu\alpha}}(x,0)$$
$$= \alpha^2 e_\alpha(x,0)\sin_{\frac{\beta}{1+\mu\alpha}}(x,0) + \alpha\beta e_\alpha(x,0)\cos_{\frac{\beta}{1+\mu\alpha}}(x,0)$$
$$+ \alpha\beta e_\alpha(x,0)\cos_{\frac{\beta}{1+\mu\alpha}}(x,0) - \beta^2 e_\alpha(x,0)\sin_{\frac{\beta}{1+\mu\alpha}}(x,0)$$
$$= (\alpha^2 - \beta^2)p(x) + 2\alpha\beta e_\alpha(x,0)\cos_{\frac{\beta}{1+\mu\alpha}}(x,0).$$

In this way, we obtain the system

$$\begin{cases} p^\Delta(x) = \alpha p(x) + \beta e_\alpha(x,0)\cos_{\frac{\beta}{1+\mu\alpha}}(x,0), \\ p^{\Delta^2}(x) = (\alpha^2 - \beta^2)p(x) + 2\alpha\beta e_\alpha(x,0)\cos_{\frac{\beta}{1+\mu\alpha}}(x,0). \end{cases}$$

Hence,

$$p^{\Delta^2}(x) - 2\alpha p^\Delta(x) = -(\alpha^2 + \beta^2)p(x),$$

or

$$p^{\Delta^2}(x) - 2\alpha p^\Delta(x) + (\alpha^2 + \beta^2)p(x) = 0.$$

Also,

$$p(0) = e_\alpha(0, 0) \sin_{\frac{\beta}{1+\mu\alpha}}(0, 0) = 0,$$
$$p^\Delta(0) = \alpha e_\alpha(0, 0) \sin_{\frac{\beta}{1+\mu\alpha}}(0, 0) + \beta e_\alpha(0, 0) \cos_{\frac{\beta}{1+\mu\alpha}}(0, 0) = \beta.$$

Consequently, we obtain the following initial value problem:

$$\begin{cases} p^{\Delta^2}(x) - 2\alpha p^\Delta(x) + (\alpha^2 + \beta^2)p(x) = 0, \\ p(0) = 0, \quad p^\Delta(0) = \beta. \end{cases} \tag{2.11}$$

Now we apply the Laplace transform to both sides of the dynamic equation (2.11), and we obtain

$$\mathscr{L}\left(p^{\Delta^2}(x) - 2\alpha p^\Delta(x) + (\alpha^2 + \beta^2)p(x)\right)(z) = 0,$$

or

$$\mathscr{L}\left(p^{\Delta^2}(x)\right)(z) - 2\alpha\mathscr{L}\left(p^\Delta(x)\right)(z) + (\alpha^2 + \beta^2)\mathscr{L}(p(x))(z) = 0,$$

or

$$z^2\mathscr{L}(p)(z) - p^\Delta(0) - zp(0) - 2\alpha z\mathscr{L}(p)(z) + 2\alpha p(0) + (\alpha^2 + \beta^2)\mathscr{L}(p)(z) = 0,$$

or

$$z^2\mathscr{L}(p)(z) - \beta - 2\alpha z\mathscr{L}(p)(z) + (\alpha^2 + \beta^2)\mathscr{L}(p)(z) = 0,$$

or

$$\left(z^2 - 2\alpha z + \alpha^2 + \beta^2\right)\mathscr{L}(p)(z) = \beta,$$

whereupon

$$\mathscr{L}(p)(z) = \frac{\beta}{(z - \alpha)^2 + \beta^2},$$

which completes the proof.

Exercise 2.5. Let $\alpha, \beta \in \mathbb{R}$ and $1 + \alpha\mu(x) \neq 0, 1 + \beta\mu(x) \neq 0$ for all $x \in \mathbb{T}$. Prove that

$$\mathscr{L}\left(e_\alpha(x, 0) \cos_{\frac{\beta}{1+\alpha\mu}}(x, 0)\right)(z) = \frac{z - \alpha}{(z - \alpha)^2 + \beta^2},$$

provided that

$$\lim_{x\to\infty} e_\alpha(x, 0) \cos_{\frac{\beta}{1+\alpha\mu}}(x, 0) = \lim_{x\to\infty} e_\alpha(x, 0) \cos^\Delta_{\frac{\beta}{1+\alpha\mu}}(x, 0) = 0.$$

Definition 2.5. Let $a \in \mathbb{T}$, $\quad a > 0$. Define the step function u_a by

$$u_a = \begin{cases} 0 & \text{if } x \in \mathbb{T} \cap (-\infty, a), \\ 1 & \text{if } x \in \mathbb{T} \cap [a, \infty). \end{cases}$$

Theorem 2.11. *Let $a \in \mathbb{T}_0$, $\quad a > 0$. Then*

$$\mathscr{L}(u_a(x))(z) = \frac{e_{\ominus z}(a, 0)}{z}$$

for all $z \in \mathscr{D}\{u_a\}$ such that

$$\lim_{x\to\infty} e_{\ominus z}(x, 0) = 0.$$

Proof. We have

$$\mathscr{L}(u_a(x))(z) = \int_0^\infty u_a(y) e_{\ominus z}(\sigma(y), 0) \Delta y$$

$$= \int_a^\infty e_{\ominus z}(\sigma(y), 0) \Delta y$$

$$= \int_a^\infty (1 + \mu \ominus z) e_{\ominus z}(y, 0) \Delta y$$

$$= \int_a^\infty \frac{1}{1 + \mu z} e_{\ominus z}(y, 0) \Delta y$$

$$= -\frac{1}{z} \int_a^\infty \frac{-z}{1 + \mu z} e_{\ominus z}(y, 0) \Delta y$$

$$= -\frac{1}{z} \int_a^\infty \ominus z e_{\ominus z}(y, 0) \Delta y$$

$$= -\frac{1}{z} \int_a^\infty e^\Delta_{\ominus z}(y, 0) \Delta y$$

$$= -\frac{1}{z} e_{\ominus z}(y, 0) \Big|_{y=a}^{y=\infty}$$

$$= -\frac{1}{z} \left(\lim_{y\to\infty} e_{\ominus z}(y, 0) - e_{\ominus z}(a, 0) \right)$$

$$= \frac{1}{z} e_{\ominus z}(a, 0).$$

This completes the proof.

Theorem 2.12. *Let $a \in \mathbb{T}_0$, $a > 0$. Assume that f is one of the functions $e_\alpha(x, 0)$,
$\cos_\alpha(x, 0)$, $\sin_\alpha(x, a)$, $\sinh_\alpha(x, 0)$, $\cosh_\alpha(x, 0)$. If $1 + z\mu(x) \neq 0$, $1 + \alpha\mu(x) \neq 0$ for
all $x \in \mathbb{T}_0$, and*

$$\lim_{x \to \infty} e_{\alpha \ominus z}(x, a) = \lim_{x \to \infty} e_{i\alpha \ominus z}(x, a) = \lim_{x \to \infty} e_{-i\alpha \ominus z}(x, a) = 0,$$

then

$$\mathscr{L}(u_a(x)f(x, a)) = e_{\ominus z}(a, 0)\mathscr{L}(f(x, a))(z).$$

Proof. 1. Let $f(x, a) = e_\alpha(x, a)$. Then

$$
\begin{aligned}
e_\alpha(x, a)e_{\ominus z}(\sigma(x), 0) &= (1 + \mu(x) \ominus z)e_\alpha(x, a)e_{\ominus z}(x, 0) \\
&= \frac{1}{1 + z\mu(x)}e_\alpha(x, a)e_{\ominus z}(x, 0)\frac{e_{\ominus z}(0, a)}{e_{\ominus z}(0, a)} \\
&= \frac{1}{1 + z\mu(x)}e_\alpha(x, a)e_{\ominus z}(x, a)\frac{1}{e_{\ominus z}(0, a)} \\
&= \frac{1}{1 + z\mu(x)}e_{\alpha \ominus z}(x, a)e_{\ominus z}(a, 0) \\
&= \frac{1}{\alpha - z}\frac{\alpha - z}{1 + z\mu(x)}e_{\alpha \ominus z}(x, a)e_{\ominus z}(a, 0) \\
&= \frac{1}{\alpha - z}(\alpha \ominus z)e_{\alpha \ominus z}(x, a)e_{\ominus z}(a, 0) \\
&= \frac{1}{\alpha - z}e_{\alpha \ominus z}^\Delta(x, a)e_{\ominus z}(a, 0).
\end{aligned}
$$

Thus

$$
\begin{aligned}
\mathscr{L}(u_f(x, a))(z) &= \int_0^\infty u_a(x)f(x, a)e_{\ominus z}(\sigma(x), 0)\Delta x \\
&= \int_a^\infty e_\alpha(x, a)e_{\ominus z}(\sigma(x), 0)\Delta x \\
&= \frac{1}{\alpha - z}e_{\ominus z}(a, 0)\int_a^\infty e_{\alpha \ominus z}^\Delta(x, a)\Delta x \\
&= \frac{1}{\alpha - z}e_{\ominus z}(a, 0)e_{\alpha \ominus z}(x, a)\Big|_{x=a}^{x=\infty} \\
&= \frac{1}{\alpha - z}e_{\ominus z}(a, 0)\left(\lim_{x \to \infty} e_{\alpha \ominus z}(x, a) - e_{\alpha \ominus z}(a, 0)\right) \\
&= \frac{1}{z - \alpha}e_{\ominus z}(a, 0) \\
&= e_{\ominus z}(a, 0)\mathscr{L}(e_\alpha(x, 0))(z).
\end{aligned}
$$

$$u_f(x, a) = u_a(x)f(x, a).$$

2. Let $f(x, a) = \cos_\alpha(x, a)$. Then

$$f(x, a) = \frac{1}{2}\left(e_{i\alpha}(x, a) + e_{-i\alpha}(x, a)\right).$$

Hence,

$$\mathcal{L}(u_a(x)f(x, a)) = \mathcal{L}\left(u_a(x)\frac{1}{2}(e_{i\alpha}(x, a) + e_{-i\alpha}(x, a))\right)(z)$$

$$= \mathcal{L}\left(\frac{1}{2}u_a(x)e_{i\alpha}(x, a) + \frac{1}{2}u_a(x)e_{-i\alpha}(x, a)\right)(z)$$

$$= \frac{1}{2}\mathcal{L}(u_a(x)e_{i\alpha}(x, a))(z) + \frac{1}{2}\mathcal{L}(u_a(x)e_{-i\alpha}(x, a))(z)$$

$$= \frac{1}{2}e_{\ominus z}(a, 0)\mathcal{L}(e_{i\alpha}(x, 0)) + \frac{1}{2}e_{\ominus z}(a, 0)\mathcal{L}(e_{-i\alpha}(x, 0))(z)$$

$$= e_{\ominus z}(a, 0)\mathcal{L}\left(\frac{1}{2}e_{i\alpha}(x, 0)\right) + e_{\ominus z}(a, 0)\mathcal{L}\left(\frac{1}{2}e_{-i\alpha}(x, 0)\right)(z)$$

$$= e_{\ominus z}(a, 0)\mathcal{L}\left(\frac{1}{2}(e_{i\alpha}(x, 0) + e_{-i\alpha}(x, 0))\right)(z)$$

$$= e_{\ominus z}(a, 0)\mathcal{L}(\cos_\alpha(x, 0))(z).$$

This completes the proof.

Definition 2.6. Let $a, b, \alpha \in \mathbb{T}$ and suppose that $f : \mathbb{T} \longmapsto \mathbb{R}$ is continuous. If $\delta_\alpha(x)$, $x \in \mathbb{T}$, satisfies the following conditions,

$$\int_a^b f(x)\delta_\alpha(x)\Delta x = \begin{cases} f(\alpha) & \text{if } \alpha \in [a, b), \\ 0 & \text{otherwise,} \end{cases}$$

then $\delta_\alpha(x)$ will be called a Dirac delta function.

Theorem 2.13. *Let* $\alpha \in \mathbb{T}_0$, $\alpha \geq 0$. *Then*

$$\mathcal{L}(\delta_\alpha(x))(z) = e_{\ominus z}^\sigma(\alpha, 0).$$

Proof. We have

$$\mathcal{L}(\delta_\alpha(x))(z) = \int_0^\infty \delta_\alpha(y)e_{\ominus z}(\sigma(y), 0)\Delta y$$

$$= e_{\ominus z}(\sigma(\alpha), 0).$$

This completes the proof.

Exercise 2.6. Prove the following relations.

1. If $\alpha \neq \beta$, then

$$e_\alpha(x, 0) \star e_\beta(x, 0) = \frac{1}{\beta - \alpha}(e_\beta(x, 0) - e_\alpha(x, 0)).$$

2.

$$e_\alpha(x, 0) \star e_\alpha(x, 0) = e_\alpha(x, 0) \int_0^x \frac{1}{1 + \alpha\mu(y)} \Delta y.$$

3. If $\alpha^2 + \beta^2 \neq 0$, then

$$e_\alpha(x, 0) \star \sin_\beta(x, 0) = \frac{\beta e_\alpha(x, 0) - \alpha \sin_\beta(x, 0) - \beta \cos_\beta(x, 0)}{\alpha^2 + \beta^2}.$$

4. If $\alpha \neq 0, \alpha \neq \beta$, then

$$\cos_\alpha(x, 0) \star \cos_\beta(x, 0) = \frac{-\beta \sin_\beta(x, 0) + \alpha \sin_\alpha(x, 0)}{\alpha^2 - \beta^2}.$$

5. If $\alpha \neq 0$, then

$$\cos_\alpha(x, 0) \star \cos_\alpha(x, 0) = \frac{1}{\alpha} \sin_\alpha(x, 0) + \frac{1}{2}x \cos_\alpha(x, 0).$$

6. If $k \geq 0$, then

$$\sin_\alpha(x, 0) \star h_k(x, 0)$$

$$= \begin{cases} (-1)^{\frac{(k+1)(k+2)}{2}} \frac{1}{\alpha^{k+1}} \cos_\alpha(x, 0) + \sum_{j=0}^{\frac{k}{2}}(-1)^j \frac{h_{k-2j}(x,0)}{\alpha^{2j+1}} & \text{if } k \text{ is even,} \\ (-1)^{\frac{(k+1)(k+2)}{2}} \frac{1}{\alpha^{k+1}} \sin_\alpha(x, 0) + \sum_{j=0}^{\frac{k-1}{2}}(-1)^j \frac{h_{k-2j}(x,0)}{\alpha^{2j+1}} & \text{if } k \text{ is odd.} \end{cases}$$

It remains an open problem to establish a formula for the inverse Laplace transform on general time scales. A formula for the inverse Laplace transform has been proved in some special cases, for instance in the case of isolated time scales.

2.2   The Laplace Transform on Isolated Time Scales

In this section we specify the Laplace transform on isolated time scales. We prove several properties of the Laplace transform in this case and establish a formula for the inverse Laplace transform. For convenience, we consider the case $n = 1$. Let $\mathbb{T}$ be a time scale defined by

$$\mathbb{T} = \{t_n : n \in \mathbb{N}_0\}$$

such that

$$\lim_{n \to \infty} t_n = \infty, \quad \sigma(t_n) = t_{n+1}, \quad \mu(t_n) = t_{n+1} - t_n, \quad n \in \mathbb{N},$$

and

$$w = \inf_{n \in \mathbb{N}_0} \mu(t_n) > 0.$$

Note that such time scales exist. For instance,

$$\mathbb{T} = \{hn : n \in \mathbb{N}_0\}, \quad h > 0,$$

and

$$\mathbb{T} = q^{\mathbb{N}_0} \bigcup \{0\}, \quad q > 1.$$

Suppose that z is a complex number such that

$$1 + z\mu(t_n) \neq 0, \quad n \in \mathbb{N}_0. \tag{2.12}$$

The solution $e_z(t_n, t_m)$ of the problem

$$y(t_{n+1}) = (1 + z\mu(t_n)) y(t_n), \quad y(t_m) = 1, \quad m, n \in \mathbb{N}_0,$$

satisfies

$$e_z(t_n, t_m) = \begin{cases} \prod_{k=m}^{n-1} (1 + \mu(t_k)z) & \text{if} \quad n \geq m, \\ \dfrac{1}{\prod_{k=n}^{m-1}(1+\mu(t_k)z)} & \text{if} \quad n \leq m. \end{cases}$$

Note that the products for $n = m$ are understood to be 1. Thus, using Definition 2.2, we make the following definition.

Definition 2.7. If $f : \mathbb{T} \to \mathbb{C}$ is a function, then the Laplace transform is defined by

$$\mathscr{L}(f)(z) = \sum_{n=0}^{\infty} \frac{\mu(t_n) f(t_n)}{\prod_{k=0}^{n} (1 + \mu(t_k) z)}$$

for all $z \in \mathbb{C}$ satisfying (2.12) for which the series converges.

We define

$$P_n(z) = \prod_{k=0}^{n} (1 + \mu(t_k) z), \quad n \in \mathbb{N}_0, \quad P_{-1}(z) = 1.$$

Theorem 2.14. *For all $n \in \mathbb{N}_0$, we have*

$$P_n(z) - P_{n-1}(z) = z\mu(t_n) P_{n-1}(z)$$

and

$$\frac{1}{P_{n-1}(z)} - \frac{1}{P_n(z)} = \frac{z\mu(t_n)}{P_n(z)}.$$

Proof. Let $n \in \mathbb{N}$. Then

$$P_n(z) - P_{n-1}(z) = \prod_{k=0}^{n}(1 + \mu(t_k)z) - \prod_{k=0}^{n-1}(1 + \mu(t_k)z)$$

$$= \left(\prod_{k=0}^{n-1}(1 + \mu(t_k)z)\right)(1 + \mu(t_n)z - 1)$$

$$= z\mu(t_n) P_{n-1}(z),$$

$$\frac{1}{P_{n-1}(z)} - \frac{1}{P_n(z)} = \frac{1}{\prod_{k=0}^{n-1}(1 + \mu(t_k)z)} - \frac{1}{\prod_{k=0}^{n}(1 + \mu(t_k)z)}$$

$$= \frac{1 + \mu(t_n)z - 1}{\prod_{k=0}^{n}(1 + \mu(t_k)z)}$$

$$= \frac{\mu(t_n)z}{P_n(z)}.$$

Let $n = 0$. Then

$$P_0(z) - P_{-1}(z) = 1 + \mu(t_0)z - 1$$

$$= \mu(t_0)z$$

$$= \mu(t_0)zP_{-1}(z),$$

$$\frac{1}{P_{-1}(z)} - \frac{1}{P_0(z)} = 1 - \frac{1}{1 + \mu(t_0)z}$$

$$= \frac{1 + \mu(t_0)z - 1}{1 + \mu(t_0)z}$$

$$= \frac{\mu(t_0)z}{P_0(z)}.$$

This completes the proof.

Observe that $\mu(t_n) \geq w$ for all $n \in \mathbb{N}_0$. Therefore, $-\frac{1}{\mu(t_n)} \in \left[-\frac{1}{w}, 0\right)$. For all $\delta > 0$, we set

$$D_\delta^n = \left\{ z \in \mathbb{C} : \left| z + \frac{1}{\mu(t_n)} \right| < \delta \right\}, \quad n \in \mathbb{N}_0,$$

$$D_\delta = \mathbb{C} \backslash \bigcup_{n=0}^{\infty} D_\delta^n.$$

Note that D_δ is a closed domain of the complex plane $\mathbb{C}$. For $z \in D_\delta$ and $a \in A$, where

$$A = \left\{ -\frac{1}{\mu(t_n)} : n \in \mathbb{N}_0 \right\},$$

we have

$$\text{dist}\,(z, a) \geq \delta \quad \text{or} \quad \left| z + \frac{1}{\mu(t_n)} \right| \geq \delta \qquad (2.13)$$

for all $n \in \mathbb{N}_0$.

Theorem 2.15. *Assume* (2.12). *Then for all* $z \in D_\delta$, *we have*

$$|P_n(z)| \geq (\delta w)^{n+1} \quad \text{and} \quad |P_n(z)| \geq \delta (\delta w)^n \mu(t_n)$$

for all $n \in \mathbb{N}_0$. *Moreover,*

$$\lim_{n \to \infty} P_n(z) = \infty$$

for all $z \in D_\delta$, *provided* $\delta > \frac{1}{w}$.

Proof. For all $z \in D_\delta$ and $n \in \mathbb{N}_0$, using (2.13), we get

$$|P_n(z)| = \left| \prod_{k=0}^{n} (1 + \mu(t_k) z) \right|$$

$$= \left| \prod_{k=0}^{n} \left(\mu(t_k) \left(\frac{1}{\mu(t_k)} + z \right) \right) \right|$$

$$= \prod_{k=0}^{n} \mu(t_k) \prod_{k=0}^{n} \left| z + \frac{1}{\mu(t_k)} \right|$$

$$= \mu\left(t_n\right) \prod_{k=0}^{n-1} \mu\left(t_k\right) \prod_{k=0}^{n} \left| z + \frac{1}{\mu\left(t_k\right)} \right|$$

$$\geq \mu\left(t_n\right) w^n \delta^{n+1}$$

$$= \delta\mu\left(t_n\right)\left(\delta w\right)^n$$

$$\geq \left(\delta w\right)^{n+1}.$$

Hence, if $z \in D_\delta$ and $\delta > \frac{1}{w}$, we have $\delta w > 1$ and

$$\lim_{n\to\infty} P_n(z) = \infty.$$

This completes the proof.

Example 2.5. We will find $\mathscr{L}(1)(z)$. Using Definition 2.7, Theorem 2.14, and Theorem 2.15, we get

$$\mathscr{L}(1)(z) = \sum_{n=0}^{\infty} \frac{\mu\left(t_n\right)}{P_n(z)}$$

$$= \sum_{n=0}^{\infty} \frac{1}{z}\left(\frac{1}{P_{n-1}(z)} - \frac{1}{P_n(z)}\right)$$

$$= \frac{1}{z} \lim_{m\to\infty} \sum_{n=0}^{m-1}\left(\frac{1}{P_{n-1}(z)} - \frac{1}{P_n(z)}\right)$$

$$= \frac{1}{z} \lim_{m\to\infty}\left(1 - \frac{1}{P_0(z)} + \frac{1}{P_0(z)} - \frac{1}{P_1(z)} + \cdots + \frac{1}{P_{m-1}(z)} - \frac{1}{P_m(z)}\right)$$

$$= \frac{1}{z} \lim_{m\to\infty}\left(1 - \frac{1}{P_m(z)}\right)$$

$$= \frac{1}{z}$$

for $z \in D_\delta$, $\delta w > 1$.

Example 2.6. Now we will find the Laplace transform of $e_\alpha\left(t_n, t_0\right)$. We have

$$e_\alpha\left(t_n, t_0\right) = P_{n-1}\left(\alpha\right), \quad n \in \mathbb{N}_0.$$

We fix $n \in \mathbb{N}_0$ and set $e_\alpha = e_\alpha(t_n, t_0)$. Then for $z \in D_\delta$, we have

$$
\begin{aligned}
\mathscr{L}(e_\alpha)(z) &= \sum_{n=0}^{\infty} \frac{\mu(t_n) e_\alpha(t_n, t_0)}{P_n(z)} \\
&= \sum_{n=0}^{\infty} \frac{\mu(t_n) P_{n-1}(\alpha)}{P_n(z)} \\
&= \sum_{n=0}^{\infty} \frac{\mu(t_n) \prod_{k=0}^{n-1}(1 + \mu(t_k)\alpha)}{\prod_{k=0}^{n}(1 + \mu(t_k) z)} \\
&= \sum_{n=0}^{\infty} \frac{\mu(t_n)}{1 + \mu(t_n) z} \prod_{k=0}^{n-1} \frac{1 + \alpha\mu(t_k)}{1 + z\mu(t_k)} \\
&= \sum_{n=0}^{\infty} \frac{\mu(t_n)}{1 + \mu(t_n) z} \prod_{k=0}^{n-1} \frac{\alpha + \dfrac{1}{\mu(t_k)}}{z + \dfrac{1}{\mu(t_k)}}.
\end{aligned}
\tag{2.14}
$$

Note that

$$
2|\alpha| + \frac{3}{\mu(t_n)} \le 2|\alpha| + \frac{3}{w}
$$

for all $n \in \mathbb{N}_0$. Therefore, there exists a sufficiently large $R_0 > 0$ such that for $|z| \ge R_0$, we have

$$
2|\alpha| + \frac{3}{\mu(t_n)} \le |z|
$$

for all $n \in \mathbb{N}_0$. Hence, for $|z| \ge R_0$, we get

$$
2|\alpha| + \frac{2}{\mu(t_n)} \le |z| - \frac{1}{\mu(t_n)},
$$

whereupon

$$
2\left|\alpha + \frac{1}{\mu(t_n)}\right| \le \left|z + \frac{1}{\mu(t_n)}\right|,
$$

or

$$
\left|\frac{\alpha + \frac{1}{\mu(t_n)}}{z + \frac{1}{\mu(t_n)}}\right| \le \frac{1}{2}.
$$

Hence for $z \in D_\delta$, $|z| \ge R_0$, we have

$$\left| \sum_{n=0}^{\infty} \frac{\mu\left(t_n\right)}{1 + z\mu\left(t_n\right)} \prod_{k=0}^{n-1} \frac{\alpha + \frac{1}{\mu(t_k)}}{z + \frac{1}{\mu(t_k)}} \right| \leq \sum_{n=0}^{\infty} \frac{1}{\left| z + \frac{1}{\mu(t_n)} \right|} \prod_{k=0}^{n-1} \left| \frac{\alpha + \frac{1}{\mu(t_k)}}{z + \frac{1}{\mu(t_k)}} \right|$$

$$\leq \sum_{n=0}^{\infty} \frac{1}{2^n \left| z + \frac{1}{\mu(t_n)} \right|}$$

$$\leq \frac{1}{\delta} \sum_{n=0}^{\infty} \frac{1}{2^n}$$

$$= \frac{2}{\delta}.$$

Therefore, for $|z| \geq R_0$, $z \in D_\delta$, (2.14) exists. Hence, for $z \in D_\delta$, $|z| \geq R_0$, using Theorem 2.14, we have

$$\mathscr{L}\left(e_\alpha\right)(z) = \sum_{n=0}^{\infty} \frac{\mu\left(t_n\right) P_{n-1}(\alpha)}{P_n(z)}$$

$$= \frac{\mu\left(t_0\right)}{P_0(z)} + \sum_{n=1}^{\infty} \frac{\mu\left(t_n\right) P_{n-1}(\alpha)}{P_n(z)}$$

$$= \frac{\mu\left(t_0\right)}{P_0(z)} + \sum_{n=1}^{\infty} \frac{1}{z} \left(\frac{P_{n-1}(\alpha)}{P_{n-1}(z)} - \frac{P_{n-1}(\alpha)}{P_n(z)} \right)$$

$$= \frac{\mu\left(t_0\right)}{P_0(z)} + \frac{1}{z} \sum_{n=1}^{\infty} \left(\frac{\left(1 + \mu\left(t_{n-1}\right)\alpha\right) P_{n-2}(\alpha)}{P_{n-1}(z)} - \frac{P_{n-1}(\alpha)}{P_n(z)} \right)$$

$$= \frac{\mu\left(t_0\right)}{P_0(z)} + \frac{1}{z} \sum_{n=1}^{\infty} \left(\frac{P_{n-2}(\alpha)}{P_{n-1}(z)} - \frac{P_{n-1}(\alpha)}{P_n(z)} \right)$$

$$+ \frac{\alpha}{z} \sum_{n=1}^{\infty} \frac{\mu\left(t_{n-1}\right) P_{n-2}(\alpha)}{P_{n-1}(z)}$$

$$= \frac{\mu\left(t_0\right)}{P_0(z)}$$

$$+ \frac{1}{z} \lim_{m \to \infty} \left(\frac{1}{P_0(z)} - \frac{P_0(\alpha)}{P_1(z)} + \frac{P_0(\alpha)}{P_1(z)} - \cdots + \frac{P_{m-2}(\alpha)}{P_{m-1}(z)} - \frac{P_{m-1}(\alpha)}{P_m(z)} \right)$$

$$+ \frac{\alpha}{z} \mathscr{L}\left(e_\alpha\right)(z)$$

$$= \frac{\mu\left(t_0\right)}{P_0(z)} + \frac{1}{zP_0(z)} + \frac{\alpha}{z} \mathscr{L}\left(e_\alpha\right)(z)$$

$$= \frac{1}{z} + \frac{\alpha}{z} \mathscr{L}\left(e_\alpha\right)(z).$$

In the last equality we have used

$$\lim_{m\to\infty}\left|\frac{P_{m-1}(\alpha)}{P_m(z)}\right| = \lim_{m\to\infty}\left|\frac{1}{1+\mu(t_m)z}\prod_{k=0}^{m-1}\frac{\alpha+\frac{1}{\mu(t_k)}}{z+\frac{1}{\mu(t_k)}}\right|$$

$$\leq \lim_{m\to\infty}\left|\frac{1}{1+\mu(t_m)z}\right|\prod_{k=0}^{m-1}\left|\frac{\alpha+\frac{1}{\mu(t_k)}}{z+\frac{1}{\mu(t_k)}}\right|$$

$$\leq \lim_{m\to\infty}\frac{1}{2^m\,|1+\mu(t_m)z|}$$

$$= 0.$$

Consequently,

$$\mathscr{L}(e_\alpha)(z) = \frac{1}{z} + \frac{\alpha}{z}\mathscr{L}(e_\alpha)(z).$$

Hence,

$$\mathscr{L}(e_\alpha)(z) = \frac{1}{z-\alpha}.$$

Theorem 2.16. *Suppose $f : \mathbb{T} \to \mathbb{C}$ satisfies the condition*

$$|f(t_n)| \leq CR^n$$

for all $n \in \mathbb{N}_0$, where C and R are some positive constants. Then the series in Definition 2.7 converges uniformly with respect to z in the region D_δ with $\delta w > R$, and hence its sum $\mathscr{L}(f)(z)$ is an analytic (holomorphic) function in D_δ, $\delta w > R$.

Proof. Using Theorem 2.15, we get

$$\left|\frac{\mu(t_n)f(t_n)}{P_n(z)}\right| \leq \frac{C\mu(t_n)R^n}{\delta(\delta w)^n\mu(t_n)}$$

$$= \frac{C}{\delta}\left(\frac{R}{\delta w}\right)^n$$

for all $n \in \mathbb{N}_0$ and $z \in D_\delta$, $\delta w > R$. Hence,

$$|\mathscr{L}(f)(z)| \leq \frac{C}{\delta}\sum_{n=0}^{\infty}\left(\frac{R}{\delta w}\right)^n$$

$$< \infty$$

for all $z \in D_\delta$, $\delta w > R$. This completes the proof.

By $\mathscr{F}_\delta$ we denote the class of all functions $f : \mathbb{T} \to \mathbb{C}$ satisfying the condition

$$\sum_{n=0}^{\infty} (\delta w)^{-n} |f(t_n)| < \infty.$$

Theorem 2.17. *For all $f \in \mathscr{F}_\delta$, the series in Definition 2.7 converges uniformly with respect to $z \in D_\delta$, and hence $\mathscr{L}(f)(z)$ is an analytic function in D_δ.*

Proof. Using Theorem 2.15, we have

$$|\mathscr{L}(f)(z)| = \left| \sum_{n=0}^{\infty} \frac{\mu(t_n) f(t_n)}{P_n(z)} \right|$$

$$\leq \sum_{n=0}^{\infty} \frac{\mu(t_n) |f(t_n)|}{|P_n(z)|}$$

$$\leq \sum_{n=0}^{\infty} \frac{\mu(t_n) |f(t_n)|}{\delta (\delta w)^n \mu(t_n)}$$

$$= \delta^{-1} \sum_{n=0}^{\infty} (\delta w)^{-n} |f(t_n)|$$

$$< \infty$$

in D_δ. This completes the proof.

Theorem 2.18. *Let $f : \mathbb{T} \to \mathbb{R}$ be a function such that $f \in \mathscr{F}_\delta$. Then $f^{\Delta^k} \in \mathscr{F}_\delta$ for all $k \in \mathbb{N}_0$ and*

$$\mathscr{L}\left(f^{\Delta^k}\right)(z) = z^k \mathscr{L}(f)(z) - \sum_{l=0}^{k-1} z^l f^{\Delta^{k-1-l}}(t_0) \qquad (2.15)$$

for all $k \in \mathbb{N}_0$ and $z \in D_\delta$.

Proof. First, we will prove that $f^{\Delta^k} \in \mathscr{F}_\delta$ for all $k \in \mathbb{N}_0$.

1. Let $k = 1$. Then

$$\sum_{n=0}^{\infty} (\delta w)^{-n} \left| f^{\Delta}(t_n) \right| = \sum_{n=0}^{\infty} (\delta w)^{-n} \frac{|f(t_{n+1}) - f(t_n)|}{\mu(t_n)}$$

$$\leq \sum_{n=0}^{\infty} \frac{(\delta w)^{-n} |f(t_{n+1})|}{\mu(t_n)} + \sum_{n=0}^{\infty} \frac{(\delta w)^{-n} |f(t_n)|}{\mu(t_n)}$$

$$\leq \delta \sum_{n=0}^{\infty} (\delta w)^{-n-1} |f(t_{n+1})| + \frac{1}{w} \sum_{n=0}^{\infty} (\delta w)^{-n} |f(t_n)|$$

$$< \infty.$$

2. Assume that $f^{\Delta^k} \in \mathscr{F}_\delta$ for some $k \in \mathbb{N}$, i.e.,

$$\sum_{n=0}^{\infty} (\delta w)^{-n} \left| f^{\Delta^k}(t_n) \right| < \infty. \tag{2.16}$$

3. We will prove that $f^{\Delta^{k+1}} \in \mathscr{F}_\delta$. Indeed, using (2.16), we get

$$\sum_{n=0}^{\infty} (\delta w)^{-n} \left| f^{\Delta^{k+1}}(t_n) \right| = \sum_{n=0}^{\infty} (\delta w)^{-n} \frac{\left| f^{\Delta^k}(t_{n+1}) - f^{\Delta^k}(t_n) \right|}{\mu(t_n)}$$

$$\leq \sum_{n=0}^{\infty} \frac{(\delta w)^{-n} \left| f^{\Delta^k}(t_{n+1}) \right|}{\mu(t_n)} + \sum_{n=0}^{\infty} \frac{(\delta w)^{-n} \left| f^{\Delta^k}(t_n) \right|}{\mu(t_n)}$$

$$\leq \delta \sum_{n=0}^{\infty} (\delta w)^{-n-1} \left| f^{\Delta^k}(t_{n+1}) \right| + \frac{1}{w} \sum_{n=0}^{\infty} (\delta w)^{-n} \left| f^{\Delta^k}(t_n) \right|$$

$$< \infty.$$

Consequently, $f^{\Delta^k} \in \mathscr{F}_\delta$ for all $k \in \mathbb{N}_0$. Now we will prove (2.15).

1. Let $k = 1$. Then

$$\mathscr{L}\left(f^\Delta\right)(z) = \sum_{n=0}^{\infty} \frac{\mu(t_n) f^\Delta(t_n)}{P_n(z)}$$

$$= \sum_{n=0}^{\infty} \frac{f(t_{n+1}) - f(t_n)}{P_n(z)}$$

$$= \sum_{n=0}^{\infty} \frac{f(t_{n+1})}{P_n(z)} - \sum_{n=0}^{\infty} \frac{f(t_n)}{P_n(z)} \tag{2.17}$$

$$= \sum_{n=0}^{\infty} \frac{f(t_{n+1})(1 + z\mu(t_{n+1}))}{P_{n+1}(z)} - \sum_{n=0}^{\infty} \frac{f(t_n)}{P_n(z)}$$

$$= z \sum_{n=0}^{\infty} \frac{f(t_{n+1})\mu(t_{n+1})}{P_{n+1}(z)} + \sum_{n=0}^{\infty} \frac{f(t_{n+1})}{P_{n+1}(z)} - \sum_{n=0}^{\infty} \frac{f(t_n)}{P_n(z)}$$

$$= z \sum_{n=1}^{\infty} \frac{f(t_n)\mu(t_n)}{P_n(z)} + \sum_{n=1}^{\infty} \frac{f(t_n)}{P_n(z)} - \sum_{n=0}^{\infty} \frac{f(t_n)}{P_n(z)}$$

$$= z \sum_{n=0}^{\infty} \frac{f(t_n)\mu(t_n)}{P_n(z)} - z \frac{f(t_0)\mu(t_0)}{P_0(z)} - \frac{f(t_0)}{P_0(z)}$$

$$= z\mathscr{L}(f)(z) - \frac{f(t_0)(1 + z\mu(t_0))}{P_0(z)}$$
$$= z\mathscr{L}(f)(z) - f(t_0), \quad z \in D_\delta.$$

2. Assume (2.15) for some $k \in \mathbb{N}$.
3. We will prove that

$$\mathscr{L}\left(f^{\Delta^{k+1}}\right)(z) = z^{k+1}\mathscr{L}(f)(z) - \sum_{l=0}^{k} z^l f^{\Delta^{k-l}}(t_0).$$

We have

$$\mathscr{L}\left(f^{\Delta^{k+1}}\right)(z) = \mathscr{L}\left(\left(f^{\Delta^k}\right)^\Delta\right)(z)$$
$$= z\mathscr{L}\left(f^{\Delta^k}\right)(z) - f^{\Delta^k}(t_0)$$
$$= z\left(z^{k-1}\mathscr{L}(f)(z) - \sum_{l=0}^{k-1} z^l f^{\Delta^{k-1-l}}(t_0)\right)$$
$$- f^{\Delta^k}(t_0)$$
$$= z^k\mathscr{L}(f)(z) - \sum_{l=1}^{k} z^l f^{\Delta^{k-l}}(t_0) - f^{\Delta^k}(t_0)$$
$$= z^k\mathscr{L}(f)(z) - \sum_{l=0}^{k} z^l f^{\Delta^{k-l}}(t_0),$$

$z \in D_\delta$. This completes the proof.

Theorem 2.19 (Initial Value and Final Value Theorem). *For all $f \in \mathscr{F}_\delta$, for some $\delta > 0$, we have*

$$f(t_0) = \lim_{z \to \infty}(z\mathscr{L}(f)(z)) \tag{2.18}$$

and

$$\lim_{n \to \infty} f(t_n) = \lim_{z \to 0}(z\mathscr{L}(f)(z)). \tag{2.19}$$

Proof. We have

$$\mathscr{L}(f)(z) = \sum_{n=0}^{\infty} \frac{\mu(t_n)f(t_n)}{\prod_{k=0}^{n}(1 + \mu(t_k)z)}$$

$$= \frac{\mu(t_0)f(t_0)}{1+\mu(t_0)z} + \frac{\mu(t_1)f(t_1)}{(1+\mu(t_0)z)(1+\mu(t_1)z)}$$
$$+ \frac{\mu(t_2)f(t_2)}{(1+\mu(t_0)z)(1+\mu(t_1)z)(1+\mu(t_2)z)} + \cdots.$$

Hence,

$$(1+\mu(t_0)z)\mathscr{L}(f)(z) = \mu(t_0)f(t_0) + \frac{\mu(t_1)f(t_1)}{(1+\mu(t_1)z)}$$
$$+ \frac{\mu(t_2)f(t_2)}{(1+\mu(t_1)z)(1+\mu(t_2)z)} + \cdots.$$

Therefore,

$$\lim_{z\to\infty}\mathscr{L}(f)(z) = 0 \quad \text{and} \quad \lim_{z\to\infty}\left((1+\mu(t_0)z)\mathscr{L}(f)(z)\right) = \mu(t_0)f(t_0),$$

whereupon

$$\mu(t_0)\lim_{z\to\infty}(z\mathscr{L}(f)(z)) = \mu(t_0)f(t_0),$$

or

$$\lim_{z\to\infty}(z\mathscr{L}(f)(z)) = f(t_0),$$

i.e., (2.18) holds. Now we will prove (2.19). By (2.17), we have

$$z\mathscr{L}(f)(z) - f(t_0) = \sum_{n=0}^{\infty}\frac{f(t_{n+1}) - f(t_n)}{P_n(z)}. \qquad (2.20)$$

Note that

$$\lim_{z\to 0}P_n(z) = 1.$$

Hence, using (2.20), we get

$$\lim_{z\to 0}(z\mathscr{L}(f)(z)) - f(t_0) = \lim_{z\to 0}\sum_{n=0}^{\infty}\frac{f(t_{n+1}) - f(t_n)}{P_n(z)}$$

$$= \sum_{n=0}^{\infty}(f(t_{n+1}) - f(t_n))$$

$$= \lim_{n\to\infty}\sum_{m=0}^{n-1}(f(t_{m+1}) - f(t_m))$$

$$
\begin{aligned}
&= \lim_{n \to \infty} (f(t_1) - f(t_0) + f(t_2) - f(t_1) \\
&\quad + \cdots + f(t_n) - f(t_{n-1})) \\
&= \lim_{n \to \infty} f(t_n) - f(t_0),
\end{aligned}
$$

whereupon we obtain (2.19). This completes the proof.

For a given function $f : \mathbb{T} \to \mathbb{C}$, we consider the shifting problem

$$
\begin{aligned}
&\mu(t_m)\left(\widehat{f}(t_{n+1}, t_{m+1}) - \widehat{f}(t_n, t_{m+1})\right) \\
&+ \mu(t_n)\left(\widehat{f}(t_n, t_{m+1}) - \widehat{f}(t_n, t_m)\right) = 0, \\
&\widehat{f}(t_n, t_0) = f(t_n), \quad n, m \in \mathbb{N}_0, \quad n \geq m.
\end{aligned}
\tag{2.21}
$$

We set

$$
\widehat{f}_{n,m} = \widehat{f}(t_n, t_m), \quad n, m \in \mathbb{N}_0, \quad n \geq m.
$$

Then the problem (2.21) can be rewritten in the form

$$
\begin{aligned}
&\mu(t_m)\left(\widehat{f}_{n+1,m+1} - \widehat{f}_{n,m+1}\right) + \mu(t_n)\left(\widehat{f}_{n,m+1} - \widehat{f}_{n,m}\right) = 0, \\
&\widehat{f}_{n,0} = f(t_n), \quad n, m \in \mathbb{N}_0, \quad n \geq m.
\end{aligned}
\tag{2.22}
$$

Theorem 2.20. *For every* $f : \mathbb{T} \to \mathbb{C}$, *the problem* (2.22) *has a unique solution.*

Proof. For $n = m$, we have

$$
\mu(t_n)\left(\widehat{f}_{n+1,n+1} - \widehat{f}_{n,n+1}\right) + \mu(t_n)\left(\widehat{f}_{n,n+1} - \widehat{f}_{n,n}\right) = 0,
$$

or

$$
\mu(t_n)\left(\widehat{f}_{n+1,n+1} - \widehat{f}_{n,n}\right) = 0,
$$

or

$$
\widehat{f}_{n+1,n+1} - \widehat{f}_{n,n} = 0, \quad n \in \mathbb{N}_0.
$$

Hence, we conclude that

$$
\widehat{f}_{n,n} = f(t_0), \quad n \in \mathbb{N}_0.
\tag{2.23}
$$

Therefore, it is enough to prove that the problem (2.22) has a unique solution satisfying the condition (2.23). For $i \in \mathbb{N}_0$, we set

$$
\mathbb{N}_i = [i, \infty) \bigcap \mathbb{N}_0.
$$

For $n \in \mathbb{N}_1$, we put $m = n - 1$ in (2.22), and we get

$$\mu(t_{n-1})\left(\widehat{f}_{n+1,n} - \widehat{f}_{n,n}\right) + \mu(t_n)\left(\widehat{f}_{n,n} - \widehat{f}_{n,n-1}\right) = 0.$$

Hence,

$$\widehat{f}_{n+1,n} = \left(1 - \frac{\mu(t_n)}{\mu(t_{n-1})}\right)f(t_0) + \frac{\mu(t_n)}{\mu(t_{n-1})}\widehat{f}_{n,n-1}, \quad n \in \mathbb{N}_1. \tag{2.24}$$

By (2.22), we get

$$\widehat{f}_{1,0} = f(t_1).$$

From this, using (2.24), we obtain $\widehat{f}_{n+1,n}$ recursively in a unique way for all $n \in \mathbb{N}_0$. Now we put $m = n - 2$, $n \in \mathbb{N}_2$, in (2.22), and we get

$$\mu(t_{n-2})\left(\widehat{f}_{n+1,n-1} - \widehat{f}_{n,n-1}\right) + \mu(t_n)\left(\widehat{f}_{n,n-1} - \widehat{f}_{n,n-2}\right) = 0,$$

or

$$\widehat{f}_{n+1,n-1} = \left(1 - \frac{\mu(t_n)}{\mu(t_{n-2})}\right)\widehat{f}_{n,n-1} + \frac{\mu(t_n)}{\mu(t_{n-2})}\widehat{f}_{n,n-2}. \tag{2.25}$$

Note that

$$\widehat{f}_{2,0} = f(t_2).$$

Therefore, using (2.25), we can determine recursively $\widehat{f}_{n+1,n-1}$ in a unique way for all $n \in \mathbb{N}_1$. Repeating this procedure, for an arbitrarily chosen $i \in \mathbb{N}$, we put $m = n - i$, $n \in \mathbb{N}_i$, in (2.22), and we obtain

$$\mu(t_{n-i})\left(\widehat{f}_{n+1,n+1-i} - \widehat{f}_{n,n+1-i}\right) + \mu(t_n)\left(\widehat{f}_{n,n+1-i} - \widehat{f}_{n,n-i}\right) = 0,$$

or

$$\widehat{f}_{n+1,n+1-i} = \left(1 - \frac{\mu(t_n)}{\mu(t_{n-i})}\right)\widehat{f}_{n,n+1-i} + \frac{\mu(t_n)}{\mu(t_{n-i})}\widehat{f}_{n,n-i}. \tag{2.26}$$

Observe that

$$\widehat{f}_{n,0} = f(t_n).$$

Since $\widehat{f}_{n,n+1-i}$ and $\widehat{f}_{n,n-i}$ are known for $n \in \mathbb{N}_i$, using (2.26), we conclude that $\widehat{f}_{n+1,n+1-i}$ can be determined recursively in a unique way for all $n \in \mathbb{N}_{i-1}$. Because $i \in \mathbb{N}$ was arbitrarily chosen, we can find $\widehat{f}_{n,m}$ in a unique way for all $m, n \in \mathbb{N}_0$, $n \geq m$. This completes the proof.

Definition 2.8. Let $f, g : \mathbb{T} \to \mathbb{C}$ be two functions and let $\widehat{f}$ be the solution satisfying the shifting problem (2.21). We define the convolution $f \star g$ of the functions f and g as follows:

$$(f \star g)(t_0) = 0,$$

$$(f \star g)(t_n) = \sum_{k=0}^{n-1} \mu(t_k)\widehat{f}(t_n, t_{k+1}) g(t_k), \quad n \in \mathbb{N}.$$

Theorem 2.21. *Let $f, g : \mathbb{T} \to \mathbb{C}$ be two functions. If $f \star g$ is identically equal to zero on $\mathbb{T}$, then at least one of the functions f and g is identically equal to zero on $\mathbb{T}$.*

Proof. Since $f \star g$ is identically equal to zero on $\mathbb{T}$, we have

$$
\begin{aligned}
(f \star g)(t_1) &= \mu(t_0)\widehat{f}(t_1, t_1) g(t_0) \\
&= \mu(t_0)\widehat{f}_{1,1} g(t_0) \\
&= 0, \\
(f \star g)(t_2) &= \mu(t_0)\widehat{f}(t_2, t_1) g(t_0) + \mu(t_1)\widehat{f}(t_2, t_2) g(t_1) \\
&= \mu(t_0)\widehat{f}_{2,1} g(t_0) + \mu(t_1)\widehat{f}_{2,2} g(t_1) \\
&= 0, \\
&\vdots \\
(f \star g)(t_n) &= \mu(t_0)\widehat{f}(t_n, t_1) g(t_0) + \mu(t_1)\widehat{f}(t_n, t_2) g(t_1) \\
&\quad + \cdots + \mu(t_{n-1})\widehat{f}(t_n, t_n) g(t_{n-1}) \\
&= \mu(t_0)\widehat{f}_{n,1} g(t_0) + \mu(t_1)\widehat{f}_{n,2} g(t_1) \\
&\quad + \cdots + \mu(t_{n-1})\widehat{f}_{n,n} g(t_{n-1}) \\
&= 0.
\end{aligned}
$$

In this way, we get the system

$$
\begin{cases}
\mu(t_0)\widehat{f}_{1,1} g(t_0) = 0, \\
\mu(t_0)\widehat{f}_{2,1} g(t_0) + \mu(t_1)\widehat{f}_{2,2} g(t_1) = 0, \\
\vdots \\
\mu(t_0)\widehat{f}_{n,1} g(t_0) + \mu(t_1)\widehat{f}_{n,2} g(t_1) + \cdots + \mu(t_{n-1})\widehat{f}_{n,n} g(t_{n-1}) = 0.
\end{cases}
\tag{2.27}
$$

Note that we can take $n \in \mathbb{N}$ as large as we wish. If f is identically equal to zero, then the assertion is proved. Assume that f is not identically equal to zero on $\mathbb{T}$. Let $f(t_m), m \in \mathbb{N}_0$, be the first value of f that is different from zero. Hence,

$$f(t_0) = \cdots = f(t_{m-1}) = 0, \quad f(t_m) \neq 0, \quad \text{if} \quad m \in \mathbb{N}. \tag{2.28}$$

If $m = 0$, then $f(t_0) \neq 0$. Consider the system (2.27) with respect to $g(t_0)$, ..., $g(t_{n-1})$. Let

$$A = \begin{pmatrix} \mu(t_0)\widehat{f}_{1,1} & 0 & \cdots & & 0 \\ \mu(t_0)\widehat{f}_{2,1} & \mu(t_1)\widehat{f}_{2,2} & \cdots & & 0 \\ \vdots & \vdots & \vdots & & \vdots \\ \mu(t_0)\widehat{f}_{n,1} & \mu(t_1)\widehat{f}_{n,2} & \cdots & \mu(t_{n-1})\widehat{f}_{n,n} \end{pmatrix}.$$

Then

$$\det A = \prod_{l=1}^{n} \left(\mu(t_{l-1})\widehat{f}_{l,l} \right)$$

$$= \prod_{l=1}^{n} (\mu(t_{l-1})f(t_0))$$

$$= (f(t_0))^n \prod_{l=1}^{n} \mu(t_{l-1}).$$

If $f(t_0) \neq 0$, we obtain that $\det A \neq 0$. From this and the system (2.27), we obtain

$$g(t_0) = \cdots = g(t_{n-1}) = 0.$$

Because $n \in \mathbb{N}$ was arbitrarily chosen, we obtain that g is identically equal to zero on $\mathbb{T}$. Let $f(t_0) = 0$ and $f(t_1) \neq 0$. Then

$$\widehat{f}_{n,n} = 0 \quad \text{for all} \quad n \in \mathbb{N}_0.$$

Let $n \in \mathbb{N}_2$. In this case, the system (2.27) takes the form

$$\begin{cases} \mu(t_0)\widehat{f}_{2,1}g(t_0) = 0, \\ \vdots \\ \mu(t_0)\widehat{f}_{n,1}g(t_0) + \cdots + \mu(t_{n-2})\widehat{f}_{n,n-1}g(t_{n-2}) = 0. \end{cases} \qquad (2.29)$$

Next, since $f(t_0) = 0$, we have

$$\widehat{f}_{n+1,n} = \frac{\mu(t_n)}{\mu(t_{n-1})}\widehat{f}_{n,n-1}$$

$$= \frac{\mu(t_n)}{\mu(t_{n-2})}\widehat{f}_{n-1,n-2}$$

$$\vdots$$

$$= \frac{\mu(t_n)}{\mu(t_0)}\widehat{f}_{1,0}.$$

$$= \frac{\mu\left(t_n\right)}{\mu\left(t_0\right)} f\left(t_1\right), \quad n \in \mathbb{N}.$$

Now we consider the system (2.29) with respect to $g\left(t_0\right), \ldots, g\left(t_{n-2}\right)$. For its matrix we have

$$A_1 = \begin{pmatrix} \mu\left(t_0\right)\widehat{f}_{2,1} & 0 & \cdots & & 0 \\ \mu\left(t_0\right)\widehat{f}_{3,1} & \mu\left(t_1\right)\widehat{f}_{3,2} & \cdots & & 0 \\ \vdots & \vdots & \vdots & & \vdots \\ \mu\left(t_0\right)\widehat{f}_{n,1} & \mu\left(t_1\right)\widehat{f}_{n,2} & \cdots & \mu\left(t_{n-2}\right)\widehat{f}_{n,n-1} \end{pmatrix}$$

and

$$\det A_1 = \prod_{l=0}^{n-2} \mu\left(t_l\right)\widehat{f}_{l+2,l+1}$$

$$= \prod_{l=0}^{n-2}\left(\mu\left(t_l\right)\frac{\mu\left(t_{l+1}\right)}{\mu\left(t_0\right)}f\left(t_1\right)\right)$$

$$= \left(f\left(t_1\right)\right)^{n-1} \prod_{l=0}^{n-2} \frac{\mu\left(t_l\right)\mu\left(t_{l+1}\right)}{\mu\left(t_0\right)}$$

$$\neq 0.$$

Therefore,

$$g\left(t_0\right) = \cdots = g\left(t_{n-2}\right) = 0,$$

and since $n \in \mathbb{N}_2$ was arbitrarily chosen, we obtain that g is identically equal to zero on $\mathbb{T}$. We see that one can discuss the case $f\left(t_m\right) \neq 0$ and argue in this way for every value $m \in \mathbb{N}_0$ in (2.28) in order to obtain that $g\left(t_n\right) = 0$ for all $n \in \mathbb{N}_0$. This completes the proof.

Theorem 2.22. *Let* $f, g : \mathbb{T} \to \mathbb{C}$ *be two functions such that* $\mathscr{L}(f)(z)$, $\mathscr{L}(g)(z)$, *and* $\mathscr{L}\left(f \star g\right)(z)$ *exist for a given* $z \in \mathbb{C}$ *satisfying* (2.12)*. Then, at the point* z,

$$\mathscr{L}\left(f \star g\right)(z) = \mathscr{L}(f)(z)\mathscr{L}(g)(z).$$

Proof. For convenience, we will use the following notation:

$$e_{n,m}(z) = e_z\left(t_n, t_m\right).$$

By the definition of the exponential function, we have the following relations:

$$e_{n,n}(z) = 1$$

for all $n \in \mathbb{N}_0$, and

$$e_{n+1,m}(z) = (1 + \mu(t_n)z) e_{n,m}(z)$$

for all $n, m \in \mathbb{N}_0, n \geq m$,

$$e_{n,m+1}(z) = \frac{e_{n,m}(z)}{1 + \mu(t_m)z}$$

for all $n, m \in \mathbb{N}, n \geq m + 1$. By the definition of the Laplace transform, we get

$$\mathscr{L}(f \star g)(z) = \sum_{n=1}^{\infty} \frac{\mu(t_n)(f \star g)(t_n)}{e_{n+1,0}(z)}$$

$$= \sum_{n=1}^{\infty} \frac{\mu(t_n)}{e_{n+1,0}(z)} \sum_{k=0}^{n-1} \mu(t_k)\widehat{f}_{n,k+1}g(t_k)$$

$$= \sum_{k=0}^{\infty} \mu(t_k)g(t_k) \sum_{n=k+1}^{\infty} \frac{\mu(t_n)\widehat{f}_{n,k+1}}{e_{n+1,0}(z)}.$$

Hence, using that

$$e_{n+1,0}(z) = e_{n+1,k+1}(z)e_{k+1,0}(z),$$

we obtain

$$\mathscr{L}(f \star g)(z) = \sum_{k=0}^{\infty} \frac{\mu(t_k)g(t_k)}{e_{k+1,0}(z)} \sum_{n=k+1}^{\infty} \frac{\mu(t_n)\widehat{f}_{n,k+1}}{e_{n+1,k+1}(z)}. \qquad (2.30)$$

We set

$$\Psi_m = \sum_{n=m}^{\infty} \frac{\mu(t_n)\widehat{f}_{n,m}}{e_{n+1,m}(z)}, \qquad m \in \mathbb{N}_0.$$

Then, using (2.22), we obtain

$$\Psi_{m+1} = \sum_{n=m+1}^{\infty} \frac{\mu(t_n)\widehat{f}_{n,m+1}}{e_{n+1,m+1}(z)}$$

$$= \sum_{n=m+1}^{\infty} \frac{(\mu(t_n) - \mu(t_m))\widehat{f}_{n,m+1} + \mu(t_m)\widehat{f}_{n,m+1}}{e_{n+1,m+1}(z)}$$

$$= \sum_{n=m+1}^{\infty} \frac{\mu(t_n)\widehat{f}_{n,m} - \mu(t_m)\widehat{f}_{n+1,m+1} + \mu(t_m)\widehat{f}_{n,m+1}}{e_{n+1,m+1}(z)}$$

$$= \sum_{n=m+1}^{\infty} \mu\,(t_n)\,\frac{\widehat{f}_{n,m}}{e_{n+1,m+1}(z)}$$

$$-\mu\,(t_m) \sum_{n=m+1}^{\infty} \left(\frac{\widehat{f}_{n+1,m+1}}{e_{n+1,m+1}(z)} - \frac{\widehat{f}_{n,m+1}}{e_{n,m+1}(z)} + \frac{\widehat{f}_{n,m+1}}{e_{n,m+1}(z)} - \frac{\widehat{f}_{n,m+1}}{e_{n+1,m+1}(z)} \right)$$

$$= \sum_{n=m+1}^{\infty} \frac{\mu\,(t_n)\widehat{f}_{n,m}\,(1 + \mu\,(t_m)\,z)}{e_{n+1,m}(z)} - \mu\,(t_m) \sum_{n=m+1}^{\infty} \frac{\widehat{f}_{n+1,m+1}}{e_{n+1,m+1}(z)}$$

$$+\mu\,(t_m)\,\frac{\widehat{f}_{m+1,m+1}}{e_{m+1,m+1}(z)} + \mu\,(t_m) \sum_{n=m+2}^{\infty} \frac{\widehat{f}_{n,m+1}}{e_{n,m+1}(z)}$$

$$-\mu\,(t_m) \sum_{n=m+1}^{\infty} \left(\frac{\widehat{f}_{n,m+1}}{e_{n,m+1}(z)} - \frac{\widehat{f}_{n,m+1}}{(1 + \mu\,(t_n)\,z)\,e_{n,m+1}(z)} \right)$$

$$= (1 + \mu\,(t_m)\,z)\,\Psi_m + \mu\,(t_m)\,\frac{\widehat{f}_{m+1,m+1}}{e_{m+1,m+1}(z)}$$

$$-\frac{\mu\,(t_m)\widehat{f}_{m,m}\,(1 + \mu\,(t_m)\,z)}{e_{m+1,m}(z)} - \mu\,(t_m) \sum_{n=m+1}^{\infty} \frac{\mu\,(t_n)\,z\widehat{f}_{n,m+1}}{e_{n,m+1}(z)}$$

$$= (1 + \mu\,(t_m)\,z)\,\Psi_m + \mu\,(t_m)\widehat{f}_{m+1,m+1}$$

$$-\mu\,(t_m)\widehat{f}_{m,m} - \mu\,(t_m)\,z\Psi_{m+1}$$

$$= (1 + \mu\,(t_m)\,z)\,\Psi_m - \mu\,(t_m)\,z\Psi_{m+1},$$

whereupon

$$(1 + \mu\,(t_m)\,z)\,\Psi_m = (1 + \mu\,(t_m)\,z)\,\Psi_{m+1},$$

or

$$\Psi_m = \Psi_{m+1}.$$

Therefore, Ψ_m is independent of $m \in \mathbb{N}_0$. From this and (2.30), we get

$$\mathscr{L}\,(f \star g)\,(z) = \sum_{k=0}^{\infty} \frac{\mu\,(t_k)\,g\,(t_k)}{e_{k+1,0}(z)} \sum_{n=0}^{\infty} \frac{\mu\,(t_n)\widehat{f}_{n,0}}{e_{n+1,0}(z)}$$

$$= \mathscr{L}(f)(z)\mathscr{L}(g)(z).$$

This completes the proof.

Theorem 2.23 (Uniqueness Theorem). *Let $f : \mathbb{T} \to \mathbb{C}$ be a function in the space $\mathscr{F}_\delta$. If $\mathscr{L}(f)(z) = 0$ for $z \in D_\delta$, then $f(t_n) = 0$ for all $n \in \mathbb{N}_0$.*

Proof. By the definition of the Laplace transform and by the condition $\mathscr{L}(f)(z) = 0$, $z \in D_\delta$, we get

$$0 = \frac{\mu(t_0)f(t_0)}{1 + \mu(t_0)z} + \frac{\mu(t_1)f(t_1)}{(1 + \mu(t_0)z)(1 + \mu(t_1)z)}$$
$$+ \frac{\mu(t_2)f(t_2)}{(1 + \mu(t_0)z)(1 + \mu(t_1)z)(1 + \mu(t_2)z)} + \cdots,$$

$z \in D_\delta$. Hence,

$$0 = \mu(t_0)f(t_0) + \frac{\mu(t_1)f(t_1)}{1 + \mu(t_1)z} + \frac{\mu(t_2)f(t_2)}{(1 + \mu(t_1)z)(1 + \mu(t_2)z)} + \cdots$$
$$\to \mu(t_0)f(t_0) \quad \text{as} \quad |z| \to \infty.$$

Therefore, $f(t_0) = 0$. Hence,

$$0 = \mu(t_1)f(t_1) + \frac{\mu(t_2)f(t_2)}{1 + \mu(t_2)z} + \cdots$$
$$\to \mu(t_1)f(t_1) \quad \text{as} \quad |z| \to \infty.$$

Therefore, $f(t_1) = 0$. Repeating this procedure, we find that $f(t_n) = 0$ for all $n \in \mathbb{N}_0$. This completes the proof.

Theorem 2.24 (Inverse Laplace Transform). *Let $f \in \mathscr{F}_\delta$ and let $\mathscr{L}(f)(z)$ be its Laplace transform. Then*

$$f(t_n) = \frac{1}{2\pi i} \int_\gamma \mathscr{L}(f)(z) \prod_{k=0}^{n-1} (1 + \mu(t_k)z)\, dz, \quad n \in \mathbb{N}_0, \tag{2.31}$$

where γ is any positively oriented closed curve in the region D_δ that encloses all the points $-\frac{1}{\mu(t_k)}$ for $k \in \mathbb{N}_0$.

Remark 2.2. Here in this book we will suppose that γ is any positively oriented closed curve in the region D_δ that encloses all the points $-\frac{1}{\mu(t_k)}$ for $k \in \mathbb{N}_0$ for which $\sqrt{1} = 1$.

Proof. Note that

$$\int_\gamma \frac{dz}{1 + \mu(t_0)z} = \frac{1}{\mu(t_0)} \int_\gamma \frac{dz}{z + \frac{1}{\mu(t_0)}}$$
$$= \frac{2\pi i}{\mu(t_0)}$$

and

$$\int_\gamma \frac{dz}{\prod_{k=0}^{n-1} (1 + \mu(t_k) z)} = 0$$

for all $n \in \mathbb{N} \setminus \{1\}$. Hence, using the uniform convergence of the series

$$\mathcal{L}(f)(z) = \frac{\mu(t_0) f(t_0)}{1 + \mu(t_0) z} + \frac{\mu(t_1) f(t_1)}{(1 + \mu(t_0) z)(1 + \mu(t_1) z)}$$

$$+ \frac{\mu(t_2) f(t_2)}{(1 + \mu(t_0) z)(1 + \mu(t_1) z)(1 + \mu(t_2) z)} + \cdots \qquad (2.32)$$

and integrating term by term, we get

$$\int_\gamma \mathcal{L}(f)(z) dz = 2\pi i f(t_0),$$

or

$$f(t_0) = \frac{1}{2\pi i} \int_\gamma \mathcal{L}(f)(z) dz.$$

Now we multiply (2.32) by $1 + \mu(t_0) z$, and we obtain

$$(1 + \mu(t_0) z) \mathcal{L}(f)(z) = \mu(t_0) f(t_0) + \frac{\mu(t_1) f(t_1)}{1 + \mu(t_1) z} + \cdots,$$

and integrating over γ with respect to z, we obtain

$$\int_\gamma (1 + \mu(t_0) z) \mathcal{L}(f)(z) dz = \mu(t_0) f(t_0) \int_\gamma dz + \mu(t_1) f(t_1) \int_\gamma \frac{1}{1 + \mu(t_1) z} dz$$

$$+ \mu(t_2) f(t_2) \int_\gamma \frac{1}{(1 + \mu(t_1) z)(1 + \mu(t_2) z)} dz + \cdots,$$

or

$$\int_\gamma (1 + \mu(t_0) z) \mathcal{L}(f)(z) dz = 2\pi i f(t_1),$$

or

$$f(t_1) = \frac{1}{2\pi i} \int_\gamma (1 + \mu(t_0) z) \mathcal{L}(f)(z) dz.$$

Repeating this procedure, we can obtain the formula (2.31) for all $n \in \mathbb{N}_0$. This completes the proof.

We will denote the inverse Laplace transform by $\mathscr{L}^{-1}$.

Example 2.7. Let $\mathbb{T} = \{t_n : n \in \mathbb{N}_0\}$ be an isolated time scale such that

$$\sigma(t_n) = t_{n+1}, \quad \mu(t_n) = t_{n+1} - t_n, \quad n \in \mathbb{N}_0, \quad w = \inf_{n \in \mathbb{N}_0} \mu(t_n) > 0.$$

Let

$$\gamma = \left\{ z \in \mathbb{C} : |z| = \frac{2}{w} \right\}.$$

Then

$$-\frac{1}{\mu(t_n)} \in \left\{ z \in \mathbb{C} : |z| < \frac{2}{w} \right\}$$

for all $n \in \mathbb{N}_0$. Hence,

$$\frac{1}{2\pi i} \int_\gamma \frac{1}{z} \prod_{k=0}^{n-1} (1 + \mu(t_k) z) \, dz$$

$$= \frac{1}{2\pi i} \int_0^{2\pi} \frac{1}{\frac{2}{w} e^{i\theta}} \prod_{k=0}^{n-1} \left(1 + \mu(t_k) \frac{2}{w} e^{i\theta} \right) \frac{2i}{w} e^{i\theta} \, d\theta$$

$$= \frac{1}{2\pi} \int_0^{2\pi} \prod_{k=0}^{n-1} \left(1 + \mu(t_k) \frac{2}{w} e^{i\theta} \right) d\theta$$

$$= 1$$

for all $n \in \mathbb{N}_0$. Therefore,

$$\mathscr{L}^{-1} \left(\frac{1}{z} \right) (t_n) = 1.$$

Example 2.8. Let $\mathbb{T} = \{t_n : n \in \mathbb{N}_0\}$ be an isolated time scale such that

$$\sigma(t_n) = t_{n+1}, \quad \mu(t_n) = t_{n+1} - t_n, \quad n \in \mathbb{N}_0, \quad w = \inf_{n \in \mathbb{N}_0} \mu(t_n) > 0.$$

Let $\alpha \in \mathbb{C}$. We choose $m \in \mathbb{N}_0$ large enough that

$$-\frac{1}{\mu(t_k)} \in \{ z \in \mathbb{C} : |z - \alpha| < m \}.$$

We take

$$\gamma = \{ z \in \mathbb{C} : |z - \alpha| = m \}.$$

Then

$$\frac{1}{2\pi i} \int_\gamma \frac{1}{z-\alpha} \prod_{k=0}^{n-1} (1 + \mu(t_k) z) \, dz$$

$$= \frac{1}{2\pi i} \int_0^{2\pi} \frac{1}{me^{i\theta}} \prod_{k=0}^{n-1} \left(1 + \mu(t_k)\left(\alpha + me^{i\theta}\right)\right) mie^{i\theta} \, d\theta$$

$$= \frac{1}{2\pi} \int_0^{2\pi} \prod_{k=0}^{n-1} \left(1 + \mu(t_k)\alpha + \mu(t_k) me^{i\theta}\right) d\theta$$

$$= \prod_{k=0}^{n-1} (1 + \mu(t_k)\alpha)$$

$$= e_\alpha(t_n, t_0)$$

for all $n \in \mathbb{N}_0$. Therefore,

$$\mathscr{L}^{-1}\left(\frac{1}{z-\alpha}\right)(t_n) = e_\alpha(t_n, t_0), \quad n \in \mathbb{N}_0.$$

Example 2.9. Let $\mathbb{T} = \{t_n : n \in \mathbb{N}_0\}$ be an isolated time scale such that

$$\sigma(t_n) = t_{n+1}, \quad \mu(t_n) = t_{n+1} - t_n, \quad n \in \mathbb{N}_0, \quad w = \inf_{n \in \mathbb{N}_0} \mu(t_n) > 0.$$

We will find

$$\mathscr{L}^{-1}\left(\frac{1}{z(z-1)(z-2)(z-3)}\right)(t_n), \quad n \in \mathbb{N}_0.$$

Note that

$$\frac{1}{z(z-1)(z-2)(z-3)} = \frac{1}{2(z-1)} - \frac{1}{2(z-2)} - \frac{1}{6z} + \frac{1}{6(z-3)}.$$

Hence,

$$\mathscr{L}^{-1}\left(\frac{1}{z(z-1)(z-2)(z-3)}\right)(t_n) = \frac{1}{2}\mathscr{L}^{-1}\left(\frac{1}{z-1}\right)(t_n) - \frac{1}{2}\mathscr{L}^{-1}\left(\frac{1}{z-2}\right)(t_n)$$

$$- \frac{1}{6}\mathscr{L}^{-1}\left(\frac{1}{z}\right)(t_n) + \frac{1}{6}\mathscr{L}^{-1}\left(\frac{1}{z-3}\right)(t_n)$$

$$= \frac{1}{2}e_1(t_n, t_0) - \frac{1}{2}e_2(t_n, t_0)$$

$$- \frac{1}{6} + \frac{1}{6}e_3(t_n, t_0).$$

Example 2.10. Let $\mathbb{T} = \{t_n : n \in \mathbb{N}_0\}$ be an isolated time scale such that

$$\sigma(t_n) = t_{n+1}, \quad \mu(t_n) = t_{n+1} - t_n, \quad n \in \mathbb{N}_0, \quad w = \inf_{n \in \mathbb{N}_0} \mu(t_n) > 0.$$

Let

$$g(z) = \prod_{k=0}^{m-1} \left(\sqrt{z} + \beta_k\right), \quad \beta_k \in \mathbb{C}, \quad k \in \{0, \ldots, m-1\}, \quad m \in \mathbb{N}.$$

We set

$$\begin{cases} b_m &= 1 \\ b_{m-1} &= \beta_0 + \beta_1 + \cdots + \beta_{m-2} + \beta_{m-1} \\ b_{m-2} &= \beta_0\beta_1 + \beta_0\beta_2 + \cdots + \beta_0\beta_{m-1} + \beta_1\beta_2 + \beta_1\beta_3 + \cdots + \beta_1\beta_{m-1} \\ & \quad + \cdots + \beta_{m-2}\beta_{m-1} \\ & \ \ \vdots \\ b_0 &= \beta_0\beta_1 \ldots \beta_{m-1}, \end{cases}$$

$$(2.33)$$

$$\begin{cases} a_n &= 1 \\ a_{n-1} &= \alpha_0 + \alpha_1 + \cdots + \alpha_{n-2} + \alpha_{n-1} \\ a_{n-2} &= \alpha_0\alpha_1 + \alpha_0\alpha_2 + \cdots + \alpha_0\alpha_{n-1} + \alpha_1\alpha_2 + \alpha_1\alpha_3 + \cdots + \alpha_1\alpha_{n-1} \\ & \quad + \cdots + \alpha_{n-2}\alpha_{n-1} \\ & \ \ \vdots \\ a_0 &= \alpha_0\alpha_1 \ldots \alpha_{n-1}, \end{cases}$$

$$(2.34)$$

where

$$\alpha_k = \frac{1}{\mu(t_k)}, \quad k \in \mathbb{N}_0.$$

Then

$$g(z) = \sum_{l=0}^{m} b_l z^{\frac{l}{2}},$$

$$\prod_{k=0}^{n-1} (1 + \mu(t_k) z) = \prod_{k=0}^{n-1} \mu(t_k) (z + \alpha_k)$$

$$= \left(\prod_{k=0}^{n-1} \mu(t_k)\right) \left(\prod_{k=0}^{n-1} (z + \alpha_k)\right)$$

$$= \left(\prod_{k=0}^{n-1} \mu(t_k) \right) \left(\sum_{l=0}^{n} a_l z^l \right).$$

Let

$$\gamma = \left\{ z \in \mathbb{C} : |z| = \frac{2}{w}, \quad \sqrt{1} = 1 \right\}.$$

Then

$$-\frac{1}{\mu(t_k)} \in \left\{ z \in \mathbb{C} : |z| \leq \frac{2}{w} \right\}$$

for all $k \in \mathbb{N}_0$. Therefore,

$$\mathscr{L}(g(z))(t_n) = \frac{1}{2\pi i} \left(\prod_{k=0}^{n-1} \mu(t_k) \right) \int_{\gamma} \left(\sum_{l=0}^{m} b_l z^{\frac{l}{2}} \right) \left(\sum_{r=0}^{n} a_r z^r \right) dz$$

$$= \frac{1}{2\pi i} \left(\prod_{k=0}^{n-1} \mu(t_k) \right) \int_{\gamma} \sum_{r=0}^{n} \sum_{l=1}^{\left[\frac{m-1}{2} \right]} a_r b_{2l+1} z^{r+l+\frac{1}{2}} dz$$

$$= \frac{1}{2\pi i} \left(\prod_{k=0}^{n-1} \mu(t_k) \right) \sum_{r=0}^{n} \sum_{l=1}^{\left[\frac{m-1}{2} \right]} a_r b_{2l+1} \left(\frac{2}{w} \right)^{r+l+\frac{3}{2}} \int_0^{2\pi} e^{i\theta \left(l+r+\frac{3}{2} \right)} d(i\theta)$$

$$= \frac{1}{2\pi} \left(\prod_{k=0}^{n-1} \mu(t_k) \right) \sum_{r=0}^{n} \sum_{l=1}^{\left[\frac{m-1}{2} \right]} (-1)^{l+r+1} \frac{a_r b_{2l+1}}{l+r+\frac{3}{2}} \left(\frac{2}{w} \right)^{r+l+\frac{3}{2}},$$

$n \in \mathbb{N}_0$.

Example 2.11. Let $\mathbb{T} = \{t_n : n \in \mathbb{N}_0\}$ be an isolated time scale such that

$$\sigma(t_n) = t_{n+1}, \quad \mu(t_n) = t_{n+1} - t_n, \quad n \in \mathbb{N}_0, \quad w = \inf_{n \in \mathbb{N}_0} \mu(t_n) > 0.$$

We will find $\mathscr{L}^{-1} \left(\frac{1}{z^k} \right) (t_n)$ for $n+1 \geq k$, $n \in \mathbb{N}_0$, and $k \in \mathbb{N}$. Let

$$\gamma = \left\{ z \in \mathbb{C} : |z| = \frac{2}{w} \right\}.$$

Note that

$$-\frac{1}{\mu(t_k)} \in \left\{ z \in \mathbb{C} : |z| < \frac{2}{w} \right\}$$

for all $k \in \mathbb{N}_0$. Then, using (2.34), we get

$$\mathscr{L}^{-1}\left(\frac{1}{z^k}\right)(t_n) = \frac{1}{2\pi i}\int_\gamma \frac{1}{z^k}\prod_{l=0}^{n-1}(1+\mu(t_l)z)\,dz$$

$$= \frac{1}{2\pi i}\int_\gamma \frac{1}{z^k}\left(\prod_{m=0}^{n-1}\mu(t_m)\right)\left(\sum_{l=0}^{n}a_l z^l\right)dz$$

$$= \left(\prod_{m=0}^{n-1}\mu(t_m)\right)\frac{1}{2\pi i}\int_\gamma \sum_{l=0}^{n}a_l z^{l-k}\,dz$$

$$= \left(\prod_{m=0}^{n-1}\mu(t_m)\right)\frac{1}{2\pi i}\int_0^{2\pi}\sum_{l=0}^{n}a_l\left(\frac{2}{w}\right)^{l-k}e^{i\theta(l-k)}\frac{2}{w}ie^{i\theta}\,d\theta$$

$$= \left(\prod_{m=0}^{n-1}\mu(t_m)\right)\frac{1}{2\pi}\int_0^{2\pi}\sum_{l=0}^{n}a_l\left(\frac{2}{w}\right)^{l-k+1}e^{i\theta(l+1-k)}\,d\theta$$

$$= a_{k-1}\left(\prod_{m=0}^{n-1}\mu(t_m)\right)$$

$$= h_{k-1}(t_n, t_0),$$

for all $n+1 \geq k$, $n \in \mathbb{N}_0$, $k \in \mathbb{N}$.

Example 2.12. Let $\mathbb{T} = \{t_n : n \in \mathbb{N}_0\}$ be an isolated time scale such that

$$\sigma(t_n) = t_{n+1}, \quad \mu(t_n) = t_{n+1} - t_n, \quad n \in \mathbb{N}_0, \quad w = \inf_{n \in \mathbb{N}_0}\mu(t_n) > 0.$$

We will find

$$\mathscr{L}^{-1}\left(\frac{1}{z^2(z^2-4)}\right)(t_n), \quad n \in \mathbb{N}.$$

We will use Example 2.11. Note that

$$\frac{1}{z^2(z^2-4)} = \frac{1}{4}\left(\frac{1}{z^2-4} - \frac{1}{z^2}\right)$$

$$= \frac{1}{4}\left(\frac{1}{(z-2)(z+2)} - \frac{1}{z^2}\right)$$

$$= \frac{1}{4} \left(\frac{1}{4} \left(\frac{1}{z-2} - \frac{1}{z+2} \right) - \frac{1}{z^2} \right)$$

$$= \frac{1}{16 \, (z-2)} - \frac{1}{16 \, (z+2)} - \frac{1}{4z^2}.$$

Hence,

$$\mathcal{L}^{-1} \left(\frac{1}{z^2 \left(z^2 - 4 \right)} \right) (t_n) = \frac{1}{16} \mathcal{L}^{-1} \left(\frac{1}{z-2} \right) - \frac{1}{16} \mathcal{L}^{-1} \left(\frac{1}{z+2} \right) - \frac{1}{4} \mathcal{L}^{-1} \left(\frac{1}{z^2} \right)$$

$$= \frac{1}{16} e_2 \, (t_n, t_0) - \frac{1}{16} e_{-2} \, (t_n, t_0) - \frac{1}{4} h_1 (t_n, t_0)$$

for all $n \in \mathbb{N}$.

Exercise 2.7. Let $\mathbb{T} = \{t_n : n \in \mathbb{N}_0\}$ be an isolated time scale such that

$$\sigma \, (t_n) = t_{n+1}, \quad \mu \, (t_n) = t_{n+1} - t_n, \quad n \in \mathbb{N}_0, \quad w = \inf_{n \in \mathbb{N}_0} \mu \, (t_n) > 0.$$

Find

$$\mathcal{L}^{-1} \left(\frac{1}{\left(z^2 - 9 \right) \left(z^2 - 4 \right)} \right) (t_n), \quad n \in \mathbb{N}_0.$$

Answer.

$$\frac{1}{30} e_3 \, (t_n, t_0) - \frac{1}{30} e_{-3} \, (t_n, t_0) - \frac{1}{20} e_2 \, (t_n, t_0) + \frac{1}{20} e_{-2} \, (t_n, t_0), \quad n \in \mathbb{N}_0.$$

2.3   Advanced Practical Problems

Problem 2.1. Let $\mathbb{T}_1 = \mathbb{T}_2 = 4^{\mathbb{N}_0} \bigcup \{0\}$ and

$$f \, (t_1, t_2) = \frac{t_1}{4t_2^4 + 1}, \quad (t_1, t_2) \in \mathbb{T}_1 \times \mathbb{T}_2.$$

Find $\mathcal{L}_2 \, (f) \, (t_1, z_2)$ for all $z_2 \in \mathbb{C}$ for which $1 + 3t_1 z_2 \neq 0$ for all $t_1 \in \mathbb{T}_1$.

Answer.

$$\mathcal{L}_2 \, (f) \, (z_1, t_2) = \frac{t_1}{1 + z_2} \left(1 + \sum_{t_2 \in [1, \infty)} \frac{3t_2}{4t_2^4 + 1} \prod_{\tau_2 \in [1, t_2]} \frac{1}{(1 + 3\tau_2 z_2)} \right).$$

Problem 2.2. Let $\mathbb{T} = \{t_n : n \in \mathbb{N}_0\}$ be an isolated time scale such that

$$\sigma\,(t_n) = t_{n+1}, \quad \mu\,(t_n) = t_{n+1} - t_n, \quad n \in \mathbb{N}_0, \quad w = \inf_{n \in \mathbb{N}_0} \mu\,(t_n) > 0.$$

Find

$$\mathscr{L}^{-1}\left(\frac{1}{z^2\,(z^2 - 16)}\right)(t_n), \quad n \in \mathbb{N}.$$

Answer.

$$\frac{1}{128}e_4\,(t_n, t_0) - \frac{1}{128}e_{-4}\,(t_n, t_+0) - \frac{1}{16}h_1(t_n, t_0).$$

Chapter 3
Convolution on Time Scales

3.1 Shifts and Convolutions

Let $\mathbb{T}$ be a time scale with forward jump operator and delta differentiation operator σ and Δ, respectively, such that $\sup \mathbb{T} = \infty$. Let also $t_0 \in \mathbb{T}$.

Definition 3.1 (Shift (Delay) of a Function). For a given function $f : [t_0, \infty) \to \mathbb{C}$, the solution of the shifting problem

$$u^{\Delta_t}(t, \sigma(s)) = -u^{\Delta_s}(t, s), \quad t, s \in \mathbb{T}, \quad t \geq s \geq t_0, \qquad (3.1)$$

$$u(t, t_0) = f(t), \quad t \in \mathbb{T}, \quad t \geq t_0, \qquad (3.2)$$

is denoted by $\hat{f}$ and is called the shift or delay of f.

Example 3.1. Let $\mathbb{T} = \mathbb{R}$. Then the problem (3.1), (3.2) takes the following form:

$$\frac{\partial u}{\partial t}(t, s) = -\frac{\partial u}{\partial s}(t, s), \quad t, s \in \mathbb{T}, \quad t \geq s \geq t_0,$$

$$u(t, t_0) = f(t), \quad t \in \mathbb{T}, \quad t \geq t_0,$$

where $f \in \mathscr{C}^1([t_0, \infty))$. Its unique solution is

$$u(t, s) = f(t - s + t_0), \quad t, s \in \mathbb{T}, \quad t \geq s \geq t_0.$$

Indeed, we have

$$u(t, t_0) = f(t - t_0 + t_0)$$

$$= f(t),$$

© Springer International Publishing AG, part of Springer Nature 2018
S. G. Georgiev, *Fractional Dynamic Calculus and Fractional Dynamic Equations on Time Scales*, https://doi.org/10.1007/978-3-319-73954-0_3

$$\frac{\partial u}{\partial t}(t, s) = f'(t - s + t_0),$$

$$\frac{\partial u}{\partial s}(t, s) = -f'(t - s + t_0), \quad t, s \in \mathbb{T}, \quad t \geq s \geq t_0.$$

Therefore,

$$\frac{\partial u}{\partial t}(t, s) = -\frac{\partial u}{\partial s}(t, s), \quad t, s \in \mathbb{T}, \quad t \geq s \geq t_0.$$

Example 3.2. Let $\mathbb{T} = \mathbb{Z}$. Then the problem (3.1), (3.2) takes the form

$$u(t + 1, s + 1) - u(t, s + 1) = -u(t, s + 1) + u(t, s), \quad t, s \in \mathbb{T}, \quad t \geq s \geq t_0,$$
$$u(t, t_0) = f(t), \quad t \in \mathbb{T}, \quad t \geq t_0.$$

Its unique solution is

$$u(t, s) = f(t - s + t_0), \quad t, s \in \mathbb{T}, \quad t \geq s \geq t_0.$$

Indeed, we have

$$u(t + 1, s + 1) = f(t + 1 - s - 1 + t_0)$$
$$= f(t - s + t_0),$$
$$u(t, s + 1) = f(t - s - 1 + t_0),$$
$$u(t + 1, s + 1) - u(t, s + 1) = f(t + s + t_0) - f(t - s - 1 + t_0)$$
$$= -u(t, s + 1) + u(t, s), \quad t, s \in \mathbb{T}, \quad t \geq s \geq t_0,$$
$$u(t, t_0) = f(t - t_0 + t_0)$$
$$= f(t), \quad t \in \mathbb{T}, \quad t \geq t_0.$$

Example 3.3. Consider the problem

$$u^{\Delta_t}(t, \sigma(s)) = -u^{\Delta_s}(t, s), \quad \forall t, s \in \mathbb{T}, \quad \text{independent of } t_0,$$
$$u(t, t_0) = e_\lambda(t, t_0), \quad \forall t \in \mathbb{T},$$

where $\lambda \in \mathbb{C}$ is a constant such that $1 + \lambda\mu(t) \neq 0$, $t \in \mathbb{T}$. We will prove that its unique solution is

$$u(t, s) = e_\lambda(t, s), \quad t, s \in \mathbb{T}, \quad \text{independent of } t_0.$$

Indeed, we have

$$u^{\Delta_t}(t, \sigma(s)) = \lambda e_\lambda(t, \sigma(s)),$$

$$u^{\Delta_s}(t, s) = e_\lambda^{\Delta_s}(t, s)$$

$$= \left(\frac{1}{e_{\lambda(s,t)}}\right)^{\Delta_s}$$

$$= -\frac{e_\lambda^{\Delta_s}(s, t)}{e_\lambda(s, t) e_\lambda(\sigma(s), t)}$$

$$= -\frac{\lambda e_\lambda(s, t)}{e_\lambda(s, t) e_\lambda(\sigma(s), t)}$$

$$= -\frac{\lambda}{e_\lambda(\sigma(s), t)}$$

$$= -\lambda e_\lambda(t, \sigma(s))$$

$$= -u^{\Delta_t}(t, \sigma(s)), \quad \forall t, s \in \mathbb{T}, \quad \text{independent of} \quad t_0,$$

$$u(t, t_0) = e_\lambda(t, t_0), \quad t \in \mathbb{T}.$$

Consequently,

$$\widehat{e_\lambda(\cdot, t_0)}(t, s) = e_\lambda(t, s), \quad \forall t, s \in \mathbb{T}, \quad \text{independent of} \quad t_0.$$

3.1.1 The Quantum Calculus Case

In this subsection, we consider $\mathbb{T} = q^{\mathbb{N}_0}, q > 1$. Calculus on this time scale is called quantum calculus. Let $t_0 = 1$. Here

$$\sigma(t) = qt, \quad \mu(t) = (q - 1)t, \quad t \in \mathbb{T}.$$

Definition 3.2. We define

$$[\alpha] = \frac{q^\alpha - 1}{q - 1}, \quad \alpha \in \mathbb{R},$$

$$[n]! = \prod_{k=1}^{n} [k],$$

$$\begin{bmatrix} m \\ n \end{bmatrix} = \frac{[m]!}{[n]![m - n]!}, \quad m, n \in \mathbb{N}, \quad m > n,$$

$$(t - s)_q^n = \prod_{k=0}^{n-1} \left(t - q^k s\right), \quad t, s \in \mathbb{T}, \quad n \in \mathbb{N}.$$

Example 3.4. Let $\alpha = 3$. Then

$$\begin{aligned}[\alpha] &= \frac{q^3 - 1}{q - 1} \\ &= \frac{(q - 1)\left(q^2 + q + 1\right)}{q - 1} \\ &= q^2 + q + 1.\end{aligned}$$

Example 3.5. Let $\alpha = -\frac{1}{3}$. Then

$$\begin{aligned}[\alpha] &= \frac{q^{-\frac{1}{3}} - 1}{q - 1} \\ &= \frac{\frac{1}{\sqrt[3]{q}} - 1}{q - 1} \\ &= -\frac{\sqrt[3]{q} - 1}{\sqrt[3]{q}\left(\sqrt[3]{q} - 1\right)\left(\sqrt[3]{q^2} + \sqrt[3]{q} + 1\right)} \\ &= -\frac{1}{\sqrt[3]{q}\left(\sqrt[3]{q^2} + \sqrt[3]{q} + 1\right)} \\ &= -\frac{1}{q + \sqrt[3]{q^2} + \sqrt[3]{q}}.\end{aligned}$$

Example 3.6. Let $\alpha = -2$. Then

$$\begin{aligned}[\alpha] &= \frac{q^{-2} - 1}{q - 1} \\ &= \frac{\frac{1}{q^2} - 1}{q - 1} \\ &= \frac{1 - q^2}{q^2(q - 1)} \\ &= -\frac{(q - 1)(q + 1)}{q^2(q - 1)} \\ &= -\frac{q + 1}{q^2} \\ &= -\frac{1}{q} - \frac{1}{q^2}.\end{aligned}$$

Exercise 3.1. Let $\alpha = -4$. Find $[\alpha]$.

Answer.

$$-\frac{1}{q} - \frac{1}{q^2} - \frac{1}{q^3} - \frac{1}{q^4}.$$

Example 3.7. Let $n = 4$. Then

$$
\begin{aligned}
[4]! &= \prod_{k=1}^{4} [k] \\
&= [1][2][3][4] \\
&= \frac{q-1}{q-1} \frac{q^2-1}{q-1} \frac{q^3-1}{q-1} \frac{q^4-1}{q-1} \\
&= \frac{(q-1)(q+1)}{q-1} \frac{(q-1)\left(q^2+q+1\right)}{q-1} \frac{(q-1)\left(q^3+q^2+q+1\right)}{q-1} \\
&= (q+1)\left(q^2+q+1\right)\left(q^3+q^2+q+1\right).
\end{aligned}
$$

Exercise 3.2. Let $n = 3$. Find $[n]!$.

Answer.

$$(q+1)\left(q^2+q+1\right).$$

Example 3.8. Let $m = 3$, $n = 2$. Then

$$
\begin{aligned}
\begin{bmatrix} 3 \\ 2 \end{bmatrix} &= \frac{[3]!}{[2]![1]!} \\
&= \frac{[1][2][3]}{[1][2][1]} \\
&= [3] \\
&= \frac{q^3-1}{q-1} \\
&= \frac{(q-1)\left(q^2+q+1\right)}{q-1} \\
&= q^2+q+1.
\end{aligned}
$$

Exercise 3.3. Let $m = 4$, $n = 2$. Find $\begin{bmatrix} m \\ n \end{bmatrix}$.

Answer.

$$\frac{\left(q^2 + q + 1\right)\left(q^3 + q^2 + q + 1\right)}{q + 1}.$$

Example 3.9. Let $n = 2$, $q = 3$. Then

$$(t - s)_q^n = \prod_{k=0}^{1} \left(t - 3^k s\right)$$

$$= (t - s)(t - 3s).$$

Exercise 3.4. Let $n = 4$, $q = 2$. Find $(t - s)_q^n$, $t, s \in \mathbb{T}$.

Answer.

$$(t - s)(t - 2s)(t - 4s)(t - 8s), \quad t, s \in \mathbb{T}.$$

Theorem 3.1. *The quantum calculus monomials are given by*

$$h_n(t, s) = \frac{(t - s)_q^n}{[n]!}, \quad n \in \mathbb{N}, \quad t, s \in \mathbb{T}. \tag{3.3}$$

Proof. Let $s, t \in \mathbb{T}$ and

$$g(t) = \frac{t^2}{q + 1} - st.$$

Then

$$g^\Delta(t) = \frac{\sigma(t) + t}{q + 1} - s$$

$$= \frac{qt + t}{q + 1} - s$$

$$= t - s$$

and

$$h_1(t, s) = \int_s^t h_0(\tau, s)\Delta\tau$$

$$= \int_s^t \Delta\tau$$

$$= t - s$$

$$= (t - s)_q^1,$$

$$h_2(t, s) = \int_s^t h_1(\tau, s) \Delta\tau$$

$$= \int_s^t (\tau - s) \Delta\tau$$

$$= \int_s^t g^\Delta(\tau) \Delta\tau$$

$$= g(\tau) \Big|_{\tau=s}^{\tau=t}$$

$$= \left(\frac{\tau^2}{q+1} - s\tau \right) \Big|_{\tau=s}^{\tau=t}$$

$$= \left(\frac{t^2}{q+1} - st \right) - \left(\frac{s^2}{q+1} - s^2 \right)$$

$$= \frac{t^2 - s^2}{q+1} - st + s^2$$

$$= \frac{(t-s)(t+s)}{q+1} - s(t-s)$$

$$= (t-s) \left(\frac{t+s}{q+1} - s \right)$$

$$= (t-s) \frac{(t-qs)}{q+1}$$

$$= \frac{(t-s)(t-qs)}{[2]!}$$

$$= \frac{\prod_{k=0}^1 \left(t - q^k s \right)}{[2]!}$$

$$= \frac{(t-s)_q^2}{[2]!}.$$

Assume (3.3) or

$$h_n(t, s) = \prod_{\nu=0}^{n-1} \frac{t - q^\nu s}{\sum_{\mu=0}^\nu q^\mu}$$

for some $n \in \mathbb{N}$. We will prove that

$$h_{n+1}(t, s) = \frac{(t-s)_q^{n+1}}{[n+1]!}$$

$$= \prod_{\nu=0}^n \frac{t - q^\nu s}{\sum_{\mu=0}^\nu q^\mu}.$$

(3.4)

We have

$$
\left(\prod_{\nu=0}^{n} \frac{t - q^{\nu} s}{\sum_{\mu=0}^{\nu} q^{\mu}} \right)^{\Delta_t} = \frac{1}{\mu(t)} \left(\prod_{\nu=0}^{n} \frac{\sigma(t) - q^{\nu} s}{\sum_{\mu=0}^{\nu} q^{\mu}} - \prod_{\nu=0}^{n} \frac{t - q^{\nu} s}{\sum_{\mu=0}^{\nu} q^{\mu}} \right)
$$

$$
= \frac{1}{\mu(t)} \left(\frac{\sigma(t) - q^n s}{\sum_{\mu=0}^{n} q^{\mu}} \prod_{\nu=0}^{n-1} \frac{\sigma(t) - q^{\nu} s}{\sum_{\mu=0}^{\nu} q^{\mu}} - \frac{t - q^n s}{\sum_{\mu=0}^{n} q^{\mu}} \prod_{\nu=0}^{n-1} \frac{t - q^{\nu} s}{\sum_{\mu=0}^{\nu} q^{\mu}} \right)
$$

$$
= \frac{1}{\mu(t)} \left(\frac{qt - q^n s}{\sum_{\mu=0}^{n} q^{\mu}} h_n(\sigma(t), s) - \frac{t - q^n s}{\sum_{\mu=0}^{n} q^{\mu}} h_n(t, s) \right)
$$

$$
= \frac{1}{\mu(t)} \left(\frac{qt - q^n s}{\sum_{\mu=0}^{n} q^{\mu}} \left(h_n(t, s) + \mu(t) h_n^{\Delta_t}(t, s) \right) \right.
$$
$$
\left. - \frac{t - q^n s}{\sum_{\mu=0}^{n} q^{\mu}} h_n(t, s) \right)
$$

$$
= \frac{1}{\mu(t)} \left(\frac{qt - q^n s}{\sum_{\mu=0}^{n} q^{\mu}} - \frac{t - q^n s}{\sum_{\mu=0}^{n} q^{\mu}} \right) h_n(t, s)
$$
$$
+ \frac{qt - q^n s}{\sum_{\mu=0}^{n} q^{\mu}} h_n^{\Delta_t}(t, s)
$$

$$
= \frac{1}{(q-1)t} \frac{(q-1)t}{\sum_{\mu=0}^{n} q^{\mu}} h_n(t, s) + \left(qt - q^n s \right) \frac{h_n^{\Delta_t}(t, s)}{\sum_{\mu=0}^{n} q^{\mu}}
$$

$$
= \frac{h_n(t, s)}{\sum_{\mu=0}^{n} q^{\mu}} + \frac{qt - q^n s}{\sum_{\mu=0}^{n} q^{\mu}} h_n^{\Delta_t}(t, s)
$$

$$
= \frac{1}{\sum_{\mu=0}^{n} q^{\mu}} \left(h_n(t, s) + \left(qt - q^n s \right) h_{n-1}(t, s) \right)
$$

$$
= \frac{1}{\sum_{\mu=0}^{n} q^{\mu}} \left(h_n(t, s) + \left(qt - q^n s \right) \prod_{\nu=0}^{n-2} \frac{t - q^{\nu} s}{\sum_{\mu=0}^{\nu} q^{\mu}} \right)
$$

$$
= \frac{1}{\sum_{\mu=0}^{n} q^{\mu}} \left(h_n(t, s) + \left(q \sum_{\mu=0}^{n-1} q^{\mu} \right) \prod_{\nu=0}^{n-1} \frac{t - q^{\nu} s}{\sum_{\mu=0}^{\nu} q^{\mu}} \right)
$$

$$
= \frac{1}{\sum_{\mu=0}^{n} q^{\mu}} \left(h_n(t, s) + \left(\sum_{\mu=1}^{n} q^{\mu} \right) h_n(t, s) \right)
$$

$$
= \frac{h_n(t, s)}{\sum_{\mu=0}^{n} q^{\mu}} \left(1 + \sum_{\mu=1}^{n} q^{\mu} \right)
$$

$$
= h_n(t, s).
$$

Hence, using that

$$h_{n+1}^{\Delta_t}(t, s) = h_n(t, s),$$

we get (3.4). Therefore, (3.3) holds for all $n \in \mathbb{N}$. This completes the proof.

Corollary 3.1. *The following formula holds:*

$$h_n\left(q^k t, t\right) = \begin{cases} \begin{bmatrix} k \\ n \end{bmatrix} (\mu(t))^n \, q^{\frac{n(n-1)}{2}} & \text{for} \quad k, n \in \mathbb{N}, \quad k > n, \\ (\mu(t))^n \, q^{\frac{n(n-1)}{2}} & \text{for} \quad k = n \in \mathbb{N}. \end{cases}$$

Proof. Let $k, n \in \mathbb{N}, k \geq n$. Then

$$\left(q^k t - t\right)_q^n = \left(q^k t - t\right)\left(q^k t - qt\right)\ldots\left(q^k t - q^{n-1} t\right)$$

$$= t^n q q^2 \ldots q^{n-1} \left(q^k - 1\right)\left(q^{k-1} - 1\right)\ldots\left(q^{k-n+1} - 1\right)$$

$$= (\mu(t))^n \, q^{\frac{n(n-1)}{2}} \frac{\left(q^k - 1\right)\left(q^{k-1} - 1\right)\ldots\left(q^{k-n+1} - 1\right)}{(q-1)^n}$$

$$= \begin{cases} (\mu(t))^n \, q^{\frac{n(n-1)}{2}} [k][k-1]\ldots[k-n+1] & \text{for} \quad k > n \\ (\mu(t))^n \, q^{\frac{n(n-1)}{2}} [n][n-1]\ldots[1] & \text{for} \quad k = n \end{cases}$$

$$= \begin{cases} [n]! \, (\mu(t))^n \, q^{\frac{n(n-1)}{2}} \begin{bmatrix} k \\ n \end{bmatrix} & \text{for} \quad k > n \\ [n]! \, (\mu(t))^n \, q^{\frac{n(n-1)}{2}} & \text{for} \quad k = n. \end{cases}$$

From this and Theorem 3.1, we get

$$h_n\left(q^k t, t\right) = \frac{\left(q^k t - t\right)_q^n}{[n]!}$$

$$= \begin{cases} (\mu(t))^n \, q^{\frac{n(n-1)}{2}} \begin{bmatrix} k \\ n \end{bmatrix} & \text{for} \quad k > n \\ (\mu(t))^n \, q^{\frac{n(n-1)}{2}} & \text{for} \quad k = n. \end{cases}$$

This completes the proof.

Lemma 3.1. *There are two q-Pascal rules:*

$$\begin{bmatrix} n \\ j \end{bmatrix} = \begin{bmatrix} n-1 \\ j-1 \end{bmatrix} + q^j \begin{bmatrix} n-1 \\ j \end{bmatrix} \tag{3.5}$$

and

$$\begin{bmatrix} n \\ j \end{bmatrix} = q^{n-j} \begin{bmatrix} n-1 \\ j-1 \end{bmatrix} + \begin{bmatrix} n-1 \\ j \end{bmatrix}, \tag{3.6}$$

for all $n \in \mathbb{N}$, $n \geq 2$, $j \in \{1, \ldots, n-1\}$.

Proof. Let $n \in \mathbb{N}$, $n \geq 2$, and $j \in \{1, \ldots, n-1\}$. Then

$$\begin{aligned}
[n] &= \frac{q^n - 1}{q - 1} \\
&= \frac{(q-1)\left(1 + q + \cdots + q^{n-1}\right)}{q-1} \\
&= 1 + q + \cdots + q^{n-1} \\
&= \left(1 + q + \cdots + q^{j-1}\right) + \left(q^j + \cdots + q^{n-1}\right) \\
&= [j] + q^j \left(1 + q + \cdots + q^{n-j-1}\right) \\
&= [j] + q^j [n-j]
\end{aligned}$$

and

$$\begin{aligned}
\begin{bmatrix} n \\ j \end{bmatrix} &= \frac{[n]!}{[j]![n-j]!} \\
&= \frac{[n][n-1]!}{[j]![n-j]!} \\
&= \frac{\left([j] + q^j[n-j]\right)[n-1]!}{[j]![n-j]!} \\
&= \frac{[j][n-1]!}{[j]![n-j]!} + \frac{q^j[n-j][n-1]!}{[j]![n-j]!} \\
&= \frac{[j][n-1]!}{[j][j-1]![n-j]!} + \frac{q^j[n-j][n-1]!}{[j]![n-j][n-j-1]!} \\
&= \frac{[n-1]!}{[j-1]![n-j]!} + \frac{q^j[n-1]!}{[j]![n-j-1]!} \\
&= \begin{bmatrix} n-1 \\ j-1 \end{bmatrix} + q^j \begin{bmatrix} n-1 \\ j \end{bmatrix},
\end{aligned}$$

i.e., (3.5) is established. Now, using (3.5), we get

$$\begin{bmatrix} n \\ j \end{bmatrix} = \frac{[n]!}{[j]![n-j]!}$$

$$= \begin{bmatrix} n \\ n-j \end{bmatrix}$$

$$= \begin{bmatrix} n-1 \\ n-j-1 \end{bmatrix} + q^{n-j}\begin{bmatrix} n-1 \\ n-j \end{bmatrix}$$

$$= \begin{bmatrix} n-1 \\ j \end{bmatrix} + q^{n-j}\begin{bmatrix} n-1 \\ j-1 \end{bmatrix},$$

i.e., (3.6) is established. This completes the proof.

Theorem 3.2. *Let* $f : \mathbb{T} \to \mathbb{R}$. *Then*

$$f^{\Delta^k}(t)\,(\mu(t))^k\,q^{\frac{k(k-1)}{2}} = \sum_{v=0}^{k}(-1)^v\begin{bmatrix} k \\ v \end{bmatrix}q^{\frac{v(v-1)}{2}}f^{\sigma^{k-v}}(t), \quad t \in \mathbb{T},$$

for all $k \in \mathbb{N}_0$.

Proof. For $k = 0$, the statement is true. Assume that the statement is true for some $k \in \mathbb{N}$. We will prove that the assertion is true for $k + 1$. We have

$$f^{\Delta^{k+1}}(t)\,(\mu(t))^{k+1}\,q^{\frac{k(k+1)}{2}} = \left(\mu(t)f^{\Delta^{k+1}}(t)\right)(\mu(t))^k\,q^{\frac{k(k+1)}{2}}$$

$$= \left(f^{\Delta^k}(\sigma(t)) - f^{\Delta^k}(t)\right)(\mu(t))^k\,q^{\frac{k(k+1)}{2}}$$

$$= f^{\Delta^k}(\sigma(t))\,(\mu(\sigma(t)))^k\,q^{-k}q^{\frac{k(k+1)}{2}} - f^{\Delta^k}(t)\,(\mu(t))^k\,q^{\frac{k(k+1)}{2}}$$

$$= f^{\Delta^k}(\sigma(t))\,(\mu(\sigma(t)))^k\,q^{\frac{k(k-1)}{2}} - f^{\Delta^k}(t)\,(\mu(t))^k\,q^{\frac{k(k-1)}{2}}q^k$$

$$= \sum_{v=0}^{k}(-1)^v\begin{bmatrix} k \\ v \end{bmatrix}q^{\frac{v(v-1)}{2}}f^{\sigma^{k-v}}(\sigma(t)) - q^k\sum_{v=0}^{k}(-1)^v\begin{bmatrix} k \\ v \end{bmatrix}q^{\frac{v(v-1)}{2}}f^{\sigma^{k-v}}(t)$$

$$= \sum_{v=0}^{k}(-1)^v\begin{bmatrix} k \\ v \end{bmatrix}q^{\frac{v(v-1)}{2}}f^{\sigma^{k-v+1}}(t) + q^k\sum_{v=1}^{k+1}(-1)^v\begin{bmatrix} k \\ v-1 \end{bmatrix}q^{\frac{(v-1)(v-2)}{2}}f^{\sigma^{k-v+1}}(t)$$

$$= f^{\sigma^{k+1}}(t) + q^k(-1)^{k+1}q^{\frac{k(k-1)}{2}}f(t)$$

$$+ \sum_{v=1}^{k}(-1)^v f^{\sigma^{k-v+1}}(t)q^{\frac{v(v-1)}{2}}\left(\begin{bmatrix} k \\ v \end{bmatrix} + q^{k-v+1}\begin{bmatrix} k \\ v-1 \end{bmatrix}\right)$$

$$= f^{\sigma^{k+1}}(t) + q^k(-1)^{k+1}q^{\frac{k(k-1)}{2}}f(t)$$

$$+ \sum_{v=1}^{k} (-1)^v f^{\sigma^{k-v+1}}(t) q^{\frac{v(v-1)}{2}} \begin{bmatrix} k+1 \\ v \end{bmatrix}$$

$$= \sum_{v=0}^{k+1} (-1)^v \begin{bmatrix} k+1 \\ v \end{bmatrix} q^{\frac{v(v-1)}{2}} f^{\sigma^{k+1-v}}(t), \quad t \in \mathbb{T}.$$

This completes the proof.

Let $n \in \mathbb{N}$ and

$$h(x) = (x + a)_q^n, \quad x \in \mathbb{T}.$$

We will expand h about $x = 0$ using Taylor's formula. We have

$$h(x) = (x + a)(x + qa) \ldots \left(x + q^{n-1}a\right),$$

$$h(0) = a(qa) \ldots \left(q^{n-1}a\right)$$

$$= q^{\frac{n(n-1)}{2}} a^n,$$

$$h^{\Delta}(x) = \frac{1}{\mu(x)} \left((\sigma(x) + a)(\sigma(x) + qa) \ldots \left(\sigma(x) + q^{n-1}a\right) \right.$$

$$\left. -(x + a)(x + qa) \ldots \left(x + q^{n-1}a\right) \right)$$

$$= \frac{1}{(q-1)x} \left((qx + a)(qx + qa) \ldots \left(qx + q^{n-1}a\right) \right.$$

$$\left. -(x + a)(x + qa) \ldots \left(x + q^{n-1}a\right) \right)$$

$$= \frac{1}{(q-1)x} \left((qx + a)q(x + a) \ldots q \left(x + q^{n-2}a\right) \right.$$

$$\left. -(x + a)(x + qa) \ldots \left(x + q^{n-2}a\right) \left(x + q^{n-1}a\right) \right)$$

$$= \frac{1}{(q-1)x} (x + a) \ldots \left(x + q^{n-2}a\right) \left(q^{n-1}(qx + a) - \left(x + q^{n-1}a\right) \right)$$

$$= \frac{1}{(q-1)x} (x + a) \ldots \left(x + q^{n-2}a\right) \left(q^n x + q^{n-1}a - x - q^{n-1}a \right)$$

$$= \frac{1}{(q-1)x} (x + a) \ldots \left(x + q^{n-2}a\right) (q^n - 1) x$$

$$= \frac{q^n - 1}{q - 1} (x + a) \ldots \left(x + q^{n-2}a\right)$$

$$= [n](x+a)\ldots \left(x+q^{n-2}a\right),$$

$$h^{\Delta}(0) = [n]a(aq)\ldots \left(aq^{n-2}\right)$$

$$= [n]a^{n-1}q^{\frac{(n-1)(n-2)}{2}},$$

$$h^{\Delta^2}(x) = \frac{1}{\mu(x)}\left(h^{\Delta}(\sigma(x)) - h^{\Delta}(x)\right)$$

$$= \frac{1}{(q-1)x}\Big([n](qx+a)(qx+qa)\ldots \left(qx+q^{n-2}a\right)$$

$$-[n](x+a)(x+qa)\ldots \left(x+q^{n-2}a\right)\Big)$$

$$= \frac{[n]}{(q-1)x}\Big((qx+a)q(x+a)\ldots q\left(x+q^{n-3}a\right)$$

$$-(x+a)\ldots \left(x+q^{n-3}a\right)\left(x+q^{n-2}a\right)\Big)$$

$$= \frac{[n]}{(q-1)x}(x+a)\ldots \left(x+q^{n-3}a\right)\left(q^{n-2}(qx+a) - x - q^{n-2}a\right)$$

$$= \frac{[n]}{(q-1)x}(x+a)\ldots \left(x+q^{n-3}a\right)\left(q^{n-1}x + q^{n-2}a - x - q^{n-2}a\right)$$

$$= \frac{[n]}{(q-1)x}(x+a)\ldots \left(x+q^{n-3}a\right)\left(q^{n-1} - 1\right)x$$

$$= [n][n-1](x+a)\ldots \left(x+q^{n-3}a\right),$$

$$h^{\Delta^2}(0) = [n][n-1]a\ldots q^{n-3}a$$

$$= [n][n-1]q^{\frac{(n-3)(n-2)}{2}}a^{n-2}.$$

Assume that

$$h^{\Delta^k}(x) = [n][n-1]\ldots[n-k+1](x+a)\ldots \left(x+q^{n-k-1}a\right)$$

for some $k < n$, $k \in \mathbb{N}$. Then

$$h^{\Delta^{k+1}}(x) = \frac{1}{\mu(x)}\left(h^{\Delta^k}(\sigma(x)) - h^{\Delta^k}(x)\right)$$

$$= \frac{1}{\mu(x)}[n][n-1]\ldots[n-k+1]\Big((\sigma(x)+a)\ldots \left(\sigma(x)+q^{n-k-1}a\right)$$

$$-(x+a)\ldots \left(x+q^{n-k-1}a\right)\Big)$$

$$= \frac{1}{(q-1)x}[n][n-1]\ldots[n-k+1]\Big((qx+a)\ldots\Big(qx+q^{n-k-1}a\Big)$$

$$-(x+a)\ldots\Big(x+q^{n-k-1}a\Big)\Big)$$

$$= \frac{1}{(q-1)x}[n][n-1]\ldots[n-k+1](x+a)\ldots\Big(x+q^{n-k-2}a\Big)$$

$$\times\Big(q^{n-k-1}(qx+a)-x-q^{n-k-1}a\Big)$$

$$= \frac{1}{(q-1)x}[n][n-1]\ldots[n-k+1](x+a)\ldots\Big(x+q^{n-k-2}a\Big)$$

$$\times\Big(q^{n-k}x+q^{n-k-1}a-x-q^{n-k-1}a\Big)$$

$$= \frac{1}{(q-1)x}[n][n-1]\ldots[n-k+1](x+a)$$

$$\ldots\Big(x+q^{n-k-2}a\Big)\Big(q^{n-k}-1\Big)x$$

$$= [n][n-1]\ldots[n-k](x+a)\ldots\Big(x+q^{n-k-2}a\Big),$$

$$h^{\Delta^{k+1}}(0) = [n][n-1]\ldots[n-k]a(qa)\ldots\Big(q^{n-k-2}a\Big)$$

$$= [n][n-1]\ldots[n-k]q^{\frac{(n-k-2)(n-k-1)}{2}}a^{n-k-1}.$$

Consequently,

$$(x+a)_q^n = \sum_{k=0}^{n}\begin{bmatrix}n\\k\end{bmatrix}q^{\frac{(n-k)(n-k-1)}{2}}a^{n-k}x^k$$

$$= \sum_{k=0}^{n}\begin{bmatrix}n\\n-k\end{bmatrix}q^{\frac{k(k-1)}{2}}a^kx^{n-k} \qquad (3.7)$$

$$= \sum_{k=0}^{n}\begin{bmatrix}n\\k\end{bmatrix}q^{\frac{k(k-1)}{2}}a^kx^{n-k}.$$

Using (3.3), we get

$$h_n(t,s) = \frac{(t-s)(t-qs)\ldots\Big(t-q^{n-1}s\Big)}{[n]!},$$

whereupon

$$h_n\Big(q^kt,t\Big) = \frac{\Big(q^kt-t\Big)\Big(q^kt-qt\Big)\ldots\Big(q^kt-q^{n-1}t\Big)}{[n]!}.$$

Therefore,

$$h_n\left(q^k t, t\right) = 0 \quad \text{for} \quad n-1 \ge k.$$

Also,

$$h_n\left(q^k t, t\right) = \frac{t^n q^{\frac{n(n-1)}{2}}\left(q^k - 1\right)\left(q^{k-1} - 1\right)\dots\left(q^{k-n+1} - 1\right)}{[n]!}$$

$$= \frac{t^n q^{\frac{n(n-1)}{2}}(q-1)^n [k][k-1]\dots[k-n+1]}{[n]!}$$

$$= \begin{bmatrix} k \\ n \end{bmatrix} q^{\frac{n(n-1)}{2}}(qt - t)^n$$

$$= \begin{bmatrix} k \\ n \end{bmatrix} q^{\frac{n(n-1)}{2}}(\mu(t))^n, \quad n < k, \quad k, n \in \mathbb{N}_0.$$

Theorem 3.3. *The shift of* $f : \mathbb{T} \to \mathbb{R}$ *is given by*

$$\hat{f}\left(q^k t, t\right) = \sum_{v=0}^{k} \begin{bmatrix} k \\ v \end{bmatrix} t^v (1 - t)_q^v f\left(q^v\right), \quad \forall k \in \mathbb{N}_0.$$

Proof. Using Theorem 3.2, we obtain

$$\hat{f}\left(q^k t, t\right) = \sum_{m=0}^{k} h_m\left(q^k t, t\right) f^{\Delta^m}(1)$$

$$= \sum_{m=0}^{k} \begin{bmatrix} k \\ m \end{bmatrix} (\mu(t))^m q^{\frac{m(m-1)}{2}} f^{\Delta^m}(1)$$

$$= \sum_{m=0}^{k} \begin{bmatrix} k \\ m \end{bmatrix} t^m f^{\Delta^m}(1) (\mu(1))^m q^{\frac{m(m-1)}{2}}$$

$$= \sum_{m=0}^{k} \begin{bmatrix} k \\ m \end{bmatrix} t^m \sum_{v=0}^{m} (-1)^v \begin{bmatrix} m \\ v \end{bmatrix} q^{\frac{v(v-1)}{2}} f^{\sigma^{m-v}}(1)$$

$$= \sum_{m=0}^{k} \sum_{v=0}^{m} \begin{bmatrix} k \\ m \end{bmatrix} t^m (-1)^{m-v} \begin{bmatrix} m \\ m-v \end{bmatrix} q^{\frac{(m-v)(m-v-1)}{2}} f\left(q^v\right)$$

$$= \sum_{v=0}^{k} \sum_{m=v}^{k} \begin{bmatrix} k \\ m \end{bmatrix} \begin{bmatrix} m \\ m-v \end{bmatrix} t^m (-1)^{m-v} q^{\frac{(m-v)(m-v-1)}{2}} f\left(q^v\right)$$

$$= \sum_{\nu=0}^{k} \sum_{m=0}^{k-\nu} \begin{bmatrix} k \\ m+\nu \end{bmatrix} \begin{bmatrix} m+\nu \\ m \end{bmatrix} t^{m+\nu} (-1)^m q^{\frac{m(m-1)}{2}} f(q^{\nu})$$

$$= \sum_{\nu=0}^{k} \sum_{m=0}^{k-\nu} \begin{bmatrix} k \\ \nu \end{bmatrix} \begin{bmatrix} k-\nu \\ m \end{bmatrix} t^{\nu} (-t)^m q^{\frac{m(m-1)}{2}} f(q^{\nu})$$

$$= \sum_{\nu=0}^{k} \begin{bmatrix} k \\ \nu \end{bmatrix} t^{\nu} \left(\sum_{m=0}^{k-\nu} \begin{bmatrix} k-\nu \\ m \end{bmatrix} (-t)^m q^{\frac{m(m-1)}{2}} \right) f(q^{\nu})$$

$$= \sum_{\nu=0}^{k} \begin{bmatrix} k \\ \nu \end{bmatrix} t^{\nu} (1-t)_q^{k-\nu} f(q^{\nu}).$$

This completes the proof.

By Theorem 3.3, it follows that the solutions of the problem (3.1), (3.2) can be represented in terms of generalized polynomials h_k, $k \in \mathbb{N}_0$. Note that for an arbitrary constant λ, the exponential function e_λ satisfies the equation (3.1). Therefore, we can construct some solutions of the problem (3.1), (3.2) in terms of the exponential function e_λ. Let $\Omega \subset \mathbb{C}$ and denote by $\mathscr{H}$ the set of functions $f : [t_0, \infty) \to \mathbb{C}$ of the form

$$f(t) = \int_{\Omega} \phi(\lambda) e_\lambda(t, t_0) dw(\lambda), \tag{3.8}$$

where w is a measure of Ω, $\phi : \Omega \to \mathbb{C}$ is a function, w and ϕ depend on f. The integral on the right-hand side of (3.8) can be understood to be a Riemann–Stieltjes or Lebesgue–Stieltjes integral, since the exponential function e_λ is well defined for all complex values of λ if $t \geq t_0$. Assume that

$$\int_{\Omega} |\lambda \phi(\lambda) e_\lambda(t, s)| \, dw(\lambda) < \infty, \quad t, s \in \mathbb{T}, \quad t \geq s \geq t_0. \tag{3.9}$$

Note that $\mathscr{H}$ is a linear space. If w is a measure concentrated on a finite set $\{\lambda_1, \ldots, \lambda_n\} \subset \mathbb{C}$ with

$$w(\{\lambda_k\}) = 1, \quad k \in \{1, \ldots, n\},$$

and if $\phi(\lambda_k) = c_k, k \in \{1, \ldots, n\}$, then (3.8) takes the form

$$f(t) = \sum_{k=1}^{n} c_k e_{\lambda_k}(t, t_0), \quad t \in \mathbb{T}, \quad t \geq t_0.$$

Consequently, $\mathscr{H}$ contains all exponential, hyperbolic, and trigonometric functions.

3.1.2 Investigation of the Shifting Problem

Suppose that $\mathbb{T}$ is an arbitrary time scale with forward jump operator and delta differentiation operator σ and Δ, respectively, such that $\sup \mathbb{T} = \infty$. Let also $t_0 \in \mathbb{T}$ be fixed.

Theorem 3.4. *Let $f \in \mathscr{P}$ be such that*

$$f(t) = \sum_{k=0}^{\infty} a_k h_k(t, t_0), \quad t \in [t_0, \infty),$$

where the coefficients a_k, $k \in \mathbb{N}_0$, satisfy

$$|a_k| \le MR^k, \quad k \in \mathbb{N}_0, \tag{3.10}$$

for some constants $M > 0$ and $R > 0$. Then the problem (3.1), (3.2) has a solution u of the form

$$u(t, s) = \sum_{k=0}^{\infty} a_k h_k(t, s), \quad t, s \in \mathbb{T}, \quad t \ge s \ge t_0. \tag{3.11}$$

This solution is unique in the class of functions u for which

$$A_k(s) = u^{\Delta_t^k}(t, s)\Big|_{t=s}, \quad k \in \mathbb{N}_0, \tag{3.12}$$

are delta differentiable with respect to $s \in \mathbb{T}$ and

$$|A_k(s)| \le A|s|^k, \quad |A_k^{\Delta}(s)| \le B|s|^k, \tag{3.13}$$

for some constants $A > 0$ and $B > 0$, for all $t, s \in \mathbb{T}$, $t \ge s \ge t_0$.

Proof. We have

$$h_k^{\Delta_t}(t, \sigma(s)) = h_{k-1}(t, \sigma(s)),$$

$$h_k^{\Delta_s}(t, s) = -h_{k-1}(t, \sigma(s)), \quad k \in \mathbb{N}, \quad t \ge s \ge t_0.$$

From (3.10), we conclude that the series (3.11) is uniformly convergent, and therefore,

$$u^{\Delta_t}(t, \sigma(s)) = \sum_{k=1}^{\infty} a_k h_{k-1}(t, \sigma(s)),$$

$$u^{\Delta_s}(t, s) = -\sum_{k=1}^{\infty} a_k h_{k-1}(t, \sigma(s)), \quad t, s \in \mathbb{T}, \quad t \ge s \ge t_0.$$

Hence, u satisfies (3.1). Also,

$$u(t, t_0) = \sum_{k=0}^{\infty} a_k h_k(t, t_0)$$

$$= f(t), \quad t \in \mathbb{T}, \quad t \geq t_0,$$

i.e., u satisfies (3.11). Now we assume that u is a solution of the problem (3.1), (3.2) that has the properties (3.11), (3.12). Then we can represent it by a Taylor series with respect to the variable t at the point $t = s$ for each fixed s:

$$u(t, s) = \sum_{k=0}^{\infty} A_k(s) h_k(t, s), \quad t, s \in \mathbb{T}, \quad t \geq s \geq t_0. \tag{3.14}$$

By the conditions (3.13), it follows that we can differentiate (3.14) term by term. Substituting (3.14) into (3.1), we get

$$\sum_{k=1}^{\infty} A_k(\sigma(s)) h_{k-1}(t, \sigma(s)) = -\sum_{k=1}^{\infty} \left(A_k^{\Delta}(s) h_k(t, s) + A_k(\sigma(s)) h_k^{\Delta_s}(t, s) \right)$$

$$= -\sum_{k=1}^{\infty} A_k^{\Delta}(s) h_k(t, s)$$

$$+ \sum_{k=1}^{\infty} A_k(\sigma(s)) h_{k-1}(t, \sigma(s)), \quad t, s \in \mathbb{T}, \quad t \geq s \geq t_0.$$

Hence,

$$\sum_{k=1}^{\infty} A_k^{\Delta}(s) h_k(t, s) = 0, \quad t, s \in \mathbb{T}, \quad t \geq s \geq t_0,$$

and then

$$A_k^{\Delta}(s) = 0, \quad k \in \mathbb{N}, \quad s \in \mathbb{T}, \quad s \geq t_0.$$

Therefore, $A_k(s)$, $k \in \mathbb{N}$, does not depend on s for $k \in \mathbb{N}_0$. If we substitute $s = t_0$ into (3.11), we obtain

$$\sum_{k=0}^{\infty} a_k h_k(t, t_0) = f(t)$$

$$= \sum_{k=0}^{\infty} A_k(t_0) h_k(t, t_0),$$

whereupon $a_k = A_k(t_0)$, $k \in \mathbb{N}_0$. Therefore, u coincides with (3.11). This completes the proof.

Theorem 3.5. *Suppose that the function $f : [t_0, \infty) \to \mathbb{C}$ has the form (3.8) and (3.9) is satisfied. Then the function*

$$u(t, s) = \int_\Omega \phi(\lambda) e_\lambda(t, s) dw(\lambda), \quad t, s \in \mathbb{T}, \quad t \geq s \geq t_0, \tag{3.15}$$

has first-order partial delta derivatives with respect to t and s for $t \geq s \geq t_0$ and satisfies (3.1), (3.2).

Proof. We have

$$u(t, t_0) = \int_\Omega \phi(\lambda) e_\lambda(t, t_0) dw(\lambda)$$
$$= f(t), \quad t \in \mathbb{T}, \quad t \geq t_0,$$

i.e., u satisfies the condition (3.2). By the condition (3.9), it follows that we can differentiate the function u, defined by (3.15), with respect to t and s, $t, s \in \mathbb{T}$, $t \geq s \geq t_0$. Then

$$u_t^{\Delta_t}(t, \sigma(s)) = \lambda \int_\Omega \phi(\lambda) e_\lambda(t, \sigma(s)) dw(\lambda),$$

$$u_s^{\Delta_s}(t, s) = -\lambda \int_\Omega \phi(\lambda) e_\lambda(t, \sigma(s)) dw(\lambda), \quad t, s \in \mathbb{T}, \quad t \geq s \geq t_0.$$

Therefore, u satisfies (3.1). This completes the proof.

By Theorem 3.5, it follows that the solution of the problem (3.1), (3.2) has the form

$$u(t, s) = \sum_{k=1}^n c_k e_{\lambda_k}(t, s), \quad t, s \in \mathbb{T}, \quad t \geq s \geq t_0.$$

Example 3.10. Let $\mathbb{T} = 2^{\mathbb{N}_0}$, $t_0 = 1$, $f(t) = t^2 + 2$, $t \in \mathbb{T}$. We will find $\hat{f}(t, s)$, $t, s \in \mathbb{T}, t \geq s \geq 1$. Consider the problem

$$u^{\Delta_t}(t, \sigma(s)) = -u^{\Delta_s}(t, s), \quad t, s \in \mathbb{T}, \quad t \geq s \geq 1,$$
$$u(t, 1) = t^2 + 2, \quad t \in \mathbb{T}, \quad t \geq 1.$$

Here

$$\sigma(t) = 2t, \quad t \in \mathbb{T}.$$

We have

$$f(1) = 3,$$
$$f^\Delta(t) = \sigma(t) + t$$
$$= 2t + t$$
$$= 3t,$$

$$f^{\Delta}(1) = 3,$$
$$f^{\Delta^2}(t) = 3,$$
$$f^{\Delta^2}(1) = 3, \quad t \in \mathbb{T}, \quad t \geq 1.$$

Hence,

$$f(t) = 3h_0(t, 1) + 3h_1(t, 1) + 3h_2(t, 1), \quad t \in \mathbb{T}, \quad t \geq 1.$$

Let

$$g(t) = \frac{t^2}{3} - st, \quad t, s \in \mathbb{T}, \quad t \geq s \geq t_0.$$

Then

$$g^{\Delta}(t) = \frac{1}{3} \left(\sigma(t) + t \right) - s$$
$$= \frac{1}{3}(2t + t) - s$$
$$= t - s.$$

From this and Theorem 3.4, we get

$$u(t, s) = 3h_0(t, s) + 3h_1(t, s) + 3h_2(t, s)$$
$$= 3 + 3(t - s) + 3 \int_s^t h_1(\tau, s) \Delta \tau$$
$$= 3 + 3(t - s) + 3 \int_s^t g^{\Delta}(\tau) \Delta \tau$$
$$= 3 + 3(t - s) + 3g(\tau) \Big|_{\tau = s}^{\tau = t}$$
$$= 3 + 3t - 3s + 3 \left(\frac{\tau^2}{3} - s\tau \right) \Big|_{\tau = s}^{\tau = t}$$
$$= 3 + 3t - 3s + 3 \left(\frac{t^2}{3} - st - \frac{s^2}{3} + s^2 \right)$$
$$= 3 + 3t - 3s + t^2 - 3st + 2s^2, \quad t, s \in \mathbb{T}, \quad t \geq s \geq 1,$$

i.e.,

$$u(t, s) = t^2 + 2s^2 - 3st + 3t - 3s + 3, \quad t, s \in \mathbb{T}, \quad t \geq s \geq 1.$$

Example 3.11. Let $\mathbb{T} = 3^{\mathbb{N}_0}$, $t_0 = 1$, $f(t) = \frac{1}{t}$, $t \in \mathbb{T}$. We will find the shift of f. Here

$$\sigma(t) = 3t, \quad t \in \mathbb{T}.$$

We have

$$f(1) = 1,$$

$$f^{\Delta}(t) = -\frac{1}{t\sigma(t)}$$

$$= -\frac{1}{3t^2},$$

$$f^{\Delta}(1) = -\frac{1}{3},$$

$$f^{\Delta^2}(t) = \frac{\sigma(t) + t}{3t^2(\sigma(t))^2}$$

$$= \frac{3t + t}{3t^2(3t)^2}$$

$$= \frac{4t}{27t^4}$$

$$= \frac{4}{27t^3},$$

$$f^{\Delta^2}(1) = \frac{4}{27},$$

$$f^{\Delta^3}(t) = -\frac{4}{27} \frac{(\sigma(t))^2 + t\sigma(t) + t^2}{t^3(\sigma(t))^3}$$

$$= -\frac{4}{27} \frac{(3t)^2 + 3t^2 + t^2}{t^3(3t)^3}$$

$$= -\frac{4}{27} \frac{9t^2 + 3t^2 + t^2}{27t^6}$$

$$= -\frac{52}{729t^4},$$

$$f^{\Delta^3}(1) = -\frac{52}{729}, \quad t \in \mathbb{T},$$

and so on. Therefore,

$$f(t) = h_0(t, 1) - \frac{1}{3}h_1(t, 1) + \frac{4}{27}h_2(t, 1) - \frac{52}{729}h_3(t, 1) + \cdots, \quad t \in \mathbb{T}, \quad t \geq 1.$$

From this and Theorem 3.4, we obtain

$$u(t, s) = h_0(t, s) - \frac{1}{3}h_1(t, s) + \frac{4}{27}h_2(t, s) - \frac{52}{729}h_3(t, s) + \cdots , \quad t, s \in \mathbb{T}, \ t \geq s \geq 1.$$

Example 3.12. Let $\mathbb{T} = \mathbb{N}_0^2$, $t_0 = 0$, $f(t) = \frac{1}{t+1}$, $t \in \mathbb{T}$. We will find the shift of f.
Here

$$\sigma(t) = \left(\sqrt{t} + 1\right)^2, \quad t \in \mathbb{T}.$$

We have

$$f(0) = 1,$$

$$f^{\Delta}(t) = -\frac{1}{(t + 1)(\sigma(t) + 1)}$$

$$= -\frac{1}{(t + 1)\left((\sqrt{t} + 1)^2 + 1\right)}$$

$$= -\frac{1}{(t + 1)\left(t + 2\sqrt{t} + 2\right)},$$

$$f^{\Delta}(0) = -\frac{1}{2},$$

$$f^{\Delta^2}(t) = -\frac{1}{\sigma(t) - t}\left(\frac{1}{(\sigma(t) + 1)\left(\sigma(t) + 2\sqrt{\sigma(t)} + 2\right)}\right.$$

$$\left. -\frac{1}{(t + 1)\left(t + 2\sqrt{t} + 2\right)}\right)$$

$$= -\frac{1}{(\sqrt{t} + 1)^2 - t}\left(\frac{1}{\left((\sqrt{t} + 1)^2 + 1\right)\left((\sqrt{t} + 1)^2 + 2(\sqrt{t} + 1) + 2\right)}\right.$$

$$\left. -\frac{1}{(t + 1)\left(t + 2\sqrt{t} + 2\right)}\right)$$

$$= -\frac{1}{1 + 2\sqrt{t}}\left(\frac{1}{\left(t + 2\sqrt{t} + 2\right)\left(t + 2\sqrt{t} + 1 + 2\sqrt{t} + 2 + 2\right)}\right.$$

$$\left. -\frac{1}{(t + 1)\left(t + 2\sqrt{t} + 2\right)}\right)$$

$$= -\frac{1}{1 + 2\sqrt{t}}\left(\frac{1}{\left(t + 2\sqrt{t} + 2\right)\left(t + 4\sqrt{t} + 5\right)} - \frac{1}{(t + 1)\left(t + 2\sqrt{t} + 2\right)}\right),$$

$$f^{\Delta^2}(0) = -\left(\frac{1}{10} - \frac{1}{2}\right)$$

$$= \frac{2}{5}, \quad t \in \mathbb{T},$$

and so on. Then

$$f(t) = h_0(t, 0) - \frac{1}{2}h_1(t, 0) + \frac{2}{5}h_2(t, 0) + \cdots, \quad t \in \mathbb{T}.$$

Therefore, using Theorem 3.4, we get

$$\hat{f}(t, s) = h_0(t, s) - \frac{1}{2}h_1(t, s) + \frac{2}{5}h_2(t, s) + \cdots, \quad t, s \in \mathbb{T}, \quad t \geq s \geq 0.$$

Exercise 3.5. Let $\mathbb{T} = 3^{\mathbb{N}_0}$, $t_0 = 1$,

$$f(t) = t^3 - 7t^2 + 4t + 5, \quad t \in \mathbb{T}.$$

Find the shift of f.

Answer.

$$\hat{f}(t, s) = 3h_0(t, s) - 11h_1(t, s) + 24h_2(t, s) + 52h_3(t, s), \quad t, s \in \mathbb{T}, \quad t \geq s \geq 1.$$

3.2 Convolutions

Suppose that $\mathbb{T}$ is an arbitrary time scale with forward jump operator and delta differentiation operator σ and Δ, respectively, and $t_0 \in \mathbb{T}$. We will begin with the following useful lemma.

Lemma 3.2. *For a given $f : \mathbb{T} \to \mathbb{C}$ we have*

$$\hat{f}(t, t) = f(t_0)$$

for all $t \in \mathbb{T}$.

Proof. Let

$$F(t) = \hat{f}(t, t), \quad t \in \mathbb{T}.$$

Then

$$F(t_0) = \hat{f}(t_0, t_0)$$
$$= f(t_0),$$
$$F^\Delta(t) = \hat{f}^{\Delta_t}(t, \sigma(t)) + \hat{f}^{\Delta_s}(t, t)$$
$$= 0, \quad t \in \mathbb{T}.$$

Therefore, $F(t) = f(t_0)$ for all $t \in \mathbb{T}$. This completes the proof.

Definition 3.3. For given functions f, $g : \mathbb{T} \to \mathbb{R}$, their convolution $f \star g$ is defined by

$$(f \star g)(t) = \int_{t_0}^{t} \hat{f}(t, \sigma(s)) g(s) \Delta s, \quad t \in \mathbb{T}, \quad t \geq t_0.$$

Example 3.13. Let $\mathbb{T} = \mathbb{Z}$, $t_0 = 0$, $f(t) = t^2$, $g(t) = t$, $t \in \mathbb{T}$. We will find $(f \star g)(t)$, $t \in \mathbb{T}$, $t \geq t_0$. We have

$$\sigma(t) = t + 1,$$
$$f(0) = 0,$$
$$f^{\Delta}(t) = \sigma(t) + t$$
$$= t + 1 + t$$
$$= 2t + 1,$$
$$f^{\Delta}(0) = 1,$$
$$f^{\Delta^2}(t) = 2,$$
$$f^{\Delta^2}(0) = 2, \quad t \in \mathbb{T}.$$

Therefore,

$$f(t) = h_1(t, 0) + 2h_2(t, 0), \quad t \in \mathbb{T}.$$

From this and Theorem 3.4, we get

$$\hat{f}(t, s) = h_1(t, s) + 2h_2(t, s), \quad t, s \in \mathbb{T}, \quad t \geq s \geq 0.$$

We set

$$g(t) = \frac{1}{2} t^2 - \frac{1}{2} t - st, \quad t \in \mathbb{T},$$

for some $s \in \mathbb{T}$. Then

$$g^{\Delta}(t) = \frac{1}{2}(\sigma(t) + t) - \frac{1}{2} - s$$
$$= \frac{1}{2}(t + 1 + t) - \frac{1}{2} - s$$
$$= \frac{1}{2}(2t + 1) - \frac{1}{2} - s$$
$$= t - s, \quad t \in \mathbb{T}.$$

Therefore,

$$h_2(t, s) = \int_s^t h_1(\tau, s)\Delta\tau$$

$$= \int_s^t (\tau - s)\Delta\tau$$

$$= \int_s^t g^\Delta(\tau)\Delta\tau$$

$$= g(\tau)\Big|_{\tau=-s}^{\tau=t}$$

$$= \left(\frac{1}{2}\tau^2 - \frac{1}{2}\tau - s\tau\right)\Big|_{\tau=s}^{\tau=t}$$

$$= \frac{1}{2}t^2 - \frac{1}{2}t - st - \frac{1}{2}s^2 + \frac{1}{2}s + s^2$$

$$= \frac{1}{2}t^2 - \frac{1}{2}t - st + \frac{1}{2}s + \frac{1}{2}s^2,$$

and

$$\hat{f}(t, s) = h_1(t, s) + 2h_2(t, s)$$

$$= t - s + 2\left(\frac{1}{2}t^2 - \frac{1}{2}t - st + \frac{1}{2}s + \frac{1}{2}s^2\right)$$

$$= t - s + t^2 - t - 2st + s + s^2$$

$$= t^2 - 2st + s^2,$$

$$\hat{f}(t, \sigma(s)) = t^2 - 2\sigma(s)t + (\sigma(s))^2$$

$$= t^2 - 2(s + 1)t + (s + 1)^2$$

$$= t^2 - 2st - 2t + s^2 + 2s + 1$$

$$= (t - 1)^2 - 2s(t - 1) + s^2, \quad t, s \in \mathbb{T}, \quad t \geq s \geq 0.$$

Let

$$l(s) = (t - 1)^2\left(\frac{1}{2}s^2 - \frac{1}{2}s\right) - 2(t - 1)\left(\frac{1}{3}s^3 - \frac{1}{2}s^2 + \frac{1}{6}s\right)$$

$$+ \frac{1}{4}s^4 - \frac{1}{2}s^3 + \frac{1}{4}s^2, \quad s \in \mathbb{T},$$

for some $t \in \mathbb{T}$. Then

$$l^{\Delta}(s) = (t-1)^2 \left(\frac{1}{2}(\sigma(s)+s) - \frac{1}{2} \right)$$

$$-2(t-1) \left(\frac{1}{3} \left((\sigma(s))^2 + s\sigma(s) + s^2 \right) - \frac{1}{2}(\sigma(s)+s) + \frac{1}{6} \right)$$

$$+\frac{1}{4} \left((\sigma(s))^3 + s(\sigma(s))^2 + s^2\sigma(s) + s^3 \right)$$

$$-\frac{1}{2} \left((\sigma(s))^2 + s\sigma(s) + s^2 \right)$$

$$+\frac{1}{4}(\sigma(s)+s)$$

$$= (t-1)^2 \left(\frac{1}{2}(s+1+s) - \frac{1}{2} \right)$$

$$-2(t-1) \left(\frac{1}{3} \left((s+1)^2 + s(s+1) + s^2 \right) - \frac{1}{2}(s+1+s) + \frac{1}{6} \right)$$

$$+\frac{1}{4} \left((s+1)^3 + s(s+1)^2 + s^2(s+1) + s^3 \right)$$

$$-\frac{1}{2} \left((s+1)^2 + s(s+1) + s^2 \right)$$

$$+\frac{1}{4}(s+1+s)$$

$$= (t-1)^2 \left(s + \frac{1}{2} - \frac{1}{2} \right)$$

$$-2(t-1) \left(\frac{1}{3} \left(s^2 + 2s + 1 + s^2 + s + s^2 \right) - s - \frac{1}{2} + \frac{1}{6} \right)$$

$$+\frac{1}{4} \left(s^3 + 3s^2 + 3s + 1 + s^3 + 2s^2 + s + s^3 + s^2 + s^3 \right)$$

$$-\frac{1}{2} \left(s^2 + 2s + 1 + s^2 + s + s^2 \right) + \frac{1}{4}(2s+1)$$

$$= (t-1)^2 s - 2(t-1) \left(\frac{1}{3} \left(3s^2 + 3s + 1 \right) - s - \frac{1}{3} \right)$$

$$+\frac{1}{4} \left(4s^3 + 6s^2 + 4s + 1 \right) - \frac{1}{2} \left(3s^2 + 3s + 1 \right) + \frac{1}{2}s + \frac{1}{4}$$

$$= (t-1)^2 s - 2(t-1) \left(s^2 + s + \frac{1}{3} - s - \frac{1}{3} \right)$$

$$+s^3 + \frac{3}{2}s^2 + s + \frac{1}{4} - \frac{3}{2}s^2 - \frac{3}{2}s - \frac{1}{2} + \frac{1}{2}s + \frac{1}{4}$$

$$= (t-1)^2 s - 2(t-1)s^2 + s^3, \quad s \in \mathbb{T},$$

for some $t \in \mathbb{T}$. Consequently,

$$
\begin{aligned}
(f \star g)(t) &= \int_0^t \hat{f}(t, \sigma(s)) g(s) \Delta s \\
&= \int_0^t \left((t-1)^2 - 2s(t-1) + s^2 \right) s \Delta s \\
&= \int_0^t \left(s(t-1)^2 - 2s^2(t-1) + s^3 \right) \Delta s \\
&= \int_0^t l^\Delta(s) \Delta s \\
&= l(s) \Big|_{s=0}^{s=t} \\
&= \left((t-1)^2 \left(\frac{1}{2}s^2 - \frac{1}{2}s \right) - 2(t-1) \left(\frac{1}{3}s^3 - \frac{1}{2}s^2 + \frac{1}{6}s \right) \right. \\
&\quad \left. + \frac{1}{4}s^4 - \frac{1}{2}s^3 + \frac{1}{4}s^2 \right) \Big|_{s=0}^{s=t} \\
&= (t-1)^2 \left(\frac{1}{2}t^2 - \frac{1}{2}t \right) - 2(t-1) \left(\frac{1}{3}t^3 - \frac{1}{2}t^2 + \frac{1}{6}t \right) \\
&\quad + \frac{1}{4}t^4 - \frac{1}{2}t^3 + \frac{1}{4}t^2 \\
&= \frac{1}{2}t(t-1)^3 - \frac{1}{3}(t-1)\left(2t^3 - 3t^2 + t \right) \\
&\quad + \frac{1}{4}t^2 \left(t^2 - 2t + 1 \right) \\
&= \frac{1}{2}t(t-1)^3 - \frac{1}{3}(t-1)\left(2t^2(t-1) - t(t-1) \right) \\
&\quad + \frac{1}{4}t^2(t-1)^2 \\
&= \frac{1}{2}t(t-1)^3 - \frac{1}{3}t(t-1)^2(2t-1) + \frac{1}{4}t^2(t-1)^2 \\
&= \frac{1}{12}t(t-1)^2 \left(6(t-1) - 4(2t-1) + 3t \right) \\
&= \frac{1}{12}t(t-1)^2 (6t - 6 - 8t + 4 + 3t) \\
&= \frac{1}{12}t(t-1)^2(t-2), \quad t \in \mathbb{T}, \quad t \geq 0.
\end{aligned}
$$

Example 3.14. Let $\mathbb{T} = 2^{\mathbb{N}_0}$, $t_0 = 1$, $f(t) = g(t) = t$, $t \in \mathbb{T}$. We will find $(f \star g)(t)$, $t \in \mathbb{T}$, $t \geq 1$. Here

$$\sigma(t) = 2t, \quad t \in \mathbb{T},$$

and

$$f(1) = 1,$$
$$f^{\Delta}(t) = 1,$$
$$f^{\Delta}(1) = 1,$$
$$f^{\Delta^k}(t) = 0, \quad k \in \mathbb{N}, \quad k \geq 2.$$

Then

$$f(t) = h_0(t, 1) + h_1(t, 1), \quad t \in \mathbb{T}, \quad t \geq 1.$$

From this and Theorem 3.4, we get

$$\hat{f}(t, s) = h_0(t, s) + h_1(t, s)$$
$$= 1 + t - s,$$
$$\hat{f}(t, \sigma(s)) = 1 + t - \sigma(s)$$
$$= 1 + t - 2s, \quad t, s \in \mathbb{T}, \quad t \geq s \geq 1.$$

Let

$$g(s) = \frac{1}{3}(1+t)s^2 - \frac{2}{7}s^3, \quad s \in \mathbb{T},$$

for some $t \in \mathbb{T}$. We have

$$g^{\Delta}(s) = \frac{1}{3}(1+t)\,(\sigma(s) + s) - \frac{2}{7}\left((\sigma(s))^2 + s\sigma(s) + s^2\right)$$
$$= \frac{1}{3}(1+t)(2s + s) - \frac{2}{7}\left((2s)^2 + s(2s) + s^2\right)$$
$$= \frac{1}{3}(1+t)(3s) - \frac{2}{7}\left(4s^2 + 2s^2 + s^2\right)$$
$$= (1+t)s - 2s^2, \quad s \in \mathbb{T},$$

for some $t \in \mathbb{T}$. Therefore,

$$(f \star g)(t) = \int_1^t \hat{f}(t, \sigma(s))g(s)\Delta s$$

$$= \int_1^t (1 + t - 2s)s\Delta s$$

$$= \int_1^t g^{\Delta}(s)\Delta s$$

$$= g(s)\Big|_{s=1}^{s=t}$$

$$= \left(\frac{1}{3}(1 + t)s^2 - \frac{2}{7}s^3\right)\Big|_{s=1}^{s=t}$$

$$= \frac{1}{3}(1 + t)t^2 - \frac{2}{7}t^3 - \frac{1}{3}(1 + t) + \frac{2}{7}$$

$$= \frac{1}{3}(1 + t)\left(t^2 - 1\right) - \frac{2}{7}\left(t^3 - 1\right)$$

$$= \frac{1}{3}(1 + t)(t - 1)(t + 1) - \frac{2}{7}(t - 1)\left(t^2 + t + 1\right)$$

$$= \frac{1}{21}(t - 1)\left(7(t + 1)^2 - 6\left(t^2 + t + 1\right)\right)$$

$$= \frac{1}{21}(t - 1)\left(7t^2 + 14t + 7 - 6t^2 - 6t - 6\right)$$

$$= \frac{1}{21}(t - 1)\left(t^2 + 8t + 1\right), \quad t \in \mathbb{T}, \quad t \geq 1.$$

Example 3.15. Let $\mathbb{T} = 3^{\mathbb{N}_0}$, $t_0 = 1$,

$$f(t) = \sum_{k=0}^{\infty} \frac{1}{k^2 + 1}h_k(t, 1), \quad g(t) = \sum_{k=0}^{\infty} \frac{k}{k^2 + k + 1}h_k(t, 1), \quad t \in \mathbb{T}.$$

We will find $(f \star g)(t)$, $t \in \mathbb{T}, t \geq 1$. Here

$$\sigma(t) = 3t, \quad t \in \mathbb{T}.$$

By Theorem 3.4, we get

$$\hat{f}(t, s) = \sum_{k=0}^{\infty} \frac{1}{k^2 + 1}h_k(t, s),$$

$$\hat{f}(t, \sigma(s)) = \sum_{k=0}^{\infty} \frac{1}{k^2 + 1}h_k(t, \sigma(s))$$

$$= \sum_{k=0}^{\infty} \frac{1}{k^2 + 1}h_k(t, 3s), \quad t, s \in \mathbb{T}, \quad t \geq s \geq 1.$$

Then

$$(f \star g)(t) = \int_1^t \hat{f}(t, \sigma(s))g(s)\Delta s$$

$$= \int_1^t \left(\sum_{k=0}^{\infty} \frac{1}{k^2+1} h_k(t, 3s) \right) \left(\sum_{k=0}^{\infty} \frac{k}{k^2+k+1} h_k(s, 1) \right) \Delta s$$

$$= \int_1^t \sum_{k=0}^{\infty} \sum_{l=0}^{k} \frac{1}{(k-l)^2+1} h_{k-l}(t, 3s) \frac{l}{l^2+l+1} h_l(s, 1) \Delta s$$

$$= \sum_{k=0}^{\infty} \sum_{l=0}^{k} \left(\frac{1}{(k-l)^2+1} \right) \left(\frac{l}{l^2+l+1} \right) \int_1^t h_{k-l}(t, 3s) h_l(s, 1) \Delta s$$

$$= \sum_{k=0}^{\infty} \sum_{l=0}^{k} \left(\frac{1}{(k-l)^2+1} \right) \left(\frac{l}{l^2+l+1} \right) h_{k+1}(t, 1), \quad t \in \mathbb{T}, \quad t \geq 1.$$

Exercise 3.6. Let $\mathbb{T} = 4^{\mathbb{N}_0}$, $t_0 = 4$,

$$f(t) = \sum_{k=0}^{\infty} e^{-k^2} h_k(t, 4), \quad g(t) = \sum_{k=0}^{\infty} e^{-k^3} h_k(t, 4), \quad t \in \mathbb{T}, \quad t \geq 4.$$

Find $(f \star g)(t), t \in \mathbb{T}, t \geq 4$.

Answer.

$$(f \star g)(t) = \sum_{k=0}^{\infty} \sum_{l=0}^{k} e^{-(k-l)^2-l^3} h_{k+1}(t, 4), \quad t \in \mathbb{T}, \quad t \geq 4.$$

Theorem 3.6. *Let $f, g : \mathbb{T} \to \mathbb{C}$ be given functions. The shift of the convolution $f \star g$ is given by the formula*

$$\left(\widehat{f \star g} \right)(t, s) = \int_s^t \hat{f}(t, \sigma(u)) \hat{g}(u, s) \Delta u, \quad t, s \in \mathbb{T}, \quad t \geq s \geq t_0.$$

Proof. Let

$$F(t, s) = \int_s^t \hat{f}(t, \sigma(u)) \hat{g}(u, s) \Delta u.$$

Then

$$F(t, t_0) = \int_{t_0}^{t} \hat{f}(t, \sigma(u))\hat{g}(u, t_0)\Delta u$$

$$= \int_{t_0}^{t} \hat{f}(t, \sigma(u))g(u)\Delta u$$

$$= (f \star g)(t), \quad t \in \mathbb{T}, \quad t \geq t_0,$$

and applying Lemma 3.2, we get

$$F^{\Delta_t}(t, \sigma(s)) + F^{\Delta_s}(t, s) = \int_{\sigma(s)}^{t} \hat{f}^{\Delta_t}(t, \sigma(u))\hat{g}(u, \sigma(s))\Delta u$$

$$+ \hat{f}(\sigma(t), \sigma(t))\hat{g}(t, \sigma(s))$$

$$+ \int_{s}^{t} \hat{f}(t, \sigma(u))\hat{g}^{\Delta_s}(u, s)\Delta u - \hat{f}(t, \sigma(s))\hat{g}(t, \sigma(s))$$

$$= -\int_{\sigma(s)}^{t} \hat{f}^{\Delta_u}(t, u)\hat{g}(u, \sigma(s))\Delta u + \int_{s}^{t} \hat{f}(t, \sigma(u))\hat{g}^{\Delta_s}(u, s)\Delta u$$

$$+ f(t_0)\hat{g}(t, \sigma(s)) - \hat{f}(t, \sigma(s))\hat{g}(s, \sigma(s))$$

$$= -\hat{f}(t, u)\hat{g}(u, \sigma(s))\Big|_{u=\sigma(s)}^{u=t} + \int_{\sigma(s)}^{t} \hat{f}(t, \sigma(u))\hat{g}^{\Delta_t}(u, \sigma(s))\Delta u$$

$$- \int_{s}^{t} \hat{f}(t, \sigma(u))\hat{g}^{\Delta_t}(u, \sigma(s))\,\Delta u + f(t_0)\hat{g}(t, \sigma(s)) - \hat{f}(t, \sigma(s))\hat{g}(s, \sigma(s))$$

$$= -f(t_0)\hat{g}(t, \sigma(s)) + \hat{f}(t, \sigma(s))g(t_0)$$

$$- \int_{\sigma(s)}^{s} \hat{f}(t, \sigma(u))\hat{g}^{\Delta_s}(u, s)\Delta u$$

$$+ f(t_0)\hat{g}(t, \sigma(s)) - \hat{f}(t, \sigma(s))\hat{g}(s, \sigma(s))$$

$$= \hat{f}(t, \sigma(s))g(t_0) - \hat{f}(t, \sigma(s))\hat{g}(s, \sigma(s))$$

$$+ \int_{s}^{\sigma(s)} \hat{f}(t, \sigma(u))\hat{g}^{\Delta_s}(u, s)\Delta u$$

$$= \hat{f}(t, \sigma(s))g(t_0) - \hat{f}(t, \sigma(s))\hat{g}(s, \sigma(s))$$

$$+ \mu(s)\hat{f}(t, \sigma(s))\hat{g}^{\Delta_s}(s, s)$$

$$= \hat{f}(t, \sigma(s))g(t_0) - \hat{f}(t, \sigma(s))\hat{g}(s, \sigma(s))$$

$$+ \hat{f}(t, \sigma(s))\hat{g}(s, \sigma(s)) - \hat{f}(t, \sigma(s))\hat{g}(s, s)$$

$$= 0, \qquad t, s \in \mathbb{T}, \quad t \geq s \geq t_0.$$

This completes the proof.

Theorem 3.7 (Associativity of Convolution). *Convolution is associative, i.e.,*

$$(f \star g) \star h = f \star (g \star h).$$

Proof. We have

$$
\begin{aligned}
((f \star g) \star h)(t) &= \int_{t_0}^{t} \widehat{(f \star g)}(t, \sigma(s))h(s)\Delta s \\
&= \int_{t_0}^{t} \int_{\sigma(s)}^{t} \hat{f}(t, \sigma(u))\hat{g}(u, \sigma(s))h(s)\Delta u \Delta s \\
&= \int_{t_0}^{t} \int_{t_0}^{u} \hat{f}(t, \sigma(u))\hat{g}(u, \sigma(s))h(s)\Delta s \Delta u \\
&= \int_{t_0}^{t} \hat{f}(t, \sigma(u)) \int_{t_0}^{u} \hat{g}(u, \sigma(s))h(s)\Delta s \Delta u \\
&= \int_{t_0}^{t} \hat{f}(t, \sigma(u))(g \star h)(u)\Delta u \\
&= (f \star (g \star h))(t), \quad t \in \mathbb{T}, \quad t \geq t_0.
\end{aligned}
$$

This completes the proof.

Theorem 3.8. *Let f be delta differentiable. Then*

$$(f \star g)^{\Delta} = f^{\Delta} \star g + f(t_0)g. \tag{3.16}$$

In addition, if g is delta differentiable, then

$$(f \star g)^{\Delta} = f \star g^{\Delta} + fg(t_0). \tag{3.17}$$

Proof. We have

$$
\begin{aligned}
(f \star g)^{\Delta}(t) &= \left(\int_{t_0}^{t} \hat{f}(t, \sigma(s))g(s)\Delta s \right)^{\Delta} \\
&= \hat{f}(\sigma(t), \sigma(t))g(t) + \int_{t_0}^{t} \hat{f}^{\Delta_t}(t, \sigma(s))g(s)\Delta s \\
&= f(t_0)g(t) + \int_{t_0}^{t} \hat{f}^{\Delta_t}(t, \sigma(s))g(s)\Delta s \\
&= f(t_0)g(t) + \left(f^{\Delta} \star g \right)(t), \quad t \in \mathbb{T}.
\end{aligned}
$$

If *g* is delta differentiable, using the last equation, we get

$$(f \star g)^{\Delta}(t) = f(t_0)g(t) + \left(f^{\Delta} \star g\right)(t)$$

$$= f(t_0)g(t) - \int_{t_0}^{t} \hat{f}^{\Delta_s}(t,s)g(s)\Delta s$$

$$= f(t_0)g(t)$$
$$- \int_{t_0}^{t} \left(\left(\hat{f}(t,s)g(s)\right)^{\Delta_s} - \hat{f}(t,\sigma(s))g^{\Delta}(s) \right) \Delta s$$

$$= f(t_0)g(t) - \hat{f}(t,s)g(s) \Big|_{s=t_0}^{s=t}$$
$$+ \int_{t_0}^{t} \hat{f}(t,\sigma(s))g^{\Delta}(s)\Delta s$$

$$= f(t_0)g(t) - \hat{f}(t,t)g(t) + \hat{f}(t,t_0)g(t_0)$$
$$+ \left(f \star g^{\Delta}\right)(t)$$

$$= f(t_0)g(t) - f(t_0)g(t) + f(t)g(t_0) + \left(f \star g^{\Delta}\right)(t)$$

$$= f(t)g(t_0) + \left(f \star g^{\Delta}\right)(t), \quad t \in \mathbb{T}, \quad t \geq t_0.$$

This completes the proof.

Corollary 3.2. *Let f be delta differentiable. Then*

$$\int_{t_0}^{t} \hat{f}(t,\sigma(s))\Delta s = \int_{t_0}^{t} f(s)\Delta s, \quad t \in \mathbb{T}, \quad t \geq t_0.$$

Proof. We apply Theorem 3.8 with $g = 1$, and we get

$$(f \star g)^{\Delta}(t) = f(t),$$

whereupon

$$\int_{t_0}^{t} (f \star g)^{\Delta}(\tau)\Delta\tau = \int_{t_0}^{t} f(s)\Delta s,$$

or

$$(f \star g)(\tau) \Big|_{\tau=t_0}^{\tau=t} = \int_{t_0}^{t} f(s)\Delta s,$$

or

$$(f \star g)(t) = \int_{t_0}^{t} f(s)\Delta s,$$

or

$$\int_{t_0}^{t} \hat{f}(t, \sigma(s)) \Delta s = \int_{t_0}^{t} f(s) \Delta s, \quad t \in \mathbb{T}, \quad t \geq t_0.$$

This completes the proof.

Example 3.16. Let $\mathbb{T} = \mathbb{Z}$, $t_0 = 0$, $f(t) = t^2$, $g(t) = t$, $t \in \mathbb{T}$. We will find

$$(f \star g)^{\Delta}(t), \quad t \in \mathbb{T}, \quad t \geq t_0.$$

Here

$$\sigma(t) = t + 1, \quad t \in \mathbb{T}.$$

By Example 3.13, we have

$$
\begin{aligned}
(f \star g)(t) &= \frac{1}{12}t(t-1)^2(t-2) \\
&= \frac{1}{12}t\left(t^2 - 2t + 1\right)(t-2) \\
&= \frac{1}{12}\left(t^3 - 2t^2 + t\right)(t-2) \\
&= \frac{1}{12}\left(t^4 - 2t^3 - 2t^3 + 4t^2 + t^2 - 2t\right) \\
&= \frac{1}{12}\left(t^4 - 4t^3 + 5t^2 - 2t\right), \quad t \in \mathbb{T}, \quad t \geq 0.
\end{aligned}
$$

Then

$$
\begin{aligned}
(f \star g)^{\Delta}(t) &= \frac{1}{12}\Big((\sigma(t))^3 + t(\sigma(t))^2 + t^2\sigma(t) + t^3 \\
&\quad -4\left((\sigma(t))^2 + t\sigma(t) + t^2\right) \\
&\quad +5(\sigma(t) + t) - 2\Big) \\
&= \frac{1}{12}\Big((t+1)^3 + t(t+1)^2 + t^2(t+1) + t^3 \\
&\quad -4\left((t+1)^2 + t(t+1) + t^2\right) \\
&\quad +5(t+1+t) - 2\Big) \\
&= \frac{1}{12}\Big(t^3 + 3t^2 + 3t + 1 + t\left(t^2 + 2t + 1\right) + t^3 + t^2 + t^3
\end{aligned}
$$

$$-4\left(t^2 + 2t + 1 + t^2 + t + t^2\right) + 5(2t+1) - 2\Big)$$

$$= \frac{1}{12}\left(t^3 + 3t^2 + 3t + 1 + t^3 + 2t^2 + t + 2t^3 + t^2\right.$$

$$-4\left(3t^2 + 3t + 1\right) + 10t + 5 - 2\Big)$$

$$= \frac{1}{12}\left(4t^3 + 6t^2 + 4t + 1 - 12t^2 - 12t - 4 + 10t + 3\right)$$

$$= \frac{1}{12}\left(4t^3 - 6t^2 + 2t\right)$$

$$= \frac{1}{6}\left(2t^3 - 3t^2 + t\right), \quad t \in \mathbb{T}, \quad t \geq 0.$$

Now we will compute $(f \star g)^{\Delta}(t)$ using Theorem 3.8. We have

$$f^{\Delta}(t) = \sigma(t) + t$$

$$= t + 1 + t$$

$$= 2t + 1,$$

$$f^{\Delta}(0) = 1,$$

$$f^{\Delta^2}(t) = 2,$$

$$f^{\Delta^2}(0) = 2,$$

$$f^{\Delta^k}(t) = 0, \quad k \in \mathbb{N}, \quad k \geq 3.$$

Therefore,

$$f^{\Delta}(t) = h_0(t, 0) + 2h_1(t, 0), \quad t \in \mathbb{T}, \quad t \geq 0.$$

From this and Theorem 3.4, we obtain

$$\widehat{f^{\Delta}}(t, s) = h_0(t, s) + 2h_1(t, s)$$

$$= 1 + 2(t - s)$$

$$= 2t - 2s + 1,$$

$$\widehat{f^{\Delta}}(t, \sigma(s)) = 2t - 2\sigma(s) + 1$$

$$= 2t - 2(s + 1) + 1$$

$$= 2t - 2s - 2 + 1$$

$$= 2t - 2s - 1, \quad t, s \in \mathbb{T}, \quad t \geq s \geq 0.$$

Let

$$g(s) = (2t - 1)\left(\frac{1}{2}s^2 - \frac{1}{2}s\right) - \frac{2}{3}s^3 + s^2 - \frac{1}{3}s, \quad s \in \mathbb{T},$$

for some $t \in \mathbb{T}$. Then

$$g^\Delta(s) = (2t - 1)\left(\frac{1}{2}(\sigma(s) + s) - \frac{1}{2}\right) - \frac{2}{3}\left((\sigma(s))^2 + s\sigma(s) + s^2\right)$$

$$+\sigma(s) + s - \frac{1}{3}$$

$$= (2t - 1)\left(\frac{1}{2}(s + 1 + s) - \frac{1}{2}\right) - \frac{2}{3}\left((s + 1)^2 + s(s + 1) + s^2\right)$$

$$+s + 1 + s - \frac{1}{3}$$

$$= (2t - 1)s - \frac{2}{3}\left(s^2 + 2s + 1 + s^2 + s + s^2\right) + 2s + \frac{2}{3}$$

$$= (2t - 1)s - \frac{2}{3}\left(3s^2 + 3s + 1\right) + 2s + \frac{2}{3}$$

$$= (2t - 1)s - 2s^2 - 2s - \frac{2}{3} + 2s + \frac{2}{3}$$

$$= (2t - 1)s - 2s^2$$

and

$$\left(f^\Delta \star g\right)(t) = \int_0^t f^\Delta(t, \sigma(s))g(s)\Delta s$$

$$= \int_0^t (2t - 2s - 1)s\Delta s$$

$$= \int_0^t \left((2t - 1)s - 2s^2\right)\Delta s$$

$$= \int_0^t g^\Delta(s)\Delta s$$

$$= g(s)\Big|_{s=0}^{s=t}$$

$$= \left((2t - 1)\left(\frac{1}{2}s^2 - \frac{1}{2}s\right) - \frac{2}{3}s^3 + s^2 - \frac{1}{3}s\right)\Big|_{s=0}^{s=t}$$

$$= (2t - 1)\left(\frac{1}{2}t^2 - \frac{1}{2}t\right) - \frac{2}{3}t^3 + t^2 - \frac{1}{3}t$$

$$= \frac{1}{2}(2t-1)\left(t^2 - t\right) - \frac{2}{3}t^3 + t^2 - \frac{1}{3}t$$

$$= \frac{1}{2}\left(2t^3 - 2t^2 - t^2 + t\right) - \frac{2}{3}t^3 + t^2 - \frac{1}{3}t$$

$$= \frac{1}{2}\left(2t^3 - 3t^2 + t\right) - \frac{2}{3}t^3 + t^2 - \frac{1}{3}t$$

$$= t^3 - \frac{3}{2}t^2 + \frac{1}{2}t - \frac{2}{3}t^3 + t^2 - \frac{1}{3}t$$

$$= \frac{1}{3}t^3 - \frac{1}{2}t^2 + \frac{1}{6}t$$

$$= \frac{1}{6}\left(2t^3 - 3t^2 + t\right), \quad t \in \mathbb{T}, \quad t \geq 0.$$

Hence, using that $f(0) = 0$ and (3.16), we obtain

$$(f \star g)^{\Delta}(t) = \frac{1}{6}\left(2t^3 - 3t^2 + t\right), \quad t \in \mathbb{T}, \quad t \geq 0.$$

Example 3.17. Let $\mathbb{T} = 2^{\mathbb{N}_0}, f(t) = 1, g(t) = t^2, t \in \mathbb{T}, t_0 = 1$. We will find

$$(f \star g)^{\Delta}(t), \quad t \in \mathbb{T}, \quad t \geq 1.$$

Here

$$\sigma(t) = 2t, \quad t \in \mathbb{T}.$$

Then

$$g^{\Delta}(t) = \sigma(t) + t$$

$$= 2t + t$$

$$= 3t,$$

$$f(t) = h_0(t, 1), \quad t \in \mathbb{T}, \quad t \geq 1.$$

From this and Theorem 3.4, we obtain

$$\hat{f}(t, s) = h_0(t, s), \quad t, s \in \mathbb{T}, \quad t \geq s \geq 1.$$

Therefore,

$$\left(f \star g^{\Delta}\right)(t) = \int_1^t f(t, \sigma(s))g^{\Delta}(s)\Delta s$$

$$= \int_1^t g^{\Delta}(s)\Delta s$$

$$= g(s)\Big|_{s=1}^{s=t}$$

$$= s^2\Big|_{s=1}^{s=t}$$

$$= t^2 - 1,$$

$$f(t)g(1) = 1, \quad t \in \mathbb{T}, \quad t \geq 1.$$

From this and (3.17), we get

$$(f \star g)^\Delta (t) = t^2 - 1 + 1$$

$$= t^2, \quad t \in \mathbb{T}, \quad t \geq 1.$$

Example 3.18. Let $\mathbb{T} = 3^{\mathbb{N}_0}$, $t_0 = 1, f(t) = t, g(t) = t^3, t \in \mathbb{T}$. We will find

$$(f \star g)^\Delta (t), \quad t \in \mathbb{T}, \quad t \geq 1.$$

Here

$$\sigma(t) = 3t, \quad t \in \mathbb{T}, \quad t \geq 1.$$

We have

$$g^\Delta(t) = (\sigma(t))^2 + t\sigma(t) + t^2$$

$$= (3t)^2 + t(3t) + t^2$$

$$= 9t^2 + 3t^2 + t^2$$

$$= 13t^2,$$

$$f(1) = 1,$$

$$f^\Delta(t) = 1,$$

$$f^\Delta(1) = 1, \quad t \in \mathbb{T}, \quad t \geq 1.$$

Then

$$f(t) = h_0(t, 1) + h_1(t, 1), \quad t \in \mathbb{T}.$$

From this and Theorem 3.4, we get

$$\hat{f}(t, s) = h_0(t, s) + h_1(t, s)$$

$$= 1 + t - s,$$

$$\hat{f}(t, \sigma(s)) = t - \sigma(s) + 1$$
$$= t - 3s + 1, \quad t, s \in \mathbb{T}, \quad t \geq s \geq 1.$$

Let

$$l(s) = \frac{1}{13}(1+t)s^3 - \frac{3}{40}s^4, \quad s \in \mathbb{T},$$

for some $t \in \mathbb{T}$. Then

$$l^{\Delta}(s) = \frac{1}{13}(1+t)\left((\sigma(s))^2 + s\sigma(s) + s^2\right)$$
$$- \frac{3}{40}\left((\sigma(s))^3 + s(\sigma(s))^2 + s^2\sigma(s) + s^3\right)$$
$$= \frac{1}{13}(1+t)\left((3s)^2 + s(3s) + s^2\right)$$
$$- \frac{3}{40}\left((3s)^3 + s(3s)^2 + s^2(3s) + s^3\right)$$
$$= \frac{1}{13}(1+t)\left(9s^2 + 3s^2 + s^2\right)$$
$$- \frac{3}{40}\left(27s^3 + 9s^3 + 3s^3 + s^3\right)$$
$$= (1+t)s^2 - 3s^3, \quad s \in \mathbb{T},$$

for some $t \in \mathbb{T}$. Therefore,

$$\left(f \star g^{\Delta}\right)(t) = \int_1^t \hat{f}(t, \sigma(s))g^{\Delta}(s)\Delta s$$
$$= 13 \int_1^t (1 + t - 3s)s^2 \Delta s$$
$$= 13 \int_1^t l^{\Delta}(s)\Delta s$$
$$= 13 l(s)\Big|_{s=1}^{s=t}$$
$$= 13 \left(\frac{1}{13}(1+t)s^3 - \frac{3}{40}s^4\right)\Big|_{s=1}^{s=t}$$
$$= 13 \left(\frac{(1+t)t^3}{13} - \frac{3}{40}t^4\right)$$
$$- 13 \left(\frac{1}{13}(1+t) - \frac{3}{40}\right)$$

$$= \frac{t^3(40 + 40t - 39t)}{40} - \frac{40 + 40t - 39}{40}$$

$$= \frac{t^3(40 + t) - 1 - 40t}{40}$$

$$= \frac{t^4 + 40t^3 - 40t - 1}{40}, \quad t \in \mathbb{T}, \quad t \geq 1.$$

From this and (3.17), we obtain

$$(f \star g)^\Delta (t) = \frac{t^4 + 40t^3 - 40t - 1}{40} + t$$

$$= \frac{t^4 + 40t^3 - 1}{40}, \quad t \in \mathbb{T}, \quad t \geq 1.$$

Exercise 3.7. Let $\mathbb{T} = 4^{\mathbb{N}_0}, f(t) = 1, g(t) = 3t + 7, t \in \mathbb{T}, t_0 = 1$. Find

$$(f \star g)^\Delta (t), \quad t \in \mathbb{T}, \quad t \geq 1.$$

Answer. $3t + 7, t \in \mathbb{T}, t \geq 1.$

Example 3.19. Let $\mathbb{T} = \mathbb{Z}, t_0 = 0, f(t) = t^2 - 3t, t \in \mathbb{T}$. We will compute

$$\int_0^t \hat{f}(t, \sigma(s)) \Delta s, \quad t \in \mathbb{Z}, \quad t \geq 0.$$

Here

$$\sigma(t) = t + 1, \quad t \in \mathbb{T}.$$

Let

$$g(t) = \frac{1}{3}t^3 - 2t^2 + \frac{5}{3}t, \quad t \in \mathbb{T}.$$

Then

$$g^\Delta (t) = \frac{1}{3}\left((\sigma(t))^2 + t\sigma(t) + t^2\right)$$

$$-2(\sigma(t) + t) + \frac{5}{3}$$

$$= \frac{1}{3}\left((t+1)^2 + t(t+1) + t^2\right)$$

$$-2(t + 1 + t) + \frac{5}{3}$$

$$= \frac{1}{3}\left(t^2 + 2t + 1 + t^2 + t + t^2\right)$$

$$-2(2t + 1) + \frac{5}{3}$$

$$= \frac{1}{3}\left(3t^2 + 3t + 1\right) - 4t - 2 + \frac{5}{3}$$

$$= t^2 + t + \frac{1}{3} - 4t - \frac{1}{3}$$

$$= t^2 - 3t$$

$$= f(t), \quad t \in \mathbb{T}, \quad t \geq 0.$$

Hence,

$$\int_0^t f(s)\Delta s = \int_0^t g^\Delta(s)\Delta s$$

$$= g(s)\Big|_{s=0}^{s=t}$$

$$= \left(\frac{1}{3}s^3 - 2s^2 + \frac{5}{3}s\right)\Big|_{s=0}^{s=t}$$

$$= \frac{1}{3}t^3 - 2t^2 + \frac{5}{3}t, \quad t \in \mathbb{T}, \quad t \geq 0.$$

From this and Corollary 3.2, we get that

$$\int_0^t \hat{f}(t, \sigma(s))\Delta s = \frac{1}{3}t^3 - 2t^2 + \frac{5}{3}t, \quad t \in \mathbb{T}, \quad t \geq 0.$$

Example 3.20. Let $\mathbb{T} = 2^{\mathbb{N}_0}$, $t_0 = 1$,

$$f(t) = \frac{-2t^2 - 3t + 2}{2\left(t^2 + 2\right)\left(2t^2 + 1\right)}, \quad t \in \mathbb{T}.$$

We will find

$$\int_1^t \hat{f}(t, \sigma(s))\Delta s, \quad t \in \mathbb{T}, \quad t \geq 1.$$

Here

$$\sigma(t) = 2t, \quad t \in \mathbb{T}.$$

Let

$$g(t) = \frac{t+1}{t^2+2}, \quad t \in \mathbb{T}, \quad t \geq 1.$$

Then

$$g^{\Delta}(t) = \frac{t^2 + 2 - (t+1)(\sigma(t) + t)}{(t^2 + 2)\left((\sigma(t))^2 + 2\right)}$$

$$= \frac{t^2 + 2 - (t+1)(2t+t)}{(t^2 + 2)\left((2t)^2 + 2\right)}$$

$$= \frac{t^2 + 2 - 3t(t+1)}{(t^2 + 2)\left(4t^2 + 2\right)}$$

$$= \frac{t^2 + 2 - 3t^2 - 3t}{2(t^2 + 2)\left(2t^2 + 1\right)}$$

$$= \frac{-2t^2 - 3t + 2}{2(t^2 + 2)\left(2t^2 + 1\right)}$$

$$= f(t), \quad t \in \mathbb{T}.$$

Hence,

$$\int_1^t f(s)\Delta s = \int_1^t g^{\Delta}(s)\Delta s$$

$$= g(s)\Big|_{s=1}^{s=t}$$

$$= \frac{s+1}{s^2+2}\Big|_{s=1}^{s=t}$$

$$= \frac{t+1}{t^2+2} - \frac{2}{3}$$

$$= \frac{3t + 3 - 2t^2 - 4}{3(t^2+2)}$$

$$= \frac{-2t^2 + 3t - 1}{3(t^2+2)}, \quad t \in \mathbb{T}, \quad t > 1.$$

Consequently, using Corollary 3.2, we get

$$\int_1^t \hat{f}(t, \sigma(s))\Delta s = \frac{-2t^2 + 3t - 1}{3(t^2+2)}, \quad t \in \mathbb{T}, \quad t > 1.$$

Example 3.21. Let $\mathbb{T} = 3^{\mathbb{N}_0}$, $t_0 = 1$,

$$f(t) = \frac{1}{(t+1)(3t+2)} + \sin_1(t, 1) + 3t \cos_1(t, 1), \quad t \in \mathbb{T}, \quad t > 1.$$

We will find

$$\int_1^t \hat{f}(t, \sigma(s)) \Delta s, \quad t \in \mathbb{T}, \quad t > 1.$$

Here

$$\sigma(t) = 3t, \quad t \in \mathbb{T}.$$

Let

$$g(t) = \frac{t+1}{t+2} + t \sin_1(t, 1), \quad t \in \mathbb{T}.$$

We have

$$g^\Delta(t) = \frac{t+2-(t+1)}{(t+2)(\sigma(t)+2)} + \sin_1(t, 1) + \sigma(t) \cos_1(t, 1)$$

$$= \frac{1}{(t+2)(3t+2)} + \sin_1(t, 1) + 3t \cos_1(t, 1)$$

$$= f(t), \quad t \in \mathbb{T}, \quad t > 1.$$

Then

$$\int_1^t f(s) \Delta s = \int_1^t g^\Delta(s) \Delta s$$

$$= g(s) \Big|_{s=1}^{s=t}$$

$$= \left(\frac{s+1}{s+2} + s \sin_1(s, 1) \right) \Big|_{s=1}^{s=t}$$

$$= \frac{t+1}{t+2} + t \sin_1(t, 1) - \frac{2}{3}$$

$$= \frac{3t+3-2t-4}{3(t+2)} + t \sin_1(t, 1)$$

$$= \frac{t-1}{3(t+2)} + t \sin_1(t, 1), \quad t \in \mathbb{T}, \quad t > 1.$$

Consequently, using Corollary 3.2, we get

$$\int_1^t \hat{f}(t, \sigma(s)) \Delta s = \frac{t-1}{3(t+2)} + t \sin_1(t, 1), \quad t \in \mathbb{T}, \quad t > 1.$$

Exercise 3.8. Let $\mathbb{T} = 2^{\mathbb{N}_0}$, $t_0 = 1$. Let also

$$f(t) = \frac{t^2 - 3t - 1}{t^2(t+1)(2t+1)}, \quad t \in \mathbb{T}.$$

Find

$$\int_1^t \hat{f}(t, \sigma(s)) \Delta s, \quad t \in \mathbb{T}, \quad t > 1.$$

Hint. Use the function

$$g(t) = \frac{t^2 + 2}{t^2 + t}, \quad t \in \mathbb{T}.$$

Answer.

$$\frac{-t^2 - 3t + 4}{2t(t+1)}, \quad t \in \mathbb{T}, \quad t > 1.$$

Theorem 3.9. *If f and g are infinitely Δ-differentiable, then for all $k \in \mathbb{N}_0$, we have*

$$(f \star g)^{\Delta^k} = f^{\Delta^k} \star g + \sum_{\nu=0}^{k-1} f^{\Delta^\nu}(t_0) g^{\Delta^{k-1-\nu}}$$

$$= f \star g^{\Delta^k} + \sum_{\nu=0}^{k-1} f^{\Delta^\nu} g^{\Delta^{k-1-\nu}}(t_0), \quad (3.18)$$

$$(f \star g)^{\Delta^k}(t_0) = \sum_{\nu=0}^{k-1} f^{\Delta^\nu}(t_0) g^{\Delta^{k-1-\nu}}(t_0).$$

Proof. By Theorem 3.8, it follows that the assertion is valid for $k = 1$. Assume (3.18) for some $k \in \mathbb{N}$. We will prove that

$$(f \star g)^{\Delta^{k+1}} = f^{\Delta^{k+1}} \star g + \sum_{\nu=0}^{k} f^{\Delta^\nu}(t_0) g^{\Delta^{k-\nu}}$$

$$= f \star g^{\Delta^{k+1}} + \sum_{\nu=0}^{k} f^{\Delta^\nu} g^{\Delta^{k-\nu}}(t_0),$$

$$(f \star g)^{\Delta^{k+1}}(t_0) = \sum_{\nu=0}^{k} f^{\Delta^\nu}(t_0) g^{\Delta^{k-\nu}}(t_0).$$

Indeed, using (3.16), (3.17), we get

$$
(f \star g)^{\Delta^{k+1}} = \left((f \star g)^{\Delta^k} \right)^\Delta
$$

$$
= \left(f^{\Delta^k} \star g + \sum_{\nu=0}^{k-1} f^{\Delta^\nu}(t_0) g^{\Delta^{k-1-\nu}} \right)^\Delta
$$

$$
= \left(f^{\Delta^k} \star g \right)^\Delta + \left(\sum_{\nu=0}^{k-1} f^{\Delta^\nu}(t_0) g^{\Delta^{k-1-\nu}} \right)^\Delta
$$

$$
= f^{\Delta^{k+1}} \star g + f^{\Delta^k}(t_0) g + \sum_{\nu=0}^{k-1} f^{\Delta^\nu}(t_0) g^{\Delta^{k-\nu}}
$$

$$
= f^{\Delta^{k+1}} \star g + \sum_{\nu=0}^{k} f^{\Delta^\nu}(t_0) g^{\Delta^{k-\nu}}
$$

and

$$
(f \star g)^{\Delta^{k+1}} = \left(f \star g^{\Delta^k} + \sum_{\nu=0}^{k-1} f^{\Delta^\nu} g^{\Delta^{k-1-\nu}}(t_0) \right)^\Delta
$$

$$
= \left(f \star g^{\Delta^k} \right)^\Delta + \left(\sum_{\nu=0}^{k-1} f^{\Delta^\nu} g^{\Delta^{k-1-\nu}}(t_0) \right)^\Delta
$$

$$
= f \star g^{\Delta^{k+1}} + f g^{\Delta^k}(t_0) + \sum_{\nu=0}^{k-1} f^{\Delta^{\nu+1}} g^{\Delta^{k-1-\nu}}(t_0)
$$

$$
= f \star g^{\Delta^{k+1}} + f g^{\Delta^k}(t_0) + \sum_{\nu=1}^{k} f^{\Delta^\nu} g^{\Delta^{k-\nu}}(t_0)
$$

$$
= f \star g^{\Delta^{k+1}} + \sum_{\nu=0}^{k} f^{\Delta^\nu} g^{\Delta^{k-\nu}}(t_0).
$$

From this and the principle of mathematical induction, we conclude that (3.18) is valid for all $k \in \mathbb{N}$. Note that

$$
\left(f^{\Delta^k} \star g \right)(t_0) = \int_{t_0}^{t_0} \widehat{f^{\Delta^k}}(t, \sigma(s)) g(s) \Delta s
$$

$$
= 0,
$$

$$
\left(f \star g^{\Delta^k} \right)(t_0) = \int_{t_0}^{t_0} \hat{f}(t, \sigma(s)) g^{\Delta^k}(s) \Delta s
$$

$$
= 0, \quad k \in \mathbb{N}_0.
$$

Therefore,

$$(f \star g)^{\Delta^k}(t_0) = \sum_{\nu=0}^{k-1} f^{\Delta^\nu}(t_0) g^{\Delta^{k-1-\nu}}(t_0).$$

This completes the proof.

Example 3.22. Let $\mathbb{T} = 2\mathbb{Z}$, $t_0 = 0$,

$$f(t) = t^2 + 3t + 1, \quad g(t) = t^2, \quad t \in \mathbb{T}.$$

We will find

$$(f \star g)^{\Delta^4}(t), \quad t \in \mathbb{T}, \quad t > 0.$$

Here

$$\sigma(t) = t + 2, \quad t \in \mathbb{T}.$$

We have

$$
\begin{aligned}
f(0) &= 1, \\
f^{\Delta}(t) &= \sigma(t) + t + 3 \\
&= t + 2 + t + 3 \\
&= 2t + 5, \\
f^{\Delta}(0) &= 5, \\
f^{\Delta^2}(t) &= 2, \\
f^{\Delta^2}(0) &= 2, \\
f^{\Delta^k}(t) &= 0, \\
f^{\Delta^k}(0) &= 0, \quad k \in \mathbb{N}, \quad k \geq 3, \\
g^{\Delta}(t) &= \sigma(t) + t \\
&= t + 2 + t \\
&= 2t + 2, \\
g^{\Delta^2}(t) &= 2, \\
g^{\Delta^k}(t) &= 0, \quad k \in \mathbb{N}, \quad k \geq 3.
\end{aligned}
$$

Then, using (3.18), we get

$$(f \star g)^{\Delta^4}(t) = \left(f^{\Delta^4} \star g\right)(t) + \sum_{\nu=0}^{3} f^{\Delta^\nu}(0)g^{\Delta^{3-\nu}}(t)$$

$$= f(0)f^{\Delta^3}(t) + f^{\Delta}(0)g^{\Delta^2}(t) + f^{\Delta^2}(0)g^{\Delta}(t)$$
$$+ f^{\Delta^3}(0)g(t)$$
$$= 10 + 2(2t + 2)$$
$$= 10 + 4t + 4$$
$$= 4t + 14, \quad t \in \mathbb{T}, \quad t > 0.$$

Example 3.23. Let $\mathbb{T} = 2^{\mathbb{N}_0}$, $t_0 = 1$,

$$f(t) = t^3 + t^2 + t, \quad g(t) = t^2 - 2t + 1, \quad t \in \mathbb{T}.$$

We will find

$$(f \star g)^{\Delta^3}(t), \quad t \in \mathbb{T}, \quad t > 1.$$

Here

$$\sigma(t) = 2t, \quad t \in \mathbb{T}.$$

We have

$$f(1) = 3,$$
$$f^{\Delta}(t) = (\sigma(t))^2 + t\sigma(t) + t^2 + \sigma(t) + t + 1$$
$$= (2t)^2 + t(2t) + t^2 + 2t + t + 1$$
$$= 4t^2 + 2t^2 + t^2 + 3t + 1$$
$$= 7t^2 + 3t + 1,$$
$$f^{\Delta}(1) = 11,$$
$$f^{\Delta^2}(t) = 7(\sigma(t) + t) + 3$$
$$= 7(2t + t) + 3$$
$$= 21t + 3,$$
$$f^{\Delta^2}(1) = 24,$$
$$f^{\Delta^3}(t) = 21,$$
$$f^{\Delta^3}(1) = 21,$$
$$g^{\Delta}(t) = \sigma(t) + t - 2$$

$$= 2t + t - 2$$
$$= 3t - 2,$$
$$g^{\Delta^2}(t) = 3,$$
$$g^{\Delta^3}(t) = 0, \quad t \in \mathbb{T}.$$

Also,

$$f^{\Delta^3}(t) = 21h_0(t, 1), \quad t \in \mathbb{T}.$$

From this and Theorem 3.4, we obtain

$$\widehat{f^{\Delta^3}}(t, s) = 21h_0(t, s)$$
$$= 21, \quad t, s \in \mathbb{T}, \quad t \geq s \geq 1.$$

Let

$$l(t) = \frac{1}{7}t^3 - \frac{2}{3}t^2 + t, \quad t \in \mathbb{T}.$$

Then

$$l^{\Delta}(t) = \frac{1}{7}\left((\sigma(t))^2 + t\sigma(t) + t^2\right) - \frac{2}{3}(\sigma(t) + t) + 1$$
$$= \frac{1}{7}\left((2t)^2 + t(2t) + t^2\right) - \frac{2}{3}(2t + t) + 1$$
$$= \frac{1}{7}\left(4t^2 + 2t^2 + t^2\right) - \frac{2}{3}(3t) + 1$$
$$= \frac{1}{7}(7t^2) - 2t + 1$$
$$= t^2 - 2t + 1, \quad t \in \mathbb{T}.$$

Therefore,

$$\left(f^{\Delta^3} \star g\right)(t) = \int_1^t \widehat{f^{\Delta^3}}(t, \sigma(s))g(s)\Delta s$$
$$= 21 \int_1^t g(s)\Delta s$$
$$= 21 \int_1^t l^{\Delta}(s)\Delta s$$
$$= 21l(s)\Big|_{s=1}^{s=t}$$
$$= 21\left(\frac{1}{7}s^3 - \frac{2}{3}s^2 + s\right)\Big|_{s=1}^{s=t}$$

$$= 21 \left(\frac{1}{7} t^3 - \frac{2}{3} t^2 + t - \frac{1}{7} + \frac{2}{3} - 1 \right)$$

$$= 21 \left(\frac{1}{7} t^3 - \frac{2}{3} t^2 + t - \frac{10}{21} \right)$$

$$= 3t^3 - 14t^2 + 21t - 10, \quad t \in \mathbb{T}, \quad t > 1.$$

From this and (3.18), we get

$$(f \star g)^{\Delta^3}(t) = \left(f^{\Delta^3} \star g \right)(t) + \sum_{\nu=0}^{2} f^{\Delta^\nu}(1) g^{\Delta^{2-\nu}}(t)$$

$$= 3t^3 - 14t^2 + 21t - 10$$

$$+ f(1) g^{\Delta^2}(t) + f^{\Delta}(1) g^{\Delta}(t) + f^{\Delta^2}(1) g(t)$$

$$= 3t^3 - 14t^2 + 21t - 10 + 9$$

$$+ 11(3t - 2) + 24 \left(t^2 - 2t + 1 \right)$$

$$= 3t^3 - 14t^2 + 21t - 10 + 9 + 33t - 22$$

$$+ 24t^2 - 48t + 24$$

$$= 3t^3 + 10t^2 + 6t + 1, \quad t \in \mathbb{T}, \quad t > 1.$$

Example 3.24. Let $\mathbb{T} = 3^{\mathbb{N}_0}$, $t_0 = 1$,

$$f(t) = t^2, \quad g(t) = \frac{1}{3t^2}, \quad t \in \mathbb{T}.$$

We will find

$$(f \star g)^{\Delta^2}(t), \quad t \in \mathbb{T}, \quad t > 1.$$

Here

$$\sigma(t) = 3t, \quad t \in \mathbb{T}.$$

We have

$$f(1) = 1,$$
$$f^{\Delta}(t) = \sigma(t) + t$$
$$= 3t + t$$
$$= 4t,$$
$$f^{\Delta}(1) = 4,$$

$$f^{\Delta^2}(t) = 4,$$

$$f^{\Delta^2}(1) = 4,$$

$$g^{\Delta}(t) = -\frac{\sigma(t) + t}{3t^2 (\sigma(t))^2}$$

$$= -\frac{4t}{3t^2 (3t)^2}$$

$$= -\frac{4}{3t(9t^2)}$$

$$= -\frac{4}{27t^3},$$

$$g^{\Delta^2}(t) = \frac{4}{27}\frac{(\sigma(t))^2 + t\sigma(t) + t^2}{t^3 (\sigma(t))^3}$$

$$= \frac{4}{27}\frac{(3t)^2 + 3t^2 + t^2}{t^3 27t^3}$$

$$= \frac{4}{729}\frac{9t^2 + 4t^2}{t^6}$$

$$= \frac{52}{729}\frac{1}{t^4}, \quad t \in \mathbb{T}, \quad t > 1.$$

Also,

$$f^{\Delta^2}(t) = 4h_0(t, 1), \quad t \in \mathbb{T}.$$

From this and Theorem 3.4, we get

$$\widehat{f^{\Delta^2}}(t, s) = 4h_0(t, s), \quad t, s \in \mathbb{T}, \quad t \ge s \ge 1.$$

Therefore,

$$\left(f^{\Delta^2} \star g\right)(t) = \int_1^t \widehat{f^{\Delta^2}}(t, \sigma(s))g(s)\Delta s$$

$$= 4 \int_1^t g(s)\Delta s$$

$$= 4 \left(-\frac{1}{s}\right)\Big|_{s=1}^{s=t}$$

$$= -4 \left(\frac{1}{t} - 1\right)$$

$$= -4\frac{1 - t}{t}$$

$$= \frac{4(t-1)}{t}, \quad t \in \mathbb{T}, \quad t > 1,$$

$$(f \star g)^{\Delta^2}(t) = \left(f^{\Delta^2} \star g\right)(t) + \sum_{\nu=0}^{1} f^{\Delta^\nu}(1) g^{\Delta^{1-\nu}}(t)$$

$$= \frac{4(t-1)}{t} + f(1) g^{\Delta}(t) + f^{\Delta}(1) g(t)$$

$$= \frac{4(t-1)}{t} + \left(-\frac{4}{27t^3}\right) + \frac{4}{3t^2}$$

$$= 4 \frac{27t^3 - 27t^2 - 1 + 9t}{27t^3}, \quad t \in \mathbb{T}, \quad t > 1.$$

Exercise 3.9. Let $\mathbb{T} = 2^{\mathbb{N}_0}$, $t_0 = 1$,

$$f(t) = t^2 + t + 3, \quad g(t) = t^3 - 3t^2 + t + 1, \quad t \in \mathbb{T}, \quad t > 1.$$

Find

$$(f \star g)^{\Delta^{11}}(t), \quad t \in \mathbb{T}, \quad t > 1.$$

Answer. 0.

Theorem 3.10. *If $\hat{f}$ has partial Δ-derivatives of all orders, then*

$$\hat{f}^{\Delta_t^k}(t, t) = f^{\Delta^k}(t_0) \tag{3.19}$$

for all $k \in \mathbb{N}_0$, where $\hat{f}^{\Delta_t}$ indicates the Δ-derivatives of $\hat{f}$ with respect to its first variable.

Proof. Let $k \in \mathbb{N}_0$ be arbitrarily chosen. By (3.2), we get

$$f^{\Delta^k}(t) = \hat{f}^{\Delta_t^k}(t, t_0), \quad t \in \mathbb{T}, \quad t \geq t_0.$$

Let

$$F(t) = \hat{f}^{\Delta_t^k}(t, t), \quad t \in \mathbb{T}, \quad t \geq t_0.$$

Then

$$F(t_0) = f^{\Delta^k}(t_0),$$

$$F^{\Delta}(t) = \hat{f}^{\Delta_t^k \Delta_t}(t, \sigma(t)) + \hat{f}^{\Delta_t^k \Delta_s}(t, t)$$

$$= \hat{f}^{\Delta_t \Delta_t^k}(t, \sigma(t)) + \hat{f}^{\Delta_s \Delta_t^k}(t, t)$$

$$= \left(\hat{f}^{\Delta_t}(t, \sigma(t)) + \hat{f}^{\Delta_s}(t, t)\right)^{\Delta_t^k}$$

$$= 0, \quad t \in \mathbb{T}, \quad t \geq t_0.$$

Therefore,

$$F(t) = f^{\Delta^k}(t_0), \quad t \in \mathbb{T}, \quad t \geq t_0.$$

This completes the proof.

Example 3.25. Let $\mathbb{T} = \mathbb{Z}$, $t_0 = 0$,

$$f(t) = t^3 + t^2 - 2t + 1, \quad t \in \mathbb{T}.$$

We will find

$$\hat{f}^{\Delta_t^2}(t, t), \quad t \in \mathbb{T}.$$

Here

$$\sigma(t) = t + 1, \quad t \in \mathbb{T}.$$

We have

$$
\begin{aligned}
f^{\Delta}(t) &= (\sigma(t))^2 + t\sigma(t) + t^2 + \sigma(t) + t - 2 \\
&= (t+1)^2 + t(t+1) + t^2 + t + 1 + t - 2 \\
&= t^2 + 2t + 1 + t^2 + t + t^2 + 2t - 1 \\
&= 3t^2 + 5t, \\
f^{\Delta^2}(t) &= 3(\sigma(t) + t) + 5 \\
&= 3(t + 1 + t) + 5 \\
&= 3(2t + 1) + 5 \\
&= 6t + 8, \\
f^{\Delta^2}(0) &= 8, \quad t \in \mathbb{T}.
\end{aligned}
$$

From this and (3.19), we get

$$\hat{f}^{\Delta_t^2}(t, t) = 8, \quad t \in \mathbb{T}.$$

Example 3.26. Let $\mathbb{T} = 2^{\mathbb{N}_0}$, $t_0 = 1$,

$$f(t) = \frac{t+1}{t+2}, \quad t \in \mathbb{T}.$$

We will find

$$\hat{f}^{\Delta_i^3}(t, t), \quad t \in \mathbb{T}.$$

Here

$$\sigma(t) = 2t, \quad t \in \mathbb{T}.$$

We have

$$
\begin{aligned}
f^{\Delta}(t) &= \frac{t + 2 - (t + 1)}{(t + 2)(\sigma(t) + 2)} \\
&= \frac{t + 2 - t - 1}{(t + 2)(2t + 2)} \\
&= \frac{1}{2(t + 1)(t + 2)} \\
&= \frac{1}{2\left(t^2 + 3t + 2\right)},
\end{aligned}
$$

$$
\begin{aligned}
f^{\Delta^2}(t) &= -\frac{\sigma(t) + t + 3}{2\left(t^2 + 3t + 2\right)\left((\sigma(t))^2 + 3\sigma(t) + 2\right)} \\
&= -\frac{2t + t + 3}{2(t + 1)(t + 2)\left(4t^2 + 6t + 2\right)} \\
&= -\frac{3(t + 1)}{4(t + 1)(t + 2)(t + 1)\left(t + \frac{1}{2}\right)} \\
&= -\frac{3}{2(t + 1)(t + 2)(2t + 1)} \\
&= -\frac{3}{2\left(2t^3 + 7t^2 + 7t + 2\right)},
\end{aligned}
$$

$$
\begin{aligned}
f^{\Delta^3}(t) &= \frac{3}{2} \frac{2\left((\sigma(t))^2 + t\sigma(t) + t^2\right) + 7(\sigma(t) + t) + 7}{(t + 1)(t + 2)(2t + 1)(\sigma(t) + 1)(\sigma(t) + 2)(2\sigma(t) + 1)} \\
&= \frac{3}{2} \frac{2\left(4t^2 + 2t^2 + t^2\right) + 7(3t) + 7}{(t + 1)(t + 2)(2t + 1)(2t + 1)(2t + 2)(4t + 1)} \\
&= \frac{3}{4} \frac{14t^2 + 21t + 7}{(t + 1)^2(t + 2)(2t + 1)^2(4t + 1)},
\end{aligned}
$$

$$f^{\Delta^3}(1) = \frac{7}{120}, \quad t \in \mathbb{T}.$$

From this and (3.19), we get

$$\hat{f}^{\Delta_i^3}(t, t) = \frac{7}{120}, \quad t \in \mathbb{T}.$$

Example 3.27. Let $\mathbb{T} = 3^{\mathbb{N}_0}$, $t_0 = 1$,

$$f(t) = \frac{t+2}{t^2+2t+1}, \quad t \in \mathbb{T}.$$

We will find

$$\hat{f}^{\Delta_t}(t, t), \quad t \in \mathbb{T}.$$

Here

$$\sigma(t) = 3t, \quad t \in \mathbb{T}.$$

We have

$$
\begin{aligned}
f^{\Delta}(t) &= \frac{t^2+2t+1-(t+2)(\sigma(t)+t+2)}{(t^2+2t+1)\left((\sigma(t))^2+2\sigma(t)+1\right)} \\
&= \frac{t^2+2t+1-(t+2)(3t+t+2)}{(t^2+2t+1)\left(9t^2+6t+1\right)} \\
&= \frac{t^2+2t+1-(t+2)(4t+2)}{(t+1)^2(3t+1)^2}, \quad t \in \mathbb{T},
\end{aligned}
$$

$$f^{\Delta}(1) = -\frac{7}{32}.$$

From this and (3.19), we get

$$\hat{f}^{\Delta_t}(t, t) = -\frac{7}{32}, \quad t \in \mathbb{T}.$$

Exercise 3.10. Let $\mathbb{T} = \mathbb{Z}$, $t_0 = 0$,

$$f(t) = t^2 - 11t + 12, \quad t \in \mathbb{T}.$$

Find

$$\hat{f}^{\Delta_t^2}(t, t), \quad t \in \mathbb{T}.$$

Answer. 2.

3.3 The Convolution Theorem

We will begin with the following useful theorem.

Theorem 3.11. *Let* $f : \mathbb{T} \to \mathbb{C}$ *and*

$$\psi(s) = \int_s^\infty \frac{\hat{f}(t, s)}{e_z(\sigma(t), s)} \Delta t, \quad z \in \mathscr{D}\{f\}, \quad s \in \mathbb{T}.$$

Then ψ *is a constant.*

Proof. We have

$$\psi^\Delta(s) = \int_s^\infty \frac{\hat{f}^{\Delta_s}(t, s)e_z(\sigma(t), s) - (\ominus z)(s)e_z(\sigma(t), s)\hat{f}(t, s)}{e_z(\sigma(t), s)e_z(\sigma(t), \sigma(s))} \Delta t$$

$$- \frac{\hat{f}(s, \sigma(s))}{e_z(\sigma(s), \sigma(s))}$$

$$= \int_s^\infty \left(-\frac{\hat{f}^{\Delta_t}(t, \sigma(s))}{e_z(\sigma(t), \sigma(s))} + \frac{z\hat{f}(t, s)}{(1 + \mu(s)z)\, e_z(\sigma(t), \sigma(s))} \right) \Delta t$$

$$- \hat{f}(s, \sigma(s))$$

$$= z\psi(s) - \int_s^{\sigma(s)} \frac{\hat{f}^{\Delta_t}(t, \sigma(s))}{e_z(\sigma(t), \sigma(s))} \Delta t$$

$$- \int_{\sigma(s)}^\infty \frac{\hat{f}^{\Delta_t}(t, \sigma(s))}{e_z(\sigma(t), \sigma(s))} \Delta t - \hat{f}(s, \sigma(s))$$

$$= z\psi(s) - \mu(s)\frac{\hat{f}^{\Delta_t}(s, \sigma(s))}{e_z(\sigma(s), \sigma(s))}$$

$$- \int_{\sigma(s)}^\infty \frac{\hat{f}^{\Delta_t}(t, \sigma(s))}{e_z(\sigma(t), \sigma(s))} \Delta t - \hat{f}(s, \sigma(s))$$

$$= z\psi(s) - \hat{f}(\sigma(s), \sigma(s)) + \hat{f}(s, \sigma(s))$$

$$- \int_{\sigma(s)}^\infty \hat{f}^{\Delta_t}(t, \sigma(s))e_{\ominus z}(\sigma(t), \sigma(s))\Delta t$$

$$- \hat{f}(s, \sigma(s))$$

$$= z\psi(s) - f(t_0) - \int_{\sigma(s)}^\infty \left(\hat{f}(t, \sigma(s))e_{\ominus z}(t, \sigma(s)) \right)^{\Delta_t} \Delta t$$

$$+ \int_{\sigma(s)}^\infty \hat{f}(t, \sigma(s))\, (\ominus z)\,(t)e_{\ominus z}(t, \sigma(s))\Delta t$$

$$= z\psi(s) - f(t_0) + \hat{f}(\sigma(s), \sigma(s))e_{\ominus z}(\sigma(s), \sigma(s))$$
$$-z\int_{\sigma(s)}^{\infty} \hat{f}(t, \sigma(s))\frac{1}{1 + \mu(t)z}e_{\ominus z}(t, \sigma(s))\Delta t$$
$$= z\psi(s) - f(t_0) + f(t_0)$$
$$-z\int_{\sigma(s)}^{\infty} \hat{f}(t, \sigma(s))e_{\ominus z}(\sigma(t), \sigma(s))\Delta t$$
$$= z\psi(s) - z\psi(\sigma(s))$$
$$= -z\mu(s)\psi^{\Delta}(s),$$

whereupon

$$(1 + z\mu(s))\,\psi^{\Delta}(s) = 0,$$

and therefore,

$$\psi^{\Delta}(s) = 0, \quad s \in \mathbb{T}.$$

This completes the proof.

Corollary 3.3. *Let f and ψ be as in Theorem 3.11. Then*

$$\psi(t_0) = \mathscr{L}(f)(z), \quad z \in \mathscr{D}\{f\}.$$

Proof. We have

$$\psi(t_0) = \int_{t_0}^{\infty} \frac{\hat{f}(t, t_0)}{e_z(\sigma(t), t_0)}\Delta t$$
$$= \int_{t_0}^{\infty} f(t)e_{\ominus z}(\sigma(t), t_0)\Delta t$$
$$= \mathscr{L}(f)(z), \quad z \in \mathscr{D}\{f\}.$$

This completes the proof.

Theorem 3.12 (Convolution Theorem). *Suppose $f, g : \mathbb{T} \to \mathbb{C}$ are locally Δ-integrable functions on $\mathbb{T}$. Then*

$$\mathscr{L}(f \star g)(z) = \mathscr{L}(f)(z)\mathscr{L}(g)(z), \quad z \in \mathscr{D}\{f\}\bigcap\mathscr{D}\{g\}.$$

Proof. Let $z \in \mathscr{D}\{f\}\bigcap\mathscr{D}\{g\}$. Using Theorem 3.11 and Corollary 3.3, we get

$$\mathscr{L}(f \star g)(z) = \int_{t_0}^{\infty} \frac{(f \star g)(t)}{e_z(\sigma(t), t_0)} \Delta t$$

$$= \int_{t_0}^{\infty} \frac{1}{e_z(\sigma(t), t_0)} \int_{t_0}^{t} \hat{f}(t, \sigma(s))g(s)\Delta s \Delta t$$

$$= \int_{t_0}^{\infty} \frac{g(s)}{e_z(\sigma(s), t_0)} \left(\int_{\sigma(s)}^{\infty} \frac{\hat{f}(t, \sigma(s))}{e_z(\sigma(t), \sigma(s))} \Delta t \right) \Delta s$$

$$= \int_{t_0}^{\infty} \frac{g(s)}{e_z(\sigma(s), t_0)} \psi(\sigma(s)) \Delta s$$

$$= \mathscr{L}(f)(z) \int_{t_0}^{\infty} g(s) e_{\ominus z}(\sigma(s), t_0) \Delta s$$

$$= \mathscr{L}(f)(z)\mathscr{L}(g)(z).$$

This completes the proof.

Theorem 3.13. *Let $s, t_0 \in \mathbb{T}$, $s \geq t_0$, and*

$$u_s(t) = \begin{cases} 0 & \text{if } t < s, \\ 1 & \text{if } s \geq t. \end{cases}$$

Then

$$\mathscr{L}\left(u_s \hat{f}(\cdot, s)\right)(z) = e_{\ominus z}(s, t_0)\mathscr{L}(f)(z), \quad z \in \mathscr{D}\{f\}.$$

Proof. Let $z \in \mathscr{D}\{f\}$. Using Corollary 3.3, we get

$$\mathscr{L}\left(u_s \hat{f}(\cdot, s)\right)(z) = \int_{t_0}^{\infty} u_s(t)\hat{f}(t, s)e_{\ominus z}(\sigma(t), t_0)\Delta t$$

$$= \int_{s}^{\infty} \hat{f}(t, s)e_{\ominus z}(\sigma(t), t_0)\Delta t$$

$$= e_{\ominus z}(s, t_0) \int_{s}^{\infty} \hat{f}(t, s)e_{\ominus z}(\sigma(t), s)\Delta t$$

$$= e_{\ominus z}(s, t_0)\psi(s)$$

$$= e_{\ominus z}(s, t_0)\mathscr{L}(f)(z).$$

This completes the proof.

3.4   Advanced Practical Problems

Problem 3.1. Let $\alpha = -\frac{1}{5}$. Find $[\alpha]$.

Answer.

$$-\frac{1}{k + \sqrt[5]{q^4} + \sqrt[5]{q^3} + \sqrt[5]{q^2} + \sqrt[5]{q}}.$$

Problem 3.2. Let $n = 5$. Find $[5]!$.

Answer.

$$(q+1)\left(q^2 + q + 1\right)\left(q^3 + q^2 + q + 1\right)\left(q^4 + q^3 + q^2 + q + 1\right).$$

Problem 3.3. Let $m = 5$, $n = 3$. Find $\begin{bmatrix} m \\ n \end{bmatrix}$.

Answer.

$$\frac{\left(q^3 + q^2 + q + 1\right)\left(q^4 + q^3 + q^2 + q + 1\right)}{q + 1}.$$

Problem 3.4. Let $n = 3$, $q = 4$. Find $(t - s)_q^n$, $t, s \in \mathbb{T}$.

Answer.

$$(t - s)(t - 4s)((t - 16s), \quad t, s \in \mathbb{T}.$$

Problem 3.5. Let $\mathbb{T} = \mathbb{N}_0$, $t_0 = 0$,

$$f(t) = \frac{t+1}{t+2}, \quad t \in \mathbb{T}.$$

Find the shift of f.

Answer.

$$\hat{f}(t, s) = \frac{1}{2}h_0(t, s) + \frac{1}{6}h_1(t, s) - \frac{1}{12}h_2(t, s) + \cdots, \quad t, s \in \mathbb{T}, \quad t \geq s \geq 0.$$

Problem 3.6. Let $\mathbb{T} = 2^{\mathbb{N}_0}$, $t_0 = 1$,

$$f(t) = \sum_{k=0}^{\infty} \frac{k}{2k+1}h_k(t, 1), \quad g(t) = \sum_{k=0}^{\infty} \frac{k}{k+2}h_k(t, 1), \quad t \in \mathbb{T}, \quad t \geq 1.$$

Find $(f \star g)(t)$, $t \in \mathbb{T}$, $t \geq 1$.

Answer.

$$(f \star g)(t) = \sum_{k=0}^{\infty} \sum_{l=0}^{k} \left(\frac{k-l}{2(k-l)+1} \right) \left(\frac{l}{l+2} \right) h_{k+1}(t, 1), \quad t \in \mathbb{T}, \quad t \geq 1.$$

Problem 3.7. Let $\mathbb{T} = 2^{\mathbb{N}_0}, f(t) = t^7, g(t) = 1, t \in \mathbb{T}, t_0 = 1$. Find

$$(f \star g)^{\Delta}(t), \quad t \in \mathbb{T}, \quad t \geq 1.$$

Answer. $t^7, t \in \mathbb{T}, t \geq 1$.

Problem 3.8. Let $\mathbb{T} = 2^{\mathbb{N}_0}, t_0 = 1$,

$$f(t) = \frac{-4t^2 - 3t + 4}{2(t^2 + 2)(2t^2 + 1)}, \quad t \in \mathbb{T}, \quad t > 1.$$

Find

$$\int_{t_0}^{t} \hat{f}(t, \sigma(s)) \, \Delta s, \quad t \in \mathbb{T}, \quad t > t_0.$$

Answer.

$$-\frac{(t-1)^2}{t^2 + 2}, \quad t \in \mathbb{T}, \quad t > 1.$$

Problem 3.9. Let $\mathbb{T} = 4^{\mathbb{N}_0}, t_0 = 1$,

$$f(t) = t^4 + 4t^3 + 2t^2 + t - 1, \quad g(t) = t^5 + t^4 + t^3 - 3t^2, \quad t \in \mathbb{T}, \quad t > 1.$$

Find

$$(f \star g)^{\Delta^{40}}(t), \quad t \in \mathbb{T}, \quad t > 1.$$

Answer. 0.

Problem 3.10. Let $\mathbb{T} = 2^{\mathbb{N}_0}, t_0 = 1$,

$$f(t) = t^7 + 12t^5 + t^3 + 3t^2 + t + 15, \quad t \in \mathbb{T}.$$

Find

$$\hat{f}^{\Delta_t^{20}}(t, t), \quad t \in \mathbb{T}.$$

Answer. 0.

Chapter 4
The Riemann–Liouville Fractional Δ-Integral and the Riemann–Liouville Fractional Δ-Derivative on Time Scales

In this chapter we suppose that $\mathbb{T}$ is a time scale with forward jump operator and delta differentiation operator σ and Δ, respectively, such that

$$\mathbb{T} = \{t_n : n \in \mathbb{N}_0\},$$

$$\lim_{n \to \infty} t_n = \infty,$$

$$\sigma(t_n) = t_{n+1}, \quad n \in \mathbb{N}_0,$$

$$w = \inf_{n \in \mathbb{N}_0} \mu(t_n) > 0.$$

4.1 The Δ-Power Function

Suppose that $\alpha \in \mathbb{R}$.

Definition 4.1 (Δ-Power Function). We define the generalized Δ-power function $h_\alpha(t, t_0)$ on $\mathbb{T}$ as follows:

$$h_\alpha(t, t_0) = \mathscr{L}^{-1}\left(\frac{1}{z^{\alpha+1}}\right)(t), \quad t \geq t_0, \tag{4.1}$$

for all $z \in \mathbb{C} \backslash \{0\}$ such that $\mathscr{L}^{-1}$ exists, $t \geq t_0$. The fractional generalized Δ-power function $h_\alpha(t, s)$ on $\mathbb{T}$, $t \geq s \geq t_0$, is defined as the shift of $h_\alpha(t, t_0)$, i.e.,

$$h_\alpha(t, s) = \widehat{h_\alpha(\cdot, t_0)}(t, s), \quad t, s \in \mathbb{T}, \quad t \geq s \geq t_0. \tag{4.2}$$

Let

$$\gamma = \left\{ z \in \mathbb{C} : |z| = \frac{1}{w}, \quad \sqrt{1} = 1 \right\}.$$

We can rewrite (4.1) as follows:

$$h_\alpha(t_n, t_0) = \frac{1}{2\pi i} \int_\gamma \frac{1}{z^{\alpha+1}} \prod_{k=0}^{n-1} (1 + \mu(t_k)z) \, dz, \quad n \in \mathbb{N}.$$

We have

$$h_\alpha(t_0, t_0) = \lim_{z \to \infty} \frac{z}{z^{\alpha+1}}$$

$$= \lim_{z \to \infty} \frac{1}{z^\alpha}$$

$$= \begin{cases} 0 & \text{if} \quad \alpha > 0, \\ 1 & \text{if} \quad \alpha = 0, \\ \infty & \text{if} \quad \alpha < 0, \end{cases}$$

$$\lim_{t \to \infty} h_\alpha(t, t_0) = \lim_{z \to 0} \frac{z}{z^{\alpha+1}}$$

$$= \lim_{z \to 0} \frac{1}{z^\alpha}$$

$$= \begin{cases} \infty & \text{if} \quad \alpha > 0, \\ 1 & \text{if} \quad \alpha = 0, \\ 0 & \text{if} \quad \alpha < 0. \end{cases}$$

Theorem 4.1. *Let $\alpha, \beta \in \mathbb{R}$. Then*

$$\left(h_\alpha(\cdot, t_0) \star h_\beta(\cdot, t_0) \right)(t) = h_{\alpha+\beta+1}(t, t_0), \quad t \in \mathbb{T}. \tag{4.3}$$

Proof. We have

$$\mathscr{L}\left(\left(h_\alpha(\cdot, t_0) \star h_\beta(\cdot, t_0) \right)(t) \right)(z) = \mathscr{L}\left(h_\alpha(t, t_0) \right)(z) \mathscr{L}\left(h_\beta(t, t_0) \right)(z)$$

$$= \frac{1}{z^{\alpha+1}} \frac{1}{z^{\beta+1}}$$

$$= \frac{1}{z^{\alpha+\beta+2}}$$

$$= \mathscr{L}\left(h_{\alpha+\beta+1}(t, t_0) \right)(z).$$

This completes the proof.

By (4.3), we get

$$\int_{t_0}^{t} \widehat{h_\alpha(\cdot, t_0)}(t, \sigma(u)) h_\beta(u, t_0) \Delta u = h_{\alpha+\beta+1}(t, t_0), \quad t \in \mathbb{T}.$$

Hence, for $\alpha = 0$, we get

$$\int_{t_0}^{t} h_\beta(u, t_0) \Delta u = h_{\beta+1}(t, t_0), \quad t \in \mathbb{T},$$

and

$$h_{\beta+1}^{\Delta}(t, t_0) = h_\beta(t, t_0), \quad t \in \mathbb{T}.$$

4.2 Definition of the Riemann–Liouville Fractional Δ-Integral and the Riemann–Liouville Fractional Δ-Derivative

Suppose that $\alpha \geq 0$, and Ω is a finite interval on the time scale $\mathbb{T}$. By $\overline{[-\alpha]}$ we will denote the integer part of $-\alpha$.

Definition 4.2. For a function $f : \mathbb{T} \to \mathbb{R}$, the Riemann–Liouville fractional Δ-integral of order α is defined by

$$I_{\Delta,t_0}^{0} f(t) = f(t),$$

$$\left(I_{\Delta,t_0}^{\alpha} f\right)(t) = (h_{\alpha-1}(\cdot, t_0) \star f)(t)$$

$$= \int_{t_0}^{t} \widehat{h_{\alpha-1}(\cdot, t_0)}(t, \sigma(u)) f(u) \Delta u$$

$$= \int_{t_0}^{t} h_{\alpha-1}(t, \sigma(u)) f(u) \Delta u$$

for $\alpha > 0$, $t > t_0$.

Example 4.1. Let $\mathbb{T} = \mathbb{R}$. Then

$$h_{\alpha-1}(t, t_0) = \frac{(t - t_0)^{\alpha-1}}{\Gamma(\alpha)}, \quad t \in \mathbb{T}, \quad t > t_0,$$

and

$$\left(I_{t_0+}^{\alpha} f\right)(t) = \frac{1}{\Gamma(\alpha)} \int_{t_0}^{t} (t - \tau)^{\alpha-1} f(\tau) d\tau, \quad t \in \mathbb{T}, \quad t > t_0.$$

Example 4.2. Let $\mathbb{T} = \mathbb{Z}$. Define the factorial polynomial

$$t^{(\alpha)} = \frac{\Gamma(t+1)}{\Gamma(t+1-\alpha)}, \quad t \in \mathbb{T}, \quad \alpha \in \mathbb{R}.$$

The factorial polynomial has the following properties:

1. $\Delta t^{(\alpha)} = \alpha t^{(\alpha-1)}, t \in \mathbb{T}, \alpha \in \mathbb{R}$, where Δ is the forward difference operator.

   Proof. We have

$$\Delta t^{(\alpha)} = (t+1)^{(\alpha)} - t^{(\alpha)}$$

$$= \frac{\Gamma(t+2)}{\Gamma(t+2-\alpha)} - \frac{\Gamma(t+1)}{\Gamma(t+1-\alpha)}$$

$$= \frac{(t+1)\Gamma(t+1)}{(t+1-\alpha)\Gamma(t+1-\alpha)} - \frac{\Gamma(t+1)}{\Gamma(t+1-\alpha)}$$

$$= \frac{\Gamma(t+1)}{\Gamma(t+1-\alpha)}\left(\frac{t+1}{t+1-\alpha} - 1\right)$$

$$= \frac{\Gamma(t+1)}{(t+1-\alpha)\Gamma(t+1-\alpha)}(t+1-t-1+\alpha)$$

$$= \alpha \frac{\Gamma(t+1)}{\Gamma(t+2-\alpha)}$$

$$= \alpha \frac{\Gamma(t+1)}{\Gamma(t+1-(\alpha-1))}$$

$$= \alpha t^{(\alpha-1)}, \quad t \in \mathbb{T}, \quad \alpha \in \mathbb{R}.$$

   This completes the proof.

2. $(t-\alpha)t^{(\alpha)} = t^{(\alpha+1)}, t \in \mathbb{T}, \alpha \in \mathbb{R}$.

   Proof. We have

$$(t-\alpha)t^{(\alpha)} = (t-\alpha)\frac{\Gamma(t+1)}{\Gamma(t+1-\alpha)}$$

$$= (t-\alpha)\frac{\Gamma(t+1)}{(t-\alpha)\Gamma(t-\alpha)}$$

$$= \frac{\Gamma(t+1)}{\Gamma(t+1-(\alpha+1))}$$

$$= t^{(\alpha+1)}, \quad t \in \mathbb{T}, \quad \alpha \in \mathbb{R}.$$

   This completes the proof.

3. $\alpha^{(\alpha)} = \Gamma(\alpha + 1), \alpha \in \mathbb{R}$.

Proof. We have

$$\alpha^{(\alpha)} = \frac{\Gamma(\alpha + 1)}{\Gamma(\alpha + 1 - \alpha)}$$
$$= \Gamma(\alpha + 1), \quad \alpha \in \mathbb{R}.$$

This completes the proof.

4. $t^{(\alpha+\beta)} = (t - \beta)^{(\alpha)} t^{(\beta)}, t \in \mathbb{T}, \alpha, \beta \in \mathbb{R}$.

Proof. We have

$$t^{(\alpha+\beta)} = \frac{\Gamma(t + 1)}{\Gamma(t + 1 - \alpha - \beta)}$$
$$= \left(\frac{\Gamma(t - \beta + 1)}{\Gamma(t + 1 - \alpha - \beta)} \right) \left(\frac{\Gamma(t + 1)}{\Gamma(t - \beta + 1)} \right)$$
$$= (t - \beta)^{(\alpha)} t^{(\beta)}, \quad t \in \mathbb{T}, \quad \alpha, \beta \in \mathbb{R}.$$

This completes the proof.

Using the above properties, we conclude that

$$h_{\alpha-1}(t, s) = \frac{(t - s)^{(\alpha-1)}}{\Gamma(\alpha)}, \quad t \geq s \geq t_0 - \overline{[-\alpha]},$$

and then

$$\left(I^{\alpha}_{\Delta, t_0} f \right)(t) = \int_{t_0}^{t} h_{\alpha-1}(t, \sigma(s)) f(s) \Delta s$$
$$= \frac{1}{\Gamma(\alpha)} \int_{t_0}^{t} (t - \sigma(s))^{(\alpha-1)} f(s) \Delta s$$
$$= \frac{1}{\Gamma(\alpha)} \sum_{s=t_0}^{t-1} (t - s - 1)^{(\alpha-1)} f(s), \quad t \geq t_0 - \overline{[-\alpha]}.$$

Example 4.3. Let $\mathbb{T} = \mathbb{Z}, f(t) = t, t \in \mathbb{T}, t_0 = 0, \alpha = \frac{1}{3}$. Then

$$\overline{[-\alpha]} = \overline{\left[-\frac{1}{3} \right]}$$
$$= -1,$$

$$\left(I_{\Delta,0}^{\frac{1}{3}}f\right)(t) = \frac{1}{\Gamma\left(\frac{1}{3}\right)} \sum_{s=0}^{t-1} (t-s-1)^{\left(-\frac{2}{3}\right)}s$$

$$= \frac{1}{\Gamma\left(\frac{1}{3}\right)} \sum_{s=0}^{t-1} \frac{\Gamma(t-s)}{\Gamma\left(t-s+\frac{2}{3}\right)}s, \quad t \in \mathbb{Z}, \quad t \geq 1.$$

Example 4.4. Let $\mathbb{T} = q^{\mathbb{N}_0}$, $q > 1$. Let also $\alpha \in [0,\infty)\backslash\mathbb{N}_0$. We define the q-factorial function

$$(t-s)_q^\alpha = t^\alpha \prod_{n=0}^\infty \frac{1-\frac{s}{t}q^n}{1-\frac{s}{t}q^{\alpha+n}}$$

$$= t^\alpha \prod_{n=0}^\infty \frac{t-sq^n}{t-sq^{\alpha+n}}, \quad t,s \in \mathbb{T}, \quad t \geq s. \tag{4.4}$$

Note that if $t,s \in \mathbb{T}$, $t \geq s$, then there exists $l \in \mathbb{N}_0$ such that $\frac{t}{s} = q^l$. Hence, $\alpha + n \neq l$ for all $n \in \mathbb{N}_0$, and

$$q^l \neq q^{\alpha+n} \quad \text{for any} \quad n \in \mathbb{N}_0,$$

or

$$\frac{t}{s} \neq q^{\alpha+n} \quad \text{for any} \quad n \in \mathbb{N}_0,$$

or

$$t - sq^{\alpha+n} \neq 0 \quad \text{for any} \quad n \in \mathbb{N}_0.$$

Consequently, (4.4) is well defined. We will give some properties of the q-factorial function.

1. Let $\alpha, \beta \in [0,\infty)\backslash\mathbb{N}_0$. Then

$$(t-s)_q^{\alpha+\beta} = (t-s)_q^\alpha \left(t - q^\alpha s\right)_q^\beta.$$

Proof. We have

$$(t-s)_q^{\alpha+\beta} = t^{\alpha+\beta} \prod_{n=0}^\infty \frac{t-sq^n}{t-sq^{\alpha+\beta+n}}$$

$$= t^{\alpha+\beta} \prod_{n=0}^\infty \frac{t-sq^n}{t-sq^{\alpha+n}} \frac{t-sq^{\alpha+n}}{t-sq^{\alpha+\beta+n}}$$

$$= t^{\alpha+\beta} \left(\prod_{n=0}^{\infty} \frac{t - sq^n}{t - sq^{\alpha+n}} \right) \left(\prod_{n=0}^{\infty} \frac{t - sq^{\alpha+n}}{t - sq^{\alpha+\beta+n}} \right)$$

$$= (t-s)_q^{\alpha} \left(t - q^{\alpha} s \right)_q^{\beta}, \quad t, s \in \mathbb{T}, \quad t \geq s \geq t_0.$$

This completes the proof.

2. Let $\alpha \in [0, \infty) \backslash \mathbb{N}_0$, $a \in \mathbb{R}$. Then

$$(at - as)_q^{\alpha} = a^{\alpha}(t-s)_q^{\alpha}, \quad t, s \in \mathbb{T}, \quad t \geq s \geq t_0.$$

Proof. We have

$$(at - as)_q^{\alpha} = (at)^{\alpha} \prod_{n=0}^{\infty} \frac{at - asq^n}{at - asq^{\alpha+n}}$$

$$= a^{\alpha} t^{\alpha} \prod_{n=0}^{\infty} \frac{t - sq^n}{t - sq^{\alpha+n}}$$

$$= a^{\alpha}(t-s)_q^{\alpha}, \quad t, s \in \mathbb{T}, \quad t \geq s \geq t_0.$$

This completes the proof.

3. Let $\alpha \in [0, \infty) \backslash \mathbb{N}_0$. Then

$$\left((t-s)_q^{\alpha} \right)^{\Delta_t} = \frac{q^{\alpha} - 1}{q - 1}(t-s)_q^{\alpha-1}, \quad t, s \in \mathbb{T}, \quad t \geq s \geq 1. \tag{4.5}$$

Proof. We have

$$\left((t-s)_q^{\alpha} \right)^{\Delta_t} = \frac{1}{\sigma(t) - t} \left((\sigma(t) - s)_q^{\alpha} - (t-s)_q^{\alpha} \right)$$

$$= \frac{1}{qt - t} \left((qt - s)_q^{\alpha} - (t-s)_q^{\alpha} \right)$$

$$= \frac{1}{(q-1)t} \left(q^{\alpha} t^{\alpha} \prod_{n=0}^{\infty} \frac{1 - \frac{s}{qt}q^n}{1 - \frac{s}{qt}q^{\alpha+n}} - t^{\alpha} \prod_{n=0}^{\infty} \frac{1 - \frac{s}{t}q^n}{1 - \frac{s}{t}q^{\alpha+n}} \right)$$

$$= \frac{1}{(q-1)t} \left(q^{\alpha} t^{\alpha} \prod_{n=0}^{\infty} \frac{1 - \frac{s}{t}q^{n-1}}{1 - \frac{s}{t}q^{\alpha+n-1}} - t^{\alpha} \prod_{n=0}^{\infty} \frac{1 - \frac{s}{t}q^n}{1 - \frac{s}{t}q^{\alpha+n}} \right)$$

$$= \frac{1}{(q-1)t} \left(q^{\alpha} t^{\alpha} \frac{1 - \frac{s}{qt}}{1 - \frac{sq^{\alpha}}{tq}} \prod_{n=1}^{\infty} \frac{1 - \frac{s}{t}q^{n-1}}{1 - \frac{s}{t}q^{\alpha+n-1}} - t^{\alpha} \prod_{n=0}^{\infty} \frac{1 - \frac{s}{t}q^n}{1 - \frac{s}{t}q^{\alpha+n}} \right)$$

$$= \frac{1}{(q-1)t}\left(q^\alpha t^\alpha \frac{1-\frac{s}{tq}}{1-\frac{sq^\alpha}{tq}} \prod_{n=0}^{\infty} \frac{1-\frac{s}{t}q^n}{1-\frac{s}{t}q^{\alpha+n}} - t^\alpha \prod_{n=0}^{\infty} \frac{1-\frac{s}{t}q^n}{1-\frac{s}{t}q^{\alpha+n}} \right)$$

$$= \frac{1}{(q-1)t} t^\alpha \left(\prod_{n=0}^{\infty} \frac{1-\frac{s}{t}q^n}{1-\frac{s}{t}q^{\alpha+n}} \right) \left(\frac{q^\alpha - \frac{s}{t}q^{\alpha-1}}{1-\frac{s}{t}q^{\alpha-1}} - 1 \right)$$

$$= \frac{1}{q-1} t^{\alpha-1} \left(\prod_{n=0}^{\infty} \frac{1-\frac{s}{t}q^n}{1-\frac{s}{t}q^{\alpha-1+n+1}} \right) \frac{q^\alpha - \frac{s}{t}q^{\alpha-1} - 1 + \frac{s}{t}q^{\alpha-1}}{1-\frac{s}{t}q^{\alpha-1}}$$

$$= \frac{1}{q-1} t^{\alpha-1} \left(\prod_{n=1}^{\infty} \frac{1-\frac{s}{t}q^{n-1}}{1-\frac{s}{t}q^{\alpha-1+n}} \right) \frac{(q^\alpha - 1)}{1-\frac{s}{t}q^{\alpha-1}}$$

$$= \frac{q^\alpha - 1}{q-1} t^{\alpha-1} \prod_{n=0}^{\infty} \frac{1-\frac{s}{t}q^n}{1-\frac{s}{t}q^{\alpha-1+n}}$$

$$= \frac{q^\alpha - 1}{q-1}(t-s)_q^{\alpha-1}, \quad t,s \in \mathbb{T}, \quad t \geq s \geq 1.$$

This completes the proof.

4. Let $\alpha \in [0,\infty)\backslash\mathbb{N}_0$. Then

$$\left((t-s)_q^\alpha\right)^{\Delta_s} = -\frac{q^\alpha - 1}{q-1}(t-\sigma(s))_q^{\alpha-1}, \quad t,s \in \mathbb{T}, \quad t \geq s \geq 1. \qquad (4.6)$$

Proof. We have

$$\left((t-s)_q^\alpha\right)^{\Delta_s} = \frac{1}{\sigma(s)-s}\left((t-\sigma(s))_q^\alpha - (t-s)_q^\alpha\right)$$

$$= \frac{1}{(q-1)s}\left(t^\alpha \prod_{n=0}^{\infty} \frac{1-\frac{\sigma(s)}{t}q^n}{1-\frac{\sigma(s)}{t}q^{\alpha+n}} - t^\alpha \prod_{n=0}^{\infty} \frac{1-\frac{s}{t}q^n}{1-\frac{s}{t}q^{\alpha+n}} \right)$$

$$= \frac{1}{(q-1)s}\left(t^\alpha \prod_{n=0}^{\infty} \frac{1-\frac{s}{t}q^{n+1}}{1-\frac{s}{t}q^{\alpha+n+1}} - t^\alpha \prod_{n=0}^{\infty} \frac{1-\frac{s}{t}q^n}{1-\frac{s}{t}q^{\alpha+n}} \right)$$

$$= \frac{1}{(q-1)s}\left(t^\alpha \prod_{n=1}^{\infty} \frac{1-\frac{s}{t}q^n}{1-\frac{s}{t}q^{\alpha+n}} - t^\alpha \prod_{n=0}^{\infty} \frac{1-\frac{s}{t}q^n}{1-\frac{s}{t}q^{\alpha+n}} \right)$$

$$= \frac{1}{(q-1)s} t^\alpha \prod_{n=1}^{\infty} \frac{1-\frac{s}{t}q^n}{1-\frac{s}{t}q^{\alpha+n}} \left(1 - \frac{1-\frac{s}{t}}{1-\frac{s}{t}q^\alpha} \right)$$

$$= \frac{1}{(q-1)s} t^\alpha \prod_{n=1}^{\infty} \frac{1-\frac{s}{t}q^n}{1-\frac{s}{t}q^{\alpha+n}} \left(\frac{1-\frac{s}{t}q^\alpha - 1 + \frac{s}{t}}{1-\frac{s}{t}q^\alpha} \right)$$

$$= \frac{1}{(q-1)s}t^\alpha \prod_{n=1}^{\infty} \frac{1 - \frac{s}{t}q^n}{1 - \frac{s}{t}q^{\alpha+n}} \frac{(1-q^\alpha)\frac{s}{t}}{1 - \frac{s}{t}q^\alpha}$$

$$= -\frac{q^\alpha - 1}{q - 1}t^{\alpha-1} \prod_{n=1}^{\infty} \frac{1 - \frac{s}{t}qq^{n-1}}{1 - \frac{s}{t}qq^{n-1+\alpha}} \frac{1}{1 - \frac{s}{t}q^\alpha}$$

$$= -\frac{q^\alpha - 1}{q - 1}t^{\alpha-1} \prod_{n=1}^{\infty} \frac{1 - \frac{\sigma(s)}{t}q^{n-1}}{1 - \frac{\sigma(s)}{t}q^{\alpha-1+n}} \frac{1}{1 - \frac{s}{t}qq^{\alpha-1}}$$

$$= -\frac{q^\alpha - 1}{q - 1}t^{\alpha-1} \prod_{n=1}^{\infty} \frac{1 - \frac{\sigma(s)}{t}q^{n-1}}{1 - \frac{\sigma(s)}{t}q^{\alpha-1+n}} \frac{1}{1 - \frac{\sigma(s)}{t}q^{\alpha-1}}$$

$$= -\frac{q^\alpha - 1}{q - 1}t^{\alpha-1} \prod_{n=0}^{\infty} \frac{1 - \frac{\sigma(s)}{t}q^n}{1 - \frac{\sigma(s)}{t}q^{\alpha-1+n}}$$

$$= -\frac{q^\alpha - 1}{q - 1}(t - \sigma(s))_q^{\alpha-1}, \quad t, s \in \mathbb{T}, \quad t \geq s \geq 1.$$

This completes the proof.

Define the q-gamma function $\Gamma_q : \mathbb{R}\backslash\mathbb{Z} \to \mathbb{R}$ as follows:

$$\Gamma_q\left(\frac{1}{2}\right) = 1, \quad \Gamma_q(\alpha)\frac{q^\alpha - 1}{q - 1} = \Gamma_q(\alpha - 1), \quad \alpha \in \mathbb{R}\backslash\mathbb{Z}.$$

Let

$$h_\alpha(t, s) = \Gamma_q(\alpha)(t - s)_q^\alpha, \quad t, s \in \mathbb{T}, \quad t \geq s \geq t_0.$$

By (4.5), (4.6), it follows that

$$h_\alpha^{\Delta t}(t, s) = h_{\alpha-1}(t, s),$$

$$h_\alpha^{\Delta t}(t, \sigma(s)) = -h_\alpha^{\Delta s}(t, s), \quad t, s \in \mathbb{T}, \quad t \geq s \geq 1,$$

for all $\alpha \in [0, \infty)\backslash\mathbb{N}_0$. Then

$$\left(I_{\Delta,j}^\alpha f\right)(t) = \int_1^t h_{\alpha-1}(t, \sigma(s))f(s)\Delta s$$

$$= \Gamma_q(\alpha - 1)\int_1^t (t - qs)_q^{\alpha-1}f(s)\Delta s, \quad t, s \in \mathbb{T}, \quad t \geq s \geq 1.$$

If $t = q^l$ for some $l \in \mathbb{N}$, then

$$\left(I_{\Delta,f}^\alpha\right)\left(q^l\right) = \Gamma_q(\alpha-1)q^{(\alpha-1)l}\sum_{r=0}^{l-1}\mu\left(q^r\right)\prod_{n=0}^{\infty}\frac{q^l - q^r q^{n+1}}{q^l - q^r q^{\alpha+n}}f\left(q^r\right)$$

$$= \Gamma_q(\alpha-1)q^{(\alpha-1)l}\sum_{r=0}^{l-1}(q-1)q^r\prod_{n=0}^{\infty}\frac{q^l - q^{r+n+1}}{q^l - q^{\alpha+r+n}}f\left(q^r\right).$$

Suppose that $\alpha \in [0,\infty)\backslash\mathbb{N}_0$. Let $s \in \{0,\ldots,n-1\}$ and

$$a_{n-s}(t_n,t_s) = 1,$$
$$a_{n-1-s}(t_n,t_s) = \frac{1}{\mu(t_s)} + \cdots + \frac{1}{\mu(t_{n-1})},$$
$$a_{n-2-s}(t_n,t_s) = \frac{1}{\mu(t_s)\mu(t_{s+1})} + \cdots + \frac{1}{\mu(t_s)\mu(t_{n-1})} + \cdots + \frac{1}{\mu(t_{n-2})\mu(t_{n-1})},$$
$$\vdots$$
$$a_0(t_n,t_s) = \frac{1}{\mu(t_s)\ldots\mu(t_{n-1})}.$$

$$(4.7)$$

Then

$$\prod_{k=s}^{n-1}(1+\mu(t_k)z) = \prod_{k=s}^{n-1}\mu(t_k)\left(z+\frac{1}{\mu(t_k)}\right)$$

$$= \frac{1}{a_0(t_n,t_s)}\prod_{k=s}^{n-1}\left(z+\frac{1}{\mu(t_k)}\right)$$

$$= \frac{1}{a_0(t_n,t_s)}\sum_{k=s}^{n}a_{n-k}(t_n,t_s)z^{n-k}$$

and

$$\mathscr{L}^{-1}\left(\frac{1}{z^{\alpha+1}}\right) = \frac{1}{2\pi i}\int_\gamma\frac{1}{z^{\alpha+1}}\frac{1}{a_0(t_n,t_s)}\sum_{k=s}^{n}a_{n-k}(t_n,t_s)z^{n-k}dz$$

$$= \frac{1}{2\pi i}\frac{1}{a_0(t_n,t_s)}\sum_{k=s}^{n}a_{n-k}(t_n,t_s)\int_\gamma z^{n-k-\alpha-1}dz$$

$$= \frac{1}{2\pi i}\frac{1}{a_0(t_n,t_s)}\sum_{k=s}^{n}a_{n-k}(t_n,t_s)\left(\frac{1}{w}\right)^{n-k-\alpha}\int_0^{2\pi}e^{i(n-k-\alpha)\theta}d(i\theta)$$

$$= \frac{1}{2\pi i}\frac{1}{a_0(t_n,t_s)}\sum_{k=s}^{n}a_{n-k}(t_n,t_s)\left(\frac{1}{w}\right)^{n-k-\alpha}\frac{1}{n-k-\alpha}\left(e^{2\pi(n-k-\alpha)i}-1\right),$$

i.e.,

$$h_\alpha(t_n, t_s) = \frac{1}{2\pi i} \frac{1}{a_0(t_n, t_s)} \sum_{k=s}^{n} a_{n-k}(t_n, t_s) \left(\frac{1}{w}\right)^{n-k-\alpha} \frac{1}{n-k-\alpha} \left(e^{2\pi(n-k-\alpha)i} - 1\right).$$

Therefore,

$$\left(I_{\Delta,t_0}^\alpha f\right)(t_n) = \int_{t_0}^{t_n} h_{\alpha-1}(t_n, \sigma(u)) f(u) \Delta u$$

$$= \int_{t_0}^{t_n} h_{\alpha-1}(t_n, \sigma(t_s)) f(t_s) \Delta t_s$$

$$= \int_{t_0}^{t_n} h_{\alpha-1}(t_n, t_{s+1}) f(t_s) \Delta t_s$$

$$= \sum_{r=0}^{n-1} \mu(t_r) h_{\alpha-1}(t_n, t_{r+1}) f(t_r)$$

$$= \sum_{r=0}^{n-2} \mu(t_r) f(t_r) \frac{1}{2\pi i} \frac{1}{a_0(t_n, t_{r+1})} \sum_{k=r+1}^{n-1} a_{n-k}(t_n, t_{r+1}) \left(\frac{1}{w}\right)^{n-k-\alpha}$$

$$\times \frac{1}{n-k-\alpha} \left(e^{2\pi(n-k-\alpha)i} - 1\right).$$

For a function $f : \mathbb{T} \to \mathbb{R}$ we will define

$$f^{\Delta^n} = D_\Delta^n f, \quad n \in \mathbb{N}_0.$$

Definition 4.3. Let $\alpha \geq 0$, $m = -\overline{[-\alpha]}$, $f : \mathbb{T} \to \mathbb{R}$. For $s, t \in \mathbb{T}^{\kappa^m}$, $s < t$, the Riemann–Liouville fractional Δ-derivative of order α is defined by the expression

$$D_{\Delta,s}^\alpha f(t) = D_\Delta^m I_{\Delta,s}^{m-\alpha} f(t), \quad t \in \mathbb{T},$$

if it exists. For $\alpha < 0$, we define

$$D_{\Delta,s}^\alpha f(t) = I_{\Delta,s}^{-\alpha} f(t), \quad t, s \in \mathbb{T}, \quad t > s,$$

$$I_{\Delta,s}^\alpha f(t) = D_{\Delta,s}^{-\alpha} f(t), \quad t, s \in \mathbb{T}^{\kappa^r}, \quad t > s, \quad r = \overline{[-\alpha]} + 1.$$

Suppose that $\alpha \in [0, \infty) \backslash \mathbb{N}_0$, $m = -\overline{[-\alpha]}$. Then, using the definition, we have

$$D_{\Delta,s}^\alpha f(t) = D_\Delta^m I_{\Delta,s}^{m-\alpha} f(t)$$

$$= D_\Delta^m \left(\int_{t_0}^{t} h_{m-\alpha-1}(t, \sigma(u)) f(u) \Delta u\right), \quad t \in \mathbb{T}, \quad t \geq t_0.$$

If $m = 0$, then

$$D^\alpha_{\Delta,s}f(t) = \int_{t_0}^t h_{-\alpha-1}(t, \sigma(u))f(u)\Delta u, \quad t \in \mathbb{T}, \quad t \geq t_0.$$

If $m = 1$, then

$$D^\alpha_{\Delta,s}f(t) = D_\Delta \left(\int_{t_0}^t h_{-\alpha}(t, \sigma(u))f(u)\Delta u \right)$$

$$= \int_{t_0}^t h^{\Delta_t}_{-\alpha}(t, \sigma(u))f(u)\Delta u + h_{-\alpha}(\sigma(t), \sigma(t))f(t)$$

$$= \int_{t_0}^t h_{-\alpha-1}(t, \sigma(u))f(u)\Delta u + h_{-\alpha}(\sigma(t), \sigma(t))f(t), \quad t \in \mathbb{T}, \quad t \geq t_0.$$

Example 4.5. Let $\mathbb{T} = \mathbb{Z}$, $t_0 \in \mathbb{T}$, $\alpha \in [0, \infty)\backslash\mathbb{N}_0$. Using Example 4.2, we get

$$I^{m-\alpha}_{\Delta,t_0}f(t) = \frac{1}{\Gamma(\alpha)} \sum_{s=t_0}^{t-1} (t-s-1)^{(m-\alpha-1)}f(s)$$

$$= \frac{1}{\Gamma(\alpha)} \sum_{s=t_0}^{t-1} \frac{\Gamma(t-s)}{\Gamma(t-s-1+1-m+\alpha+1)}f(s)$$

$$= \frac{1}{\Gamma(\alpha)} \sum_{s=t_0}^{t-1} \frac{\Gamma(t-s)}{\Gamma(t-s-m+\alpha+1)}f(s), \quad t \in \mathbb{T}, \quad t \geq t_0.$$

Then

$$D^\alpha_{\Delta,s}f(t) = D^m_\Delta \left(I^{m-\alpha}_{\Delta,t_0}f(t) \right)$$

$$= D^m_\Delta \left(\frac{1}{\Gamma(\alpha)} \sum_{s=t_0}^{t-1} \frac{\Gamma(t-s)}{\Gamma(t-s-m+\alpha+1)}f(s) \right)$$

$$= \sum_{k=0}^m \binom{m}{k} (-1)^k \left(\frac{1}{\Gamma(\alpha)} \sum_{s=t_0}^{t+m-k-1} \frac{\Gamma(t+m-k-s)}{\Gamma(t+m-k-s-m+\alpha+1)}f(s) \right)$$

$$= \sum_{k=0}^m \binom{m}{k} (-1)^k \left(\frac{1}{\Gamma(\alpha)} \sum_{s=t_0}^{t+m-k-1} \frac{\Gamma(t+m-k-s)}{\Gamma(t-s-k+\alpha+1)}f(s) \right),$$

$$t \in \mathbb{T}, \quad t > t_0.$$

Definition 4.4. The Δ-Mittag-Leffler function is defined by

$$\Delta F_{\alpha,\beta}(\lambda, t, t_0) = \sum_{j=0}^{\infty} \lambda^j h_{j\alpha+\beta-1}(t, t_0), \tag{4.8}$$

provided that the right-hand side is convergent, where $\alpha, \beta > 0, \lambda \in \mathbb{R}$.

Remark 4.1. When $0 < \lambda < 1, \alpha > 0, \beta > 0$, and

$$\left| h_{j\alpha+\beta-1}(t, t_0) \right| < M, \quad t \in \mathbb{T}, \quad t > t_0,$$

for all $j \in \mathbb{N}_0$, then the series (4.8) is convergent.

Remark 4.2. When $\mathbb{T} = \mathbb{R}, \alpha > 0, \beta > 0$, then

$$\Delta F_{\alpha,\beta}(\lambda, t, t_0) = \sum_{j=0}^{\infty} \lambda^j \frac{(t - t_0)^{j\alpha+\beta-1}}{\Gamma(j\alpha + \beta)},$$

which is a convergent series for all $t \geq t_0$. That is, $\Delta F_{\alpha,\beta}(\lambda, t, t_0)$ is defined as $t \geq t_0$.

Example 4.6. Let $\mathbb{T} = \mathbb{Z}, \alpha > 0, \beta > 0$. Then

$$
\begin{aligned}
h_{j\alpha+\beta-1}(t, t_0) &= \frac{(t - t_0)^{(j\alpha+\beta-1)}}{\Gamma(j\alpha + \beta)} \\[2mm]
&= \frac{\Gamma(t - t_0 + 1)}{\Gamma(t - t_0 + 1 - j\alpha - \beta + 1)\Gamma(j\alpha + \beta)} \\[2mm]
&= \frac{\Gamma(t - t_0 + 1)}{\Gamma(t - t_0 - j\alpha - \beta + 2)\Gamma(j\alpha + \beta)}, \quad j \in \mathbb{N}_0.
\end{aligned}
$$

Hence,

$$\Delta F_{\alpha,\beta}(\lambda, t, t_0) = \sum_{j=0}^{\infty} \lambda^j \frac{\Gamma(t - t_0 + 1)}{\Gamma(t - t_0 - j\alpha - \beta + 2)\Gamma(j\alpha + \beta)}.$$

Example 4.7. Let $\mathbb{T} = q^{\mathbb{N}_0}, q > 1, \alpha > 0, \beta > 0$. Then

$$h_{j\alpha+\beta-1}(t, t_0) = \Gamma_q(j\alpha + \beta - 1)t^{j\alpha+\beta-1}\prod_{n=0}^{\infty} \frac{t - t_0 q^n}{t - t_0 q^{j\alpha+\beta-1+n}}, \quad j \in \mathbb{N}_0,$$

and

$$\varDelta F_{\alpha,\beta}(\lambda, t, t_0) = \sum_{j=0}^{\infty} \lambda^j h_{j\alpha+\beta-1}(t, t_0)$$

$$= \sum_{j=0}^{\infty} \lambda^j \Gamma_q(j\alpha + \beta - 1) t^{j\alpha+\beta-1} \prod_{n=0}^{\infty} \frac{t - t_0 q^n}{t - t_0 q^{j\alpha+\beta-1+n}},$$

$$t, t_0 \in \mathbb{T}, \quad t > t_0.$$

Theorem 4.2. *We have*

$$\mathscr{L}\left(\varDelta F_{\alpha,\beta}(\lambda, t, t_0)\right)(z, t_0) = \frac{z^{\alpha-\beta}}{z^\alpha - \lambda}, \quad |\lambda| < |z|^\alpha, \tag{4.9}$$

$\alpha, \beta > 0.$

Proof. By the definition of the Laplace transform, we get

$$\mathscr{L}\left(\varDelta F_{\alpha,\beta}(\lambda, t, t_0)\right)(z, t_0) = \int_{t_0}^{\infty} \varDelta F_{\alpha,\beta}(\lambda, t, t_0) e_{\ominus z}(\sigma(t), t_0) \Delta t$$

$$= \int_{t_0}^{\infty} \sum_{j=0}^{\infty} \lambda^j h_{j\alpha+\beta-1}(t, t_0) e_{\ominus z}(\sigma(t), t_0) \Delta t$$

$$= \sum_{j=0}^{\infty} \lambda^j \int_{t_0}^{\infty} h_{j\alpha+\beta-1}(t, t_0) e_{\ominus z}(\sigma(t), t_0) \Delta t$$

$$= \sum_{j=0}^{\infty} \lambda^j \mathscr{L}\left(h_{j\alpha+\beta-1}(t, t_0)\right)(z, t_0)$$

$$= \sum_{j=0}^{\infty} \lambda^j \frac{1}{z^{j\alpha+\beta}}$$

$$= z^{-\beta} \sum_{j=0}^{\infty} \left(\lambda z^{-\alpha}\right)^j$$

$$= \frac{z^{-\beta}}{1 - \lambda z^{-\alpha}}$$

$$= \frac{z^{\alpha-\beta}}{z^\alpha - \lambda}, \quad |\lambda| < |z|^\alpha.$$

This completes the proof.

Suppose that all conditions of Theorem 4.2 hold. We differentiate the equality (4.9) k times, $k \in \mathbb{N}$, with respect to λ, and we get

$$\mathscr{L}\left(\frac{\partial^k}{\partial \lambda^k}\Delta F_{\alpha,\beta}(\lambda, t, t_0)\right)(z, t_0) = \frac{k!z^{\alpha-\beta}}{(z^\alpha - \lambda)^{k+1}}, \quad |\lambda| < |z|^\alpha.$$

4.3 Properties of the Riemann–Liouville Fractional Δ-Integral and the Riemann–Liouville Fractional Δ-Derivative on Time Scales

Theorem 4.3. *Let* $\alpha > 0$, $m = -\overline{[-\alpha]}$, $\beta \in \mathbb{R}$. *Then*

$$I^\alpha_{\Delta,t_0}h_{\beta-1}(t, t_0) = h_{\beta+\alpha-1}(t, t_0), \quad t \in \mathbb{T}, \quad t \geq t_0.$$

Proof. Using Theorem 4.1, we derive

$$\left(I^\alpha_{\Delta,t_0}h_{\beta-1}\right)(t, t_0) = \int_{t_0}^t h_{\alpha-1}(t, \sigma(u))h_{\beta-1}(u, t_0)\Delta u$$

$$= \left(h_{\alpha-1}(\cdot, t_0) \star h_{\beta-1}(\cdot, t_0)\right)(t)$$

$$= h_{\alpha+\beta-1}(t, t_0), \quad t \in \mathbb{T}, \quad t \geq t_0.$$

This completes the proof.

Example 4.8. Let $\alpha = \beta = \frac{1}{2}$. Then

$$I^{\frac{1}{2}}_{\Delta,t_0}h_{-\frac{1}{2}}(t, t_0) = h_{\frac{1}{2}+\frac{1}{2}-1}(t, t_0)$$

$$= h_0(t, t_0)$$

$$= 1, \quad t \in \mathbb{T}, \quad t > t_0.$$

Example 4.9. Let $\alpha = \frac{1}{3}$, $\beta = \frac{5}{3}$. Then

$$I^{\frac{1}{3}}_{\Delta,t_0}h_{\frac{5}{3}-1}(t, t_0) = I^{\frac{1}{3}}_{\Delta,t_0}h_{\frac{2}{3}}(t, t_0)$$

$$= h_{\frac{1}{3}+\frac{5}{3}-1}(t, t_0)$$

$$= h_1(t, t_0)$$

$$= t - t_0, \quad t \in \mathbb{T}, \quad t > t_0.$$

Example 4.10. Let $\mathbb{T} = 2^{\mathbb{N}_0}$, $t_0 = 1$, $\alpha = \frac{4}{3}$, $\beta = \frac{5}{3}$. Then

$$I_{\Delta,1}^{\frac{4}{3}}h_{\frac{2}{3}}(t, 1) = h_{\frac{4}{3}+\frac{5}{3}-1}(t, 1)$$

$$= h_2(t, 1)$$

$$= \frac{(t - 1)_{\frac{2}{2}}}{[2]!}$$

$$= \frac{(t - 1)(t - 2)}{\frac{2^2-1}{2-1}}$$

$$= \frac{(t - 1)(t - 2)}{3}, \quad t \in \mathbb{T}, \quad t > 1.$$

Exercise 4.1. Let $\mathbb{T} = \mathbb{Z}$, $t_0 = 0$, $\alpha = \frac{1}{2}$, $\beta = \frac{5}{2}$. Find

$$I_{\Delta,0}^{\alpha}h_{\beta-1}(t, 0), \quad t \in \mathbb{T}, \quad t > 0.$$

Answer. $\frac{t(t-1)}{2}, t \in \mathbb{T}, t > 0.$

Theorem 4.4. *Let* $\alpha > 0$, $m = -\overline{[-\alpha]}$, $\beta \in \mathbb{R}$. *Then*

$$D_{\Delta,t_0}^{\alpha}h_{\beta-1}(t, t_0) = h_{\beta-\alpha-1}(t, t_0), \quad t \in \mathbb{T}, \quad t \geq t_0.$$

Proof. Using Theorem 4.3, we derive

$$D_{\Delta,t_0}^{\alpha}h_{\beta-1}(t, t_0) = D_{\Delta}^{m}I_{\Delta,t_0}^{m-\alpha}h_{\beta-1}(t, t_0)$$

$$= D_{\Delta}^{m}h_{m-\alpha+\beta-1}(t, t_0)$$

$$= h_{\beta-\alpha-1}(t, t_0), \quad t \in \mathbb{T}, \quad t \geq t_0.$$

This completes the proof.

Example 4.11. Let $\alpha = \frac{1}{2}$, $\beta = \frac{3}{2}$. Then

$$D_{\Delta,t_0}^{\frac{1}{2}}h_{\frac{3}{2}-1}(t, t_0) = D_{\Delta,t_0}^{\frac{1}{2}}h_{\frac{1}{2}}(t, t_0)$$

$$= h_{\frac{3}{2}-\frac{1}{2}-1}(t, t_0)$$

$$= h_0(t, t_0)$$

$$= 1, \quad t \in \mathbb{T}, \quad t \geq t_0.$$

Example 4.12. Let $\alpha = \frac{1}{3}$, $\beta = \frac{7}{3}$. Then

$$D_{\Delta,t_0}^{\frac{1}{3}}h_{\frac{7}{3}-1}(t, t_0) = D_{\Delta,t_0}^{\frac{1}{3}}h_{\frac{4}{3}}(t, t_0)$$

$$= h_{\frac{7}{3}-\frac{1}{3}-1}(t, t_0)$$

$$= h_1(t, t_0)$$

$$= t - t_0, \quad t \in \mathbb{T}, \quad t > t_0.$$

Example 4.13. Let $\mathbb{T} = 4^{\mathbb{N}_0}$, $\alpha = \frac{1}{4}$, $\beta = \frac{13}{4}$. Then

$$D_{\Delta,1}^{\frac{1}{4}} h_{\frac{13}{4}-1}(t, 1) = D_{\Delta,1}^{\frac{1}{4}} h_{\frac{9}{4}}(t, 1)$$

$$= h_{\frac{13}{4}-\frac{1}{4}-1}(t, 1)$$

$$= h_2(t, 1)$$

$$= \frac{(t-1)(t-4)}{5}, \quad t \in \mathbb{T}, \quad t > 1.$$

Exercise 4.2. Let $\mathbb{T} = 2\mathbb{Z}$, $t_0 = 0$, $\alpha = \frac{1}{7}$, $\beta = \frac{22}{7}$. Find

$$D_{\Delta,0}^{\frac{1}{7}} h_{\frac{15}{7}}(t, 0), \quad t \in \mathbb{T}, \quad t > 0.$$

Answer. $\frac{1}{2}t^2 - t, t \in \mathbb{T}, t > 0.$

Example 4.14. Let $\alpha = \frac{1}{3}$, $\beta = \frac{2}{3}$. We will find

$$\mathscr{L}\left(D_{\Delta,t_0}^{\frac{1}{3}} h_{-\frac{1}{3}}(\cdot, t_0)\right)(z, t_0).$$

We have

$$D_{\Delta,t_0}^{\alpha} h_{\beta-1}(t, t_0) = D_{\Delta,t_0}^{\frac{1}{3}} h_{\frac{2}{3}-1}(t, t_0)$$

$$= D_{\Delta,t_0}^{\frac{1}{3}} h_{-\frac{1}{3}}(t, t_0)$$

$$= h_{\frac{2}{3}-\frac{1}{3}-1}(t, t_0)$$

$$= h_{-\frac{2}{3}}(t, t_0), \quad t \in \mathbb{T}, \quad t > t_0.$$

Hence,

$$\mathscr{L}\left(D_{\Delta,t_0}^{\frac{1}{3}} h_{-\frac{1}{3}}(\cdot, t_0)\right)(z, t_0) = \mathscr{L}\left(h_{-\frac{2}{3}}(\cdot, t_0)\right)(z, t_0)$$

$$= \frac{1}{z^{1-\frac{2}{3}}}$$

$$= \frac{1}{z^{\frac{1}{3}}}.$$

Example 4.15. Let $\alpha = \frac{1}{7}$, $\beta = \frac{2}{3}$. We will find

$$\mathscr{L}\left(I^{\frac{1}{7}}_{\Delta,t_0} h_{-\frac{1}{3}}(\cdot, t_0)\right)(z, t_0).$$

We have

$$I^{\alpha}_{\Delta,t_0} h_{\beta-1}(t, t_0) = I^{\frac{1}{7}}_{\Delta,t_0} h_{\frac{2}{3}-1}(t, t_0)$$

$$= I^{\frac{1}{7}}_{\Delta,t_0} h_{-\frac{1}{3}}(t, t_0)$$

$$= h_{\frac{1}{7}+\frac{2}{3}-1}(t, t_0)$$

$$= h_{-\frac{4}{21}}(t, t_0), \quad t \in \mathbb{T}, \quad t > t_0.$$

Therefore,

$$\mathscr{L}\left(I^{\alpha}_{\Delta,t_0} h_{\beta-1}(\cdot, t_0)\right)(z, t_0) = \mathscr{L}\left(h_{-\frac{4}{21}}(\cdot, t_0)\right)(z, t_0)$$

$$= \frac{1}{z^{-\frac{4}{21}+1}}$$

$$= \frac{1}{z^{\frac{17}{21}}}.$$

Example 4.16. Let $\alpha = \frac{1}{2}$, $\beta = \frac{3}{2}$. We will find

$$\mathscr{L}\left(I^{\frac{1}{2}}_{\Delta,t_0} h_{\frac{1}{2}}(\cdot, t_0) + h_4(\cdot, t_0) \star D^{\frac{1}{2}}_{\Delta,t_0} h_{\frac{1}{2}}(\cdot, t_0)\right)(z, t_0).$$

We have

$$I^{\alpha}_{\Delta,t_0} h_{\beta-1}(t, t_0) = I^{\frac{1}{2}}_{\Delta,t_0} h_{\frac{3}{2}-1}(t, t_0)$$

$$= I^{\frac{1}{2}}_{\Delta,t_0} h_{\frac{1}{2}}(t, t_0)$$

$$= h_{\frac{1}{2}+\frac{3}{2}-1}(t, t_0)$$

$$= h_1(t, t_0), \quad t \in \mathbb{T}, \quad t > t_0,$$

$$\mathscr{L}\left(I^{\frac{1}{2}}_{\Delta,t_0} h_{\frac{1}{2}}(\cdot, t_0)\right)(z, t_0) = \mathscr{L}\left(h_1(\cdot, t_0)\right)(z, t_0)$$

$$= \frac{1}{z^2},$$

$$D^{\alpha}_{\Delta,t_0} h_{\beta-1}(t, t_0) = D^{\frac{1}{2}}_{\Delta,t_0} h_{\frac{3}{2}-1}(t, t_0)$$

$$= D^{\frac{1}{2}}_{\Delta,t_0} h_{\frac{1}{2}}(t, t_0)$$

$$= h_{\frac{3}{2}-\frac{1}{2}-1}(t, t_0)$$

$$= h_0(t, t_0),$$

$$\left(h_4(\cdot, t_0) \star D^{\frac{1}{2}}_{\Delta,t_0} h_{\frac{1}{2}}(\cdot, t_0) \right) = (h_4(\cdot, t_0) \star h_0(\cdot, t_0))(t, t_0)$$

$$= h_{4+0+1}(t, t_0)$$

$$= h_5(t, t_0), \quad t \in \mathbb{T}, \quad t > t_0,$$

$$\mathscr{L}\left(h_4(\cdot, t_0) \star D^{\frac{1}{2}}_{\Delta,t_0} h_{\frac{1}{2}}(\cdot, t_0) \right)(z, t_0) = \mathscr{L}(h_5(\cdot, t_0))(z, t_0)$$

$$= \frac{1}{z^6}.$$

Consequently,

$$\mathscr{L}\left(I^{\frac{1}{2}}_{\Delta,t_0} h_{\frac{1}{2}}(\cdot, t_0) + h_4(\cdot, t_0) \star D^{\frac{1}{2}}_{\Delta,t_0} h_{\frac{1}{2}}(\cdot, t_0) \right)(z, t_0) = \mathscr{L}\left(I^{\frac{1}{2}}_{\Delta,t_0} h_{\frac{1}{2}}(\cdot, t_0) \right)(z, t_0)$$

$$+ \mathscr{L}\left(h_4(\cdot, t_0) \star D^{\frac{1}{2}}_{\Delta,t_0} h_{\frac{1}{2}}(\cdot, t_0) \right)(z, t_0)$$

$$= \frac{1}{z^2} + \frac{1}{z^6}$$

$$= \frac{z^4 + 1}{z^6}.$$

Exercise 4.3. Let $\alpha = \frac{1}{3}$, $\beta = \frac{5}{3}$. Find

$$\mathscr{L}\left(h_{\frac{1}{2}}(\cdot, t_0) \star I^{\frac{1}{3}}_{\Delta,t_0} h_{\frac{2}{3}}(\cdot, t_0) - 2h_2(\cdot, t_0) \star D^{\frac{1}{3}}_{\Delta,t_0} h_{\frac{2}{3}}(\cdot, t_0) \right)(t, t_0).$$

Answer.

$$\frac{1}{z^{\frac{7}{2}}} - \frac{2}{z^{\frac{13}{3}}}.$$

Corollary 4.1. *Let* $0 < \alpha < 1$. *Then*

$$D^{\alpha}_{\Delta,t_0} 1 = h_{-\alpha}(t, t_0), \quad t \in \mathbb{T}, \quad t \geq t_0. \tag{4.10}$$

Proof. We apply Theorem 4.4 with $\beta = 1$, and we get (4.10). This completes the proof.

Example 4.17. Let $\alpha = \frac{1}{3}$, $\beta = \frac{2}{3}$. We will find

$$f(z) = \mathscr{L}\left(h_1(\cdot, t_0) \star \left(D_{\Delta, t_0}^{\frac{1}{3}} \left(I_{\Delta, t_0}^{\frac{1}{3}} h_{-\frac{1}{3}} \right) \right)(\cdot, t_0) - 4h_2(\cdot, t_0) \star h_{\frac{1}{3}}(\cdot, t_0) \right)(z, t_0).$$

We have

$$I_{\Delta, t_0}^{\alpha} h_{\beta-1}(t, t_0) = I_{\Delta, t_0}^{\frac{1}{3}} h_{\frac{2}{3}-1}(t, t_0)$$

$$= I_{\Delta, t_0}^{\frac{1}{3}} h_{-\frac{1}{3}}(t, t_0)$$

$$= h_{\frac{1}{3}+\frac{2}{3}-1}(t, t_0)$$

$$= h_0(t, t_0)$$

$$= 1,$$

$$D_{\Delta, t_0}^{\frac{1}{3}} \left(I_{\Delta, t_0}^{\frac{1}{3}} h_{-\frac{1}{3}}(t, t_0) \right) = D_{\Delta, t_0}^{\frac{1}{3}} 1$$

$$= h_{-\frac{1}{3}}(t, t_0), \quad t \in \mathbb{T}, \quad t > t_0,$$

$$h_1(\cdot, t_0) \star \left(D_{\Delta, t_0}^{\frac{1}{3}} \left(I_{\Delta, t_0}^{\frac{1}{3}} h_{-\frac{1}{3}} \right) \right)(\cdot, t_0) = h_1(\cdot, t_0) \star h_{-\frac{1}{3}}(\cdot, t_0)$$

$$= h_{\frac{5}{3}}(\cdot, t_0),$$

$$\mathscr{L}\left(h_1(\cdot, t_0) \star \left(D_{\Delta, t_0}^{\frac{1}{3}} \left(I_{\Delta, t_0}^{\frac{1}{3}} h_{-\frac{1}{3}} \right) \right)(\cdot, t_0) \right)(z, t_0) = \mathscr{L}\left(h_{\frac{5}{3}}(\cdot, t_0) \right)(z, t_0)$$

$$= \frac{1}{z^{\frac{5}{3}+1}}$$

$$= \frac{1}{z^{\frac{8}{3}}},$$

$$h_2(\cdot, t_0) \star h_{\frac{1}{3}}(\cdot, t_0) = h_{2+\frac{1}{3}+1}(\cdot, t_0)$$

$$= h_{\frac{10}{3}}(\cdot, t_0),$$

$$\mathscr{L}\left(h_2(\cdot, t_0) \star h_{\frac{1}{3}}(\cdot, t_0) \right)(z, t_0) = \mathscr{L}\left(h_{\frac{10}{3}}(\cdot, t_0) \right)(z, t_0)$$

$$= \frac{1}{z^{\frac{10}{3}+1}}$$

$$= \frac{1}{z^{\frac{13}{3}}}.$$

Consequently,

$$f(z) = \mathscr{L}\left(h_1(\cdot, t_0) \star \left(D^{\frac{1}{3}}_{\Delta, t_0}\left(I^{\frac{1}{3}}_{\Delta, t_0} h_{-\frac{1}{3}}\right)\right)(\cdot, t_0)\right)(z, t_0)$$
$$-4\mathscr{L}\left(h_2(\cdot, t_0) \star h_{\frac{1}{3}}(\cdot, t_0)\right)(z, t_0)$$
$$= \frac{1}{z^{\frac{8}{3}}} - \frac{4}{z^{\frac{13}{3}}}.$$

Example 4.18. Let $\alpha \in (0, 1)$, $\beta = 1 - \alpha$. We will prove that

$$h_{\alpha-1}(\cdot, t_0) \star \left(D^\alpha_{\Delta, t_0}\left(I^\alpha_{\Delta, t_0} h_{\beta-1}\right)\right)(\cdot, t_0) = 1.$$

Indeed, we have

$$I^\alpha_{\Delta, t_0} h_{\beta-1}(t, t_0) = h_{\beta+\alpha-1}(t, t_0)$$
$$= h_{1-\alpha+\alpha-1}(t, t_0)$$
$$= h_0(t, t_0)$$
$$= 1,$$

$$D^\alpha_{\Delta, t_0}\left(I^\alpha_{\Delta, t_0} h_{\beta-1}(t, t_0)\right) = D^\alpha_{\Delta, t_0} 1$$
$$= h_{-\alpha}(t, t_0),$$

$$h_{\alpha-1}(\cdot, t_0) \star D^\alpha_{\Delta, t_0}\left(I^\alpha_{\Delta, t_0} h_{\beta-1}(t, t_0)\right)(\cdot, t_0) = h_{\alpha-1}(\cdot, t_0) \star h_{-\alpha}(\cdot, t_0)$$
$$= h_{\alpha+1-\alpha-1}(\cdot, t_0)$$
$$= h_0(\cdot, t_0)$$
$$= 1.$$

Corollary 4.2. *Let $\alpha > 0$, $m = -\overline{[-\alpha]}$. Then*

$$D^\alpha_{\Delta, t_0} h_{\alpha-j}(t, s) = 0, \quad t, s \in \mathbb{T}, \quad t \geq s \geq t_0,$$

for all $j \in \{1, \ldots, m\}$.

Proof. Let $j \in \{1, \ldots, m\}$ be arbitrarily chosen. We apply Theorem 4.3 with $\beta = \alpha - j$, and we get

$$D^\alpha_{\Delta, t_0} h_{\alpha-j}(t, s) = D^m_\Delta I^{m-\alpha}_{\Delta, t_0} h_{\alpha-j}(t, s)$$
$$= D^m_\Delta I^{m-\alpha}_{\Delta, t_0} h_{\alpha+1-j-1}(t, s)$$
$$= D^m_\Delta h_{m-j}(t, s)$$
$$= 0, \quad t, s \in \mathbb{T}, \quad t \geq s \geq t_0.$$

This completes the proof.

Corollary 4.3. *Let* $\mathbb{T} = \mathbb{R}$, $\alpha \geq 0$, $\beta > 0$, $t_0 \in \mathbb{R}$. *Then*

$$\left(I_{t_0+}^{\alpha}(t - t_0)^{\beta-1}\right)(x) = \frac{\Gamma(\beta)}{\Gamma(\beta + \alpha)}(x - t_0)^{\beta + \alpha - 1}, \quad \alpha > 0, \quad x \in \mathbb{R}, \quad x \geq t_0,$$

$$\left(D_{t_0+}^{\alpha}(t - t_0)^{\beta-1}\right)(x) = \frac{\Gamma(\beta)}{\Gamma(\beta - \alpha)}(x - t_0)^{\beta - \alpha - 1}, \quad \alpha \geq 0, \quad x \in \mathbb{R}, \quad x \geq t_0.$$

In particular, if $\beta = 1$ *and* $\alpha \geq 0$, *then the Riemann–Liouville fractional derivatives of a constant are, in general, not equal to zero,*

$$\left(D_{t_0+}^{\alpha} 1\right)(x) = \frac{(x - t_0)^{-\alpha}}{\Gamma(1 - \alpha)}, \quad 0 < \alpha < 1, \quad x \in \mathbb{T}, \quad x \geq t_0.$$

On the other hand, for $j \in \{1, \ldots, -\overline{[-\alpha]}\}$, *we have*

$$\left(D_{t_0^+}^{\alpha}(t - t_0)^{\alpha - j}\right)(x) = 0, \quad x \in \mathbb{T}, \quad x \geq t_0.$$

Proof. Let $\alpha > 0$. By Theorem 4.3, we get

$$\left(I_{t_0+}^{\alpha} h_{\beta-1}(t, t_0)\right)(x) = h_{\alpha+\beta-1}(x, t_0),$$

or

$$\left(I_{t_0+}^{\alpha} \frac{(t - t_0)^{\beta-1}}{\Gamma(\beta)}\right)(x) = \frac{(x - t_0)^{\alpha+\beta-1}}{\Gamma(\alpha + \beta)},$$

or

$$\left(I_{t_0+}^{\alpha}(t - t_0)^{\beta-1}\right)(x) = \frac{\Gamma(\beta)}{\Gamma(\alpha + \beta)}(x - t_0)^{\alpha+\beta-1}, \quad x \in \mathbb{R}, \quad x \geq t_0.$$

Let now $\alpha \geq 0$. Then, using Theorem 4.4, we obtain

$$\left(D_{t_0+}^{\alpha} h_{\beta-1}(t, t_0)\right)(x) = h_{\beta-\alpha-1}(x, t_0),$$

or

$$\left(D_{t_0+}^{\alpha} \frac{(t - t_0)^{\beta-1}}{\Gamma(\beta)}\right)(x) = \frac{(x - t_0)^{\beta-\alpha-1}}{\Gamma(\beta - \alpha)},$$

or

$$\left(D_{t_0+}^{\alpha}(t - t_0)^{\beta-1}\right)(x) = \frac{\Gamma(\beta)}{\Gamma(\beta - \alpha)}(x - t_0)^{\beta-\alpha-1}, \quad x \in \mathbb{R}, \quad x \geq t_0.$$

Now we suppose that $0 < \alpha < 1$. Then by Corollary 4.1, we obtain

$$\left(D_{t_0+}^{\alpha} 1\right)(x) = h_{-\alpha}(x, t_0)$$

$$= \frac{(x - t_0)^{-\alpha}}{\Gamma(1 - \alpha)}, \quad x \in \mathbb{R}, \quad x \geq t_0.$$

Suppose that $\alpha \geq 0$. Then for all $j \in \{1, \ldots, -\overline{[-\alpha]}\}$, using Corollary 4.2, we obtain

$$0 = \left(D_{t_0+}^{\alpha} h_{\alpha-j}(t, t_0)\right)(x)$$

$$= \frac{1}{\Gamma(\alpha - j + 1)} D_{t_0+}^{\alpha} \left((t - t_0)^{\alpha-j}\right)(x),$$

i.e.,

$$\left(D_{t_0+}^{\alpha}(t - t_0)^{\alpha-j}\right)(x) = 0, \quad x \in \mathbb{R}, \quad x \geq t_0.$$

This completes the proof.

Corollary 4.4. *Let* $\mathbb{T} = \mathbb{Z}$, $\alpha \geq 0$, $\beta \in \mathbb{R} \backslash \{\ldots, -2, -1\}$. *Then*

$$D_{\Delta, t_0}^{-\alpha}(t - t_0)^{(\beta)} = \beta^{(-\alpha)}(t - t_0)^{(\alpha+\beta)}, \quad t \geq t_0 - \overline{[-\beta]}.$$

Proof. By Theorem 4.4, we get

$$D_{\Delta, t_0}^{-\alpha} h_\beta(t, t_0) = h_{\beta+\alpha}(t, t_0),$$

or

$$D_{\Delta, t_0}^{-\alpha} \frac{(t - t_0)^{(\beta)}}{\Gamma(\beta + 1)} = \frac{(t - t_0)^{(\beta+\alpha)}}{\Gamma(\alpha + \beta + 1)},$$

or

$$D_{\Delta, t_0}^{-\alpha}(t - t_0)^{(\beta)} = \frac{\Gamma(\beta + 1)}{\Gamma(\alpha + \beta + 1)}(t - t_0)^{(\alpha+\beta)}$$

$$= \beta^{(-\alpha)}(t - t_0)^{(\alpha+\beta)}, \quad t, t_0 \in \mathbb{T}, \quad t \geq t_0 - \overline{[-\beta]}.$$

This completes the proof.

Theorem 4.5. *Let* $\alpha > 0$, $m = -\overline{[-\alpha]}$, *and let* f *be a function that is m-times Δ-differentiable on* $\mathbb{T}^{\kappa^m}$ *with* f^{Δ^m} *rd-continuous over* $\mathbb{T}$. *Then*

$$I^{\alpha}_{\Delta,t_0}f(t) = \sum_{k=0}^{m-1} h_{k+\alpha}(t, t_0)f^{\Delta^k}(t_0) + \left(h_{m+\alpha-1}(\cdot, t_0) \star f^{\Delta^m}\right)(t), \quad t \in \mathbb{T}^{\kappa^m}.$$

Proof. By Taylor's formula, Theorem 1.60, we have

$$f(t) = \sum_{k=0}^{m-1} h_k(t, t_0)f^{\Delta^k}(t_0) + \int_{t_0}^{t} h_{m-1}(t, \sigma(\tau))f^{\Delta^m}(\tau)\Delta\tau. \quad t \in \mathbb{T}^{\kappa^m}.$$

Hence, applying Theorem 1.56, we get

$$\left(I^{\alpha}_{\Delta,t_0}f\right)(t) = \int_{t_0}^{t} h_{\alpha-1}(t, \sigma(u))f(u)\Delta u$$

$$= \int_{t_0}^{t} h_{\alpha-1}(t, \sigma(u))\left(\sum_{k=0}^{m-1} h_k(u, t_0)f^{\Delta^k}(t_0) + \int_{t_0}^{u} h_{m-1}(u, \sigma(\tau))f^{\Delta^m}(\tau)\Delta\tau\right)$$

$$= \int_{t_0}^{t} h_{\alpha-1}(t, \sigma(u))\left(\sum_{k=0}^{m-1} h_k(u, t_0)f^{\Delta^k}(t_0)\right)\Delta u$$

$$+ \int_{t_0}^{t} h_{\alpha-1}(t, \sigma(u))\left(\int_{t_0}^{u} h_{m-1}(u, \sigma(\tau))f^{\Delta^m}(\tau)\Delta\tau\right)\Delta u$$

$$= \sum_{k=0}^{m-1} f^{\Delta^k}(t_0)\int_{t_0}^{t} h_{\alpha-1}(t, \sigma(u))h_k(u, t_0)\Delta u$$

$$+ \int_{t_0}^{t} h_{\alpha-1}(t, \sigma(u))\left(h_{m-1}(\cdot, t_0) \star f^{\Delta^m}\right)(u)\Delta u$$

$$= \sum_{k=0}^{m-1} f^{\Delta^k}(t_0)h_{k+\alpha}(t, t_0) + \left(h_{\alpha-1}(\cdot, t_0) \star \left(h_{m-1}(\cdot, t_0) \star f^{\Delta^m}\right)\right)(t)$$

$$= \sum_{k=0}^{m-1} f^{\Delta^k}(t_0)h_{k+\alpha}(t, t_0) + \left((h_{\alpha-1}(\cdot, t_0) \star h_{m-1}(\cdot, t_0)) \star f^{\Delta^m}\right)(t)$$

$$= \sum_{k=0}^{m-1} f^{\Delta^k}(t_0)h_{k+\alpha}(t, t_0) + \left(h_{m+\alpha-1}(\cdot, t_0) \star f^{\Delta^m}\right)(t), \quad t \in \mathbb{T}^{\kappa^m}.$$

This completes the proof.

Example 4.19. Let $\alpha = \frac{1}{2}, f(t) = h_2(t, t_0), t \in \mathbb{T}, t > t_0$. Then $m = 1$ and

$$I^{\frac{1}{2}}_{\Delta,t_0} h_2(t, t_0) = \sum_{k=0}^{0} h_{k+\frac{1}{2}}(t, t_0) h_2^{\Delta^k}(t_0, t_0)$$

$$+ \left(h_{\frac{1}{2}}(\cdot, t_0) \star h_2^{\Delta}(\cdot, t_0) \right)(t, t_0)$$

$$= h_{\frac{1}{2}}(t, t_0) h_2(t_0, t_0)$$

$$+ \left(h_{\frac{1}{2}}(\cdot, t_0) \star h_1(\cdot, t_0) \right)(t, t_0)$$

$$= h_{\frac{5}{2}}(t, t_0), \quad t \in \mathbb{T}, \quad t > t_0.$$

Example 4.20. Let $\alpha = \frac{9}{2}, f(t) = h_3(t, t_0), t \in \mathbb{T}, t > t_0.$ Then

$$m = \overline{\left[\frac{9}{2} \right]} + 1$$

$$= 4 + 1$$

$$= 5$$

and

$$I^{\frac{9}{2}}_{\Delta,t_0} h_3(t, t_0) = \sum_{k=0}^{4} h_{k+\frac{9}{2}}(t, t_0) h_3^{\Delta^k}(t_0, t_0)$$

$$+ \left(h_{5+\frac{9}{2}-1}(\cdot, t_0) \star h_3^{\Delta^5}(\cdot, t_0) \right)(t)$$

$$= h_{\frac{9}{2}}(t, t_0) h_3(t_0, t_0)$$

$$+ h_{\frac{11}{2}}(t, t_0) h_3^{\Delta^2}(t_0, t_0)$$

$$+ h_{\frac{13}{2}}(t, t_0) h_3^{\Delta^5}(t_0, t_0)$$

$$+ h_{\frac{15}{2}}(t, t_0) h_3^{\Delta^3}(t_0, t_0)$$

$$+ h_{\frac{17}{2}}(t, t_0) h_3^{\Delta^4}(t_0, t_0)$$

$$= h_{\frac{15}{2}}(t, t_0), \quad t \in \mathbb{T}, \quad t > t_0.$$

Example 4.21. Let $\alpha = \frac{5}{3}, f(t) = e_a(t, t_0), t \in \mathbb{T}, t \geq t_0,$ and let $a \in \mathbb{C}$ be a given constant. We will find

$$\mathscr{L}\left(I^{\frac{5}{3}}_{\Delta,t_0} e_a(\cdot, t_0) \right)(z, t_0).$$

Note that

$$m = \left\lceil \frac{5}{3} \right\rceil + 1$$
$$= 1 + 1$$
$$= 2$$

and

$$e_a(t_0, t_0) = 1,$$
$$e_a^{\Delta}(t, t_0) = a e_a(t, t_0),$$
$$e_a^{\Delta}(t_0, t_0) = a,$$
$$e_a^{\Delta^2}(t, t_0) = a^2 e_a(t, t_0),$$
$$e_a^{\Delta^2}(t_0, t_0) = a^2.$$

We have

$$I_{\Delta,t_0}^{\frac{5}{3}} e_a(t, t_0) = \sum_{k=0}^{1} h_{k+\frac{5}{3}}(t, t_0) e_a^{\Delta^k}(t_0, t_0)$$
$$+ \left(h_{2+\frac{5}{3}-1}(\cdot, t_0) \star e_a^{\Delta^2}(\cdot, t_0) \right)(t, t_0)$$
$$= h_{\frac{5}{3}}(t, t_0) e_a(t_0, t_0)$$
$$+ h_{\frac{8}{3}}(t, t_0) e_a^{\Delta}(t_0, t_0)$$
$$+ \left(h_{\frac{8}{3}}(\cdot, t_0) \star \left(a^2 e_a(\cdot, t_0) \right) \right)(t, t_0)$$
$$= h_{\frac{5}{3}}(t, t_0) + a h_{\frac{8}{3}}(t, t_0)$$
$$+ a^2 \left(h_{\frac{8}{3}}(\cdot, t_0) \star e_a(\cdot, t_0) \right)(t), \quad t \in \mathbb{T}, \quad t > t_0.$$

Therefore,

$$\mathscr{L} \left(I_{\Delta,t_0}^{\frac{5}{3}} e_a(t, t_0) \right)(z, t_0) = \mathscr{L} \left(h_{\frac{5}{3}}(\cdot, t_0) \right)(z, t_0)$$
$$+ a \mathscr{L} \left(h_{\frac{8}{3}}(\cdot, t_0) \right)(z, t_0)$$
$$+ a^2 \mathscr{L} \left(h_{\frac{8}{3}}(\cdot, t_0) \star e_a(\cdot, t_0) \right)(z, t_0)$$

$$= \frac{1}{z^{\frac{5}{3}+1}} + a\frac{1}{z^{\frac{8}{3}+1}}$$

$$+ a^2 \mathscr{L}\left(h_{\frac{8}{3}}(\cdot, t_0)\right)(z, t_0)\mathscr{L}\left(e_a(\cdot, t_0)\right)(z, t_0)$$

$$= \frac{1}{z^{\frac{8}{3}}} + \frac{a}{z^{\frac{11}{3}}} + \frac{a^2}{z^{\frac{8}{3}+1}}\frac{1}{z - a}$$

$$= \frac{1}{z^{\frac{8}{3}}} + \frac{a}{z^{\frac{11}{3}}} + \frac{a^2}{(z - a)z^{\frac{11}{3}}}.$$

Exercise 4.4. Let $\alpha = \frac{3}{2}, f(t) = \cos_2(t, t_0), t \in \mathbb{T}, t \geq t_0$. Find

$$\mathscr{L}\left(I_{\Delta, t_0}^{\frac{3}{2}}\cos_2(\cdot, t_0)\right)(z, t_0).$$

Answer. $\dfrac{1}{z^{\frac{1}{2}}\left(z^2+4\right)}$.

Theorem 4.6. *Let $\alpha > 0$, $m = -\overline{[-\alpha]}$, let f be a function that is m-times Δ-differentiable on $\mathbb{T}^{\kappa^m}$ with f^{Δ^m} rd-continuous over $\mathbb{T}$, and suppose that $D_{\Delta, t_0}^{\alpha} f$ exists almost everywhere on $\mathbb{T}$. Then*

$$D_{\Delta, t_0}^{\alpha} f(t) = \sum_{k=0}^{m-1} h_{k-\alpha}(t, t_0)f^{\Delta^k}(t_0) + \left(h_{m-\alpha-1}(\cdot, t_0) \star f^{\Delta^m}\right)(t), \quad t \in \mathbb{T}^{\kappa^m}.$$

Proof. Using Theorem 4.5, we obtain

$$D_{\Delta, t_0}^{\alpha} f(t) = D_{\Delta}^m I_{\Delta, t_0}^{m-\alpha} f(t)$$

$$= D_{\Delta}^m \left(\sum_{k=0}^{m-1} h_{k+m-\alpha}(t, t_0)f^{\Delta^k}(t_0)\right.$$

$$\left. + \left(h_{2m-\alpha-1}(\cdot, t_0) \star f^{\Delta^m}\right)(t)\right)$$

$$= \sum_{k=0}^{m-1} h_{k-\alpha}(t, t_0)f^{\Delta^k}(t_0) + \left(h_{m-\alpha-1}(\cdot, t_0) \star f^{\Delta^m}\right)(t)$$

$$+ \sum_{k=0}^{m-1} h_{2m-\alpha-k-1}(t_0, t_0)f^{\Delta^{m-k}}(t)$$

$$= \sum_{k=0}^{m-1} h_{k-\alpha}(t, t_0)f^{\Delta^k}(t_0) + \left(h_{m-\alpha-1}(\cdot, t_0) \star f^{\Delta^m}\right)(t),$$

$t \in \mathbb{T}^{\kappa^m}$. This completes the proof.

Example 4.22. Let $\alpha > 0$ be arbitrarily chosen, $m = -\overline{[-\alpha]}$. If $r \in \mathbb{N}$, $r < m$, we get

$$
\begin{aligned}
D^\alpha_{\Delta,t_0} h_r(t, t_0) &= \sum_{k=0}^{m-1} h_{k-\alpha}(t, t_0) h_r^{\Delta^k}(t_0, t_0) \\
&\quad + \left(h_{m-\alpha-1}(\cdot, t_0) \star h_r^{\Delta^m}(\cdot, t_0) \right)(t, t_0) \\
&= h_{r-\alpha}(t, t_0), \quad t \in \mathbb{T}, \quad t > t_0.
\end{aligned}
$$

If $r \in \mathbb{N}$, $r \geq m$, then

$$
\begin{aligned}
D^\alpha_{\Delta,t_0} h_r(t, t_0) &= \sum_{k=0}^{m-1} h_{k-\alpha}(t, t_0) h_r^{\Delta^k}(t_0, t_0) \\
&\quad + \left(h_{m-\alpha-1}(\cdot, t_0) \star h_r^{\Delta^m}(\cdot, t_0) \right)(t, t_0) \\
&= \left(h_{m-\alpha-1}(\cdot, t_0) \star h_{r-m}(\cdot, t_0) \right)(t, t_0) \\
&= h_{r-\alpha}(t, t_0), \quad t \in \mathbb{T}, \quad t > t_0.
\end{aligned}
$$

Example 4.23. Let $\alpha > 0$ be arbitrarily chosen, $m = -\overline{[-\alpha]}$, $a \in \mathbb{C}$. We will find

$$
\mathscr{L}\left(D^\alpha_{\Delta,t_0} e_a(\cdot, t_0) \right)(z, t_0).
$$

We have

$$
\begin{aligned}
e_a^{\Delta^k}(t, t_0) &= a^k e_a(t, t_0), \\
e_a^{\Delta^k}(t_0, t_0) &= a^k, \quad t \in \mathbb{T}, \quad t \geq t_0.
\end{aligned}
$$

Then

$$
\begin{aligned}
D^\alpha_{\Delta,t_0} e_a(t, t_0) &= \sum_{k=0}^{m-1} h_{k-\alpha}(t, t_0) e_a^{\Delta^k}(t_0, t_0) \\
&\quad + \left(h_{m-\alpha-1}(\cdot, t_0) \star e_a^{\Delta^m}(\cdot, t_0) \right)(t, t_0) \\
&= \sum_{k=0}^{m-1} a^k h_{k-\alpha}(t, t_0) \\
&\quad + a^m \left(h_{m-\alpha-1}(\cdot, t_0) \star e_a(\cdot, t_0) \right)(t, t_0), \quad t \in \mathbb{T}, \quad t \geq t_0.
\end{aligned}
$$

Hence,

$$
\mathcal{L}\left(D^{\alpha}_{\Delta,t_0} e_a(\cdot,t_0)\right)(z,t_0) = \sum_{k=0}^{m-1} a^k \mathcal{L}\left(h_{k-\alpha}(\cdot,t_0)\right)(z,t_0)
$$

$$
+ a^m \mathcal{L}\left(h_{m-\alpha-1}(\cdot,t_0) \star e_a(\cdot,t_0)\right)(z,t_0)
$$

$$
= \sum_{k=0}^{m-1} \frac{a^k}{z^{k-\alpha+1}}
$$

$$
+ a^m \mathcal{L}\left(h_{m-\alpha-1}(\cdot,t_0)\right)(z,t_0) \mathcal{L}\left(e_a(\cdot,t_0)\right)(z,t_0)
$$

$$
= \sum_{k=0}^{m-1} \frac{a^k}{z^{k-\alpha+1}} + \frac{a^m}{z^{m-\alpha}(z-a)}.
$$

Example 4.24. Let $\alpha = \frac{5}{4}, f(t) = \sin_1(t,t_0), t \in \mathbb{T}, t \geq t_0$. We will find

$$
\mathcal{L}\left(D^{\frac{5}{4}}_{\Delta,t_0} \sin_1(\cdot,t_0)\right)(z,t_0).
$$

We have

$$
m = \left\lceil \frac{5}{4} \right\rceil + 1
$$

$$
= 1 + 1
$$

$$
= 2
$$

and

$$
D^{\frac{5}{4}}_{\Delta,t_0} \sin_1(t,t_0) = \sum_{k=0}^{1} h_{k-\frac{5}{4}}(t,t_0) \sin_1^{\Delta^k}(t_0,t_0)
$$

$$
+ \left(h_{2-\frac{5}{4}-1}(\cdot,t_0) \star \sin_1^{\Delta^2}(\cdot,t_0)\right)(z,t_0)
$$

$$
= h_{-\frac{5}{4}}(t,t_0) \sin_1(t_0,t_0)
$$

$$
+ h_{1-\frac{5}{4}}(t,t_0) \cos_1(t_0,t_0)
$$

$$
- \left(h_{-\frac{1}{4}}(\cdot,t_0) \star \sin_1(\cdot,t_0)\right)(t,t_0)
$$

$$
= h_{-\frac{1}{4}}(t,t_0)
$$

$$
- \left(h_{-\frac{1}{4}}(\cdot,t_0) \star \sin_1(\cdot,t_0)\right)(t,t_0), \quad t \in \mathbb{T}, \quad t \geq t_0.
$$

Hence,

$$\mathscr{L}\left(D_{\Delta,t_0}^{\frac{5}{4}}\sin_1(t,t_0)\right)(z,t_0) = \mathscr{L}\left(h_{-\frac{1}{4}}(\cdot,t_0)\right)(z,t_0)$$

$$-\mathscr{L}\left(h_{-\frac{1}{4}}(\cdot,t_0)\star\sin_1(\cdot,t_0)\right)(z,t_0)$$

$$= \frac{1}{z^{-\frac{1}{4}+1}}$$

$$-\mathscr{L}\left(h_{-\frac{1}{4}}(\cdot,t_0)\right)(z,t_0)\mathscr{L}\left(\sin_1(\cdot,t_0)\right)(z,t_0)$$

$$= \frac{1}{z^{\frac{3}{4}}} - \frac{1}{z^{-\frac{1}{4}+1}}\frac{1}{z^2+1}$$

$$= \frac{z^2+1-1}{z^{\frac{3}{4}}\left(z^2+1\right)}$$

$$= \frac{z^{\frac{5}{4}}}{z^2+1}.$$

Exercise 4.5. Let $\alpha = \frac{1}{2}, f(t) = \sin_1(t,t_0), t \in \mathbb{T}, t \geq t_0$. Find

$$\mathscr{L}\left(D_{\Delta,t_0}^{\frac{1}{2}}\sin_1(\cdot,t_0)\right)(z,t_0).$$

Answer. $\frac{z^{\frac{1}{2}}}{z^2+1}$.

Theorem 4.7. *Let $\alpha > 0, \beta > 0$. Then*

$$\left(I_{\Delta,t_0}^{\alpha}I_{\Delta,t_0}^{\beta}f\right)(t) = I_{\Delta,t_0}^{\alpha+\beta}f(t), \quad t \in \mathbb{T}, \quad t > t_0.$$

Proof. Using Theorem 4.1, we have

$$\left(I_{\Delta,t_0}^{\alpha}I_{\Delta,t_0}^{\beta}f\right)(t) = I_{\Delta,t_0}^{\alpha}\left(I_{\Delta,t_0}^{\beta}f\right)(t)$$

$$= I_{\Delta,t_0}^{\alpha}\left(h_{\beta-1}\star f\right)(t)$$

$$= \left(h_{\alpha-1}\star\left(h_{\beta-1}\star f\right)\right)(t)$$

$$= \left(\left(h_{\alpha-1}\star h_{\beta-1}\right)\star f\right)(t)$$

$$= \left(h_{\alpha+\beta-1}\star f\right)(t)$$

$$= I_{\Delta,t_0}^{\alpha+\beta}f(t), \quad t \in \mathbb{T}, \quad t > t_0.$$

This completes the proof.

Theorem 4.8. *Let* $\alpha > 0$, $m = -\overline{[-\alpha]}$, $n \in \mathbb{N}$, *and suppose that* f *is* m-*times* Δ-*differentiable and* f^{Δ^m} *is rd-continuous on* $\mathbb{T}^{\kappa^m}$. *Then*

$$D_\Delta^n D_{\Delta,t_0}^\alpha f(t) = D_{\Delta,t_0}^{n+\alpha} f(t), \quad t \in \mathbb{T}^{\kappa^m}.$$

Proof. Using Theorem 4.6, we get

$$D_\Delta^n D_{\Delta,t_0}^\alpha f(t) = D_\Delta^n \left(D_{\Delta,t_0}^\alpha f(t) \right)$$

$$= D_\Delta^n \left(\sum_{k=0}^{m-1} h_{k-\alpha}(t,t_0) f^{\Delta^k}(t_0) + \left(h_{m-\alpha-1}(\cdot,t_0) \star f^{\Delta^m} \right)(t) \right)$$

$$= \sum_{k=0}^{m-1} D_\Delta^n \left(h_{k-\alpha}(t,t_0) f^{\Delta^k}(t_0) \right)$$

$$+ D_\Delta^n \left(h_{m-\alpha-1}(\cdot,t_0) \star f^{\Delta^m} \right)(t)$$

$$= \sum_{k=0}^{m-1} h_{k-\alpha-n}(t,t_0) f^{\Delta^k}(t_0)$$

$$+ \left(h_{m-n-\alpha-1}(\cdot,t_0) \star f^{\Delta^m} \right)(t)$$

$$= D_{\Delta,t_0}^{n+\alpha} f(t), \quad t \in \mathbb{T}^{\kappa^m}.$$

This completes the proof.

Theorem 4.9. *Let* $\alpha > 0$, $m = -\overline{[-\alpha]}$, $n \in \mathbb{N}$, *and suppose that* f *is* m-*times* Δ-*differentiable and* f^{Δ^m} *is rd-continuous on* $\mathbb{T}^{\kappa^m}$. *Then*

$$D_\Delta^n I_{\Delta,t_0}^\alpha f(t) = I_{\Delta,t_0}^{\alpha-n} f(t), \quad t \in \mathbb{T}^{\kappa^m}.$$

Proof. Using Theorem 4.5, we get

$$D_\Delta^n I_{\Delta,t_0}^\alpha f(t) = D_\Delta^n \left(I_{\Delta,t_0}^\alpha f(t) \right)$$

$$= D_\Delta^n \left(\sum_{k=0}^{m-1} h_{k+\alpha}(t,t_0) f^{\Delta^k}(t_0) + \left(h_{m+\alpha-1}(\cdot,t_0) \star f^{\Delta^m} \right)(t) \right)$$

$$= \sum_{k=0}^{m-1} h_{k+\alpha-n}(t,t_0) f^{\Delta^k}(t_0)$$

$$+ D_\Delta^n \left(\left(h_{m+\alpha-1}(\cdot,t_0) \star f^{\Delta^m} \right)(t) \right)$$

$$= \sum_{k=0}^{m-1} h_{k+\alpha-n}(t, t_0) f^{\Delta^k}(t_0)$$

$$+ \left(h_{m+\alpha-n-1}(\cdot, t_0) \star f^{\Delta^m} \right)(t)$$

$$= I_{\Delta, t_0}^{\alpha-n} f(t), \quad t \in \mathbb{T}^{\kappa^m}.$$

This completes the proof.

Theorem 4.10. *Let* $\alpha > 0$, $m = -\overline{[-\alpha]}$, $n \in \mathbb{N}$, *and suppose that* f *is* $(n+m)$-*times* Δ-*differentiable and* $f^{\Delta^{n+m}}$ *is rd-continuous over* $\mathbb{T}$. *Then*

$$D_{\Delta, t_0}^{n+\alpha} f(t) = D_{\Delta, t_0}^{\alpha} D_{\Delta}^n f(t) + \sum_{k=0}^{n-1} h_{k-\alpha-n}(t, t_0) f^{\Delta^k}(t_0), \quad t \in \mathbb{T}^{\kappa^{n+m}}.$$

Proof. By Theorem 4.8, we have

$$D_{\Delta}^n D_{\Delta, t_0}^{\alpha} f(t) = D_{\Delta, t_0}^{n+\alpha} f(t)$$

$$= \sum_{k=0}^{n+m-1} h_{k-\alpha-n}(t, t_0) f^{\Delta^k}(t_0) \tag{4.11}$$

$$+ \left(h_{m-\alpha-1}(\cdot, t_0) \star f^{\Delta^{n+m}} \right)(t), \quad t \in \mathbb{T}^{\kappa^{n+m}}.$$

On the other hand,

$$D_{\Delta, t_0}^{\alpha} D_{\Delta}^n f(t) = D_{\Delta, t_0}^{\alpha} \left(D_{\Delta}^n f(t) \right)$$

$$= D_{\Delta, t_0}^{\alpha} f^{\Delta^n}(t)$$

$$= \sum_{k=0}^{m-1} h_{k-\alpha}(t, t_0) f^{\Delta^{n+k}}(t_0)$$

$$+ \left(h_{m-\alpha-1}(\cdot, t_0) \star f^{\Delta^{m+n}} \right)(t), \quad t \in \mathbb{T}^{\kappa^{m+n}}.$$

Hence,

$$\left(h_{m-\alpha-1}(\cdot, t_0) \star f^{\Delta^{m+n}} \right)(t) = D_{\Delta, t_0}^{\alpha} D_{\Delta}^n f(t)$$

$$- \sum_{k=0}^{m-1} h_{k-\alpha}(t, t_0) f^{\Delta^{n+k}}(t_0)$$

$$= D_{\Delta, t_0}^{\alpha} D_{\Delta}^n f(t)$$

$$- \sum_{r=n}^{m+n-1} h_{r-\alpha-n}(t, t_0) f^{\Delta^r}(t_0), \quad t \in \mathbb{T}^{\kappa^{m+n}}.$$

From this and (4.11), we get

$$
\begin{aligned}
D_{\Delta,t_0}^{n+\alpha} f(t) &= D_{\Delta}^n D_{\Delta,t_0}^{\alpha} f(t) \\
&= \sum_{k=0}^{n+m-1} h_{k-\alpha-n}(t,t_0) f^{\Delta^k}(t_0) \\
&\quad + D_{\Delta,t_0}^{\alpha} D_{\Delta}^n f(t) \\
&\quad - \sum_{k=n}^{m+n-1} h_{k-\alpha-n}(t,t_0) f^{\Delta^k}(t_0) \\
&= \sum_{k=0}^{n-1} h_{k-\alpha-n}(t,t_0) f^{\Delta^k}(t_0) \\
&\quad + D_{\Delta,t_0}^{\alpha} D_{\Delta}^n f(t), \quad t \in \mathbb{T}^{\kappa^{n+m}}.
\end{aligned}
$$

This completes the proof.

Theorem 4.11. Let $\alpha > 0$, $m = -\overline{[-\alpha]}$, $n \in \mathbb{N}$, and suppose that f is $(n+m)$-times Δ-differentiable and $f^{\Delta^{n+m}}$ is rd-continuous over $\mathbb{T}$. Then

$$
D_{\Delta}^n I_{\Delta,t_0}^{\alpha} f(t) = I_{\Delta,t_0}^{\alpha} D_{\Delta}^n f(t) + \sum_{k=0}^{n-1} h_{k+\alpha-n}(t,t_0) f^{\Delta^k}(t_0), \quad t \in \mathbb{T}^{\kappa^{n+m}}.
$$

Proof. By Theorem 4.5, we have

$$
\begin{aligned}
I_{\Delta,t_0}^{\alpha} D_{\Delta}^n f(t) &= I_{\Delta,t_0}^{\alpha} \left(D_{\Delta}^n f(t) \right) \\
&= I_{\Delta,t_0}^{\alpha} \left(f^{\Delta^n}(t) \right) \\
&= \sum_{k=0}^{m-1} h_{k+\alpha}(t,t_0) f^{\Delta^{k+n}}(t_0) \\
&\quad + \left(h_{m+\alpha-1}(\cdot,t_0) \star f^{\Delta^{n+m}} \right)(t) \\
&= \sum_{k=n}^{n+m-1} h_{k-n+\alpha}(t,t_0) f^{\Delta^k}(t_0) \\
&\quad + \left(h_{m+\alpha-1}(\cdot,t_0) \star f^{\Delta^{n+m}} \right)(t), \quad t \in \mathbb{T}^{\kappa^{n+m}}.
\end{aligned}
$$

Therefore,

$$\left(h_{m+\alpha-1}(\cdot, t_0) \star f^{\Delta^{n+m}}\right)(t) = I^{\alpha}_{\Delta,t_0} D^n_\Delta f(t) - \sum_{k=n}^{n+m-1} h_{k-n+\alpha}(t, t_0) f^{\Delta^k}(t_0), \quad t \in \mathbb{T}^{\kappa^{n+m}}.$$

Hence,

$$
\begin{aligned}
D^n_\Delta I^{\alpha}_{\Delta,t_0} f(t) &= D^n_\Delta \left(I^{\alpha}_{\Delta,t_0} f(t)\right) \\
&= \sum_{k=0}^{m-1} h_{k+\alpha-n}(t, t_0) f^{\Delta^k}(t_0) \\
&\quad + \left(h_{m+\alpha-1}(\cdot, t_0) \star f^{\Delta^{m+n}}\right)(t) \\
&\quad + \sum_{v=0}^{n-1} h_{m+\alpha-1-v}(t, t_0) f^{\Delta^{m+n-1-v}}(t_0) \\
&= \sum_{k=0}^{m-1} h_{k+\alpha-n}(t, t_0) f^{\Delta^k}(t_0) \\
&\quad + \sum_{k=m}^{m+n-1} h_{k+\alpha-n}(t, t_0) f^{\Delta^k}(t_0) \\
&\quad + \left(h_{m+\alpha-1}(\cdot, t_0) \star f^{\Delta^{m+n}}\right)(t) \\
&= \sum_{k=0}^{n+m-1} h_{k+\alpha-n}(t, t_0) f^{\Delta^k}(t_0) \\
&\quad + \left(h_{m+\alpha-1}(\cdot, t_0) \star f^{\Delta^{m+n}}\right)(t) \\
&= \sum_{k=0}^{n+m-1} h_{k+\alpha-n}(t, t_0) f^{\Delta^k}(t_0) \\
&\quad + I^{\alpha}_{\Delta,t_0} D^n_\Delta f(t) \\
&\quad - \sum_{k=n}^{n+m-1} h_{k-n+\alpha}(t, t_0) f^{\Delta^k}(t_0) \\
&= I^{\alpha}_{\Delta,t_0} D^n_\Delta f(t) + \sum_{k=0}^{n-1} h_{k-n+\alpha}(t, t_0) f^{\Delta^k}(t_0), \quad t \in \mathbb{T}^{\kappa^{n+m}}.
\end{aligned}
$$

This completes the proof.

Theorem 4.12. *Let $\alpha > 0$, $\beta > 0$, $M = -\overline{[-\beta]}$, and let f be Δ-differentiable and f^{Δ^M} rd-continuous over $\mathbb{T}$. Then*

$$\left(I_{\Delta,t_0}^\alpha D_{\Delta,t_0}^\beta f \right)(t) = D_{\Delta,t_0}^{\beta-\alpha} f(t) - \sum_{k=1}^{M} h_{\alpha-k}(t, t_0) D_{\Delta,t_0}^{\beta-k} f(t_0), \quad t \in \mathbb{T}^{\kappa^M}.$$

Proof. Using Theorem 4.11, we get

$$\left(I_{\Delta,t_0}^\alpha D_{\Delta,t_0}^\beta f \right)(t) = \left(I_{\Delta,t_0}^\alpha D_{\Delta,t_0}^M I_{\Delta,t_0}^{M-\beta} f \right)(t)$$

$$= \left(D_\Delta^M I_{\Delta,t_0}^\alpha I_{\Delta,t_0}^{M-\beta} f \right)(t)$$

$$\qquad - \sum_{k=0}^{M-1} h_{k+\alpha-M}(t, t_0) D_\Delta^k \left(I_{\Delta,t_0}^{M-\beta} f \right)(t_0)$$

$$= \left(I_{\Delta,t_0}^{\alpha-M} I_{\Delta,t_0}^{M-\beta} f \right)(t)$$

$$\qquad - \sum_{k=0}^{M-1} h_{k+\alpha-M}(t, t_0) I_{\Delta,t_0}^{M-\beta-k} f(t_0)$$

$$= I_{\Delta,t_0}^{\alpha-\beta} f(t) - \sum_{k=0}^{M-1} h_{k+\alpha-M}(t, t_0) D_{\Delta,t_0}^{\beta+k-M} f(t_0)$$

$$= I_{\Delta,t_0}^{\alpha-\beta} f(t) - \sum_{k=1}^{M} h_{\alpha-k}(t, t_0) D_{\Delta,t_0}^{\beta-k} f(t_0), \quad t \in \mathbb{T}^{\kappa^M}.$$

This completes the proof.

Theorem 4.13. *Let $\alpha > 0$, $\beta > 0$, $M = -\overline{[-\beta]}$, and suppose that f is Δ-differentiable and f^{Δ^M} is rd-continuous over $\mathbb{T}$. Then*

$$\left(D_{\Delta,t_0}^\beta I_{\Delta,t_0}^\alpha f \right)(t) = I_{\Delta,t_0}^{\alpha-\beta} f(t), \quad t \in \mathbb{T}^{\kappa^M}.$$

Proof. Using Theorem 4.7, we get

$$\left(D_{\Delta,t_0}^\beta I_{\Delta,t_0}^\alpha f \right)(t) = \left(D_\Delta^M I_{\Delta,t_0}^{M-\beta} I_{\Delta,t_0}^\alpha f \right)(t)$$

$$= \left(D_\Delta^M I_{\Delta,t_0}^{M+\alpha-\beta} f \right)(t)$$

$$= D_\Delta^M \left(\left(h_{M+\alpha-\beta-1}(\cdot, t_0) \star f \right)(t) \right)$$

$$= \left(h_{\alpha-\beta-1}(\cdot, t_0) \star f\right)(t)$$
$$= I_{\Delta,t_0}^{\alpha-\beta} f(t), \quad t \in \mathbb{T}^{\kappa^M}.$$

This completes the proof.

Theorem 4.14. *Let $\alpha > 0$, and let $f : \mathbb{T} \to \mathbb{R}$ be locally Δ-integrable. For $s, t \in \mathbb{T}$, $t \geq s \geq t_0$, one has*

$$\mathscr{L}\left(I_{\Delta,t_0}^{\alpha} f(t)\right)(z, t_0) = \frac{1}{z^{\alpha}} \mathscr{L}\left(f(t)\right)(z, t_0).$$

Proof. Let $t, s \in \mathbb{T}, t \geq s \geq t_0$. Then

$$\mathscr{L}\left(I_{\Delta,t_0}^{\alpha} f(t)\right)(z, t_0) = \mathscr{L}\left((h_{\alpha-1}(\cdot, t_0) \star f)(t)\right)(z, t_0)$$
$$= \mathscr{L}\left(h_{\alpha-1}(\cdot, t_0)\right)(z, t_0)\mathscr{L}(f(t))(z)$$
$$= \frac{1}{z^{\alpha}} \mathscr{L}(f(t))(z).$$

This completes the proof.

Example 4.25. Let $\alpha > 0$ and $\beta \in \mathbb{R}$ be arbitrarily chosen. Then

$$\mathscr{L}\left(I_{\Delta,t_0}^{\alpha} h_{\beta}(\cdot, t_0)\right)(z, t_0) = \frac{1}{z^{\alpha}} \mathscr{L}\left(h_{\beta}(\cdot, t_0)\right)(z, t_0)$$
$$= \frac{1}{z^{\alpha}} \frac{1}{z^{\beta+1}}$$
$$= \frac{1}{z^{\alpha+\beta+1}}.$$

Example 4.26. Let $\alpha > 0$ and $a \in \mathbb{C}$ be arbitrarily chosen. Then

$$\mathscr{L}\left(I_{\Delta,t_0}^{\alpha} e_a(\cdot, t_0)\right)(z, t_0) = \frac{1}{z^{\alpha}} \mathscr{L}\left(e_a(\cdot, t_0)\right)(z, t_0)$$
$$= \frac{1}{z^{\alpha}(z-a)}.$$

Example 4.27. Let $\alpha > 0$ and $a \in \mathbb{C}$ be arbitrarily chosen. Then

$$\mathscr{L}\left(I_{\Delta,t_0}^{\alpha} \cos_a(\cdot, t_0)\right)(z, t_0) = \frac{1}{z^{\alpha}} \mathscr{L}\left(\cos_a(\cdot, t_0)\right)(z, t_0)$$
$$= \frac{z}{z^{\alpha}\left(z^2 + a^2\right)}$$
$$= \frac{z^{1-\alpha}}{z^2 + a^2}.$$

Exercise 4.6. Let $\alpha > 0$ and $a \in \mathbb{C}$ be arbitrarily chosen. Find

$$\mathscr{L}\left(I^\alpha_{\Delta,t_0} \sin_a(\cdot, t_0)\right)(z, t_0).$$

Answer. $\frac{a}{z^\alpha(z^2+a^2)}$.

Theorem 4.15. *Let $\alpha > 0$, $m = -\overline{[-\alpha]}$, and let $f : \mathbb{T} \to \mathbb{R}$ be locally Δ-integrable. For $s, t \in \mathbb{T}$, $t \geq s \geq t_0$, one has*

$$\mathscr{L}\left(D^\alpha_{\Delta,t_0} f(t)\right)(z, t_0) = z^\alpha \mathscr{L}(f(t))(z, t_0) - \sum_{j=1}^m z^{j-1} D^{\alpha-j}_{\Delta,t_0} f(t_0).$$

Proof. Let $t, s \in \mathbb{T}$, $t \geq s \geq t_0$. Then

$$\mathscr{L}\left(D^\alpha_{\Delta,t_0} f(t)\right)(z, t_0) = \mathscr{L}\left(D^m_{\Delta,t_0} I^{m-\alpha}_{\Delta,t_0} f(t)\right)(z, t_0)$$

$$= z^m \mathscr{L}\left(I^{m-\alpha}_{\Delta,t_0} f(t)\right)(z, t_0)$$

$$- \sum_{j=0}^{m-1} z^{m-j-1} D^j_\Delta I^{m-\alpha}_{\Delta,t_0} f(t_0)$$

$$= z^m \frac{1}{z^{m-\alpha}} \mathscr{L}\left(f(t)\right)(z)$$

$$- \sum_{j=0}^{m-1} z^{m-j-1} D^{j-m+\alpha}_{\Delta,t_0} f(t_0)$$

$$= z^\alpha \mathscr{L}\left(f(t)\right)(z, t_0) - \sum_{j=1}^m z^{j-1} D^{\alpha-j}_{\Delta,t_0} f(t_0).$$

This completes the proof.

4.4 Advanced Practical Problems

Problem 4.1. Let $\mathbb{T} = 3^{\mathbb{N}_0}$, $t_0 = 1$, $\alpha = \frac{1}{4}$, $\beta = \frac{11}{4}$. Find

$$I^\alpha_{\Delta,1} h_{\beta-1}(t, 1), \quad t \in \mathbb{T}, \quad t > 1.$$

Answer. $\frac{(t-1)(t-3)}{4}$, $t \in \mathbb{T}, t > 1$.

Problem 4.2. Let $\mathbb{T} = 3^{\mathbb{N}_0}$, $t_0 = 1$, $\alpha = \frac{1}{2}$, $\beta = \frac{9}{2}$. Find

$$D_{\Delta,1}^{\frac{1}{2}} h_{\frac{7}{2}}(t, 1), \quad t \in \mathbb{T}, \quad t > 1.$$

Answer.

$$\frac{(t-1)(t-3)(t-9)}{52}, \quad t \in \mathbb{T}, \quad t > 1.$$

Problem 4.3. Let $\alpha = \frac{1}{5}$, $\beta = \frac{7}{5}$. Find

$$\mathscr{L}\left(h_{\frac{3}{5}}(\cdot, t_0) \star I_{\Delta,t_0}^{\frac{1}{2}} h_{\frac{3}{5}}(\cdot, t_0) - 3D_{\Delta,t_0}^{\frac{1}{5}} h_{\frac{3}{5}}(\cdot, t_0) \right)(z, t_0).$$

Answer. $-\frac{3}{z^{\frac{6}{5}}} + \frac{1}{z^{\frac{7}{2}}}$.

Problem 4.4. Let $\alpha \in (0, 1)$, $\beta = 1 + \alpha$. Prove that

$$\left(h_{\beta-2}(\cdot, t_0) \star D_{\Delta,t_0}^{\alpha}\left(D_{\Delta,t_0}^{\alpha} h_{\beta-1}\right)(\cdot, t_0)\right)(t, t_0) = 1.$$

Problem 4.5. Let $\alpha = \frac{4}{3}$, $f(t) = \sin_3(t, t_0)$, $t \in \mathbb{T}$, $t \geq t_0$. Find

$$\mathscr{L}\left(I_{\Delta,t_0}^{\frac{4}{3}} \sin_3(\cdot, t_0) \right)(z, t_0).$$

Answer. $\dfrac{3}{z^{\frac{4}{3}}(z^2+9)}$.

Problem 4.6. Let $\alpha = \frac{1}{3}$, $f(t) = \sin_3(t, t_0)$, $t \in \mathbb{T}$, $t \geq t_0$. Find

$$\mathscr{L}\left(D_{\Delta,t_0}^{\frac{1}{3}} \sin_3(\cdot, t_0) \right)(z, t_0).$$

Answer. $3\dfrac{z^{\frac{1}{3}}}{z^2+9} - I_{\Delta,t_0}^{\frac{2}{3}} \sin_3(t_0, t_0)$.

Problem 4.7. Let $\alpha > 0$ be arbitrarily chosen and let $f : \mathbb{T} \to \mathbb{R}$ be a continuous function on $\mathbb{T}$. Prove that

$$\mathscr{L}\left(I_{\Delta,t_0}^{\alpha} \int_{t_0}^{t} f(y)\Delta y \right)(z, t_0) = \frac{1}{z^{\alpha+1}}\mathscr{L}(f)(z).$$

Chapter 5
Cauchy-Type Problems with the Riemann–Liouville Fractional Δ-Derivative

In this chapter we suppose that $\mathbb{T}$ is a time scale with forward jump operator and delta differentiation operator σ and Δ, respectively, such that

$$\mathbb{T} = \{t_n : n \in \mathbb{N}_0\},$$

$$\lim_{n \to \infty} t_n = \infty,$$

$$\sigma(t_n) = t_{n+1}, \quad n \in \mathbb{N}_0,$$

$$w = \inf_{n \in \mathbb{N}_0} \mu(t_n) > 0.$$

5.1 Existence and Uniqueness of Solutions

Let $\alpha > 0$ and $m = -\overline{[-\alpha]}$. Consider the Cauchy problem

$$D_{\Delta, t_0}^{\alpha} y(t) = f(t, y(t)), \quad t > t_0, \tag{5.1}$$

$$D_{\Delta, t_0}^{\alpha-k} y(t_0) = b_k, \quad k \in \{1, \ldots, m\}, \tag{5.2}$$

where $f : \mathbb{T} \times \mathbb{R} \to \mathbb{R}$ is a given function, b_k, $k \in \{1, \ldots, m\}$, are given constants. Let $a \in \mathbb{T}$, $t_0 < a$. By $L_{\Delta}[t_0, a)$ we will denote the space of Δ-Lebesgue summable functions in $[t_0, a)$. We define the space

$$L_{\Delta}^{\alpha}[t_0, a) = \left\{ y \in L_{\Delta}[t_0, a) : D_{\Delta, t_0}^{\alpha} y \in L_{\Delta}[t_0, a) \right\}.$$

Theorem 5.1. *Let G be an open set in $\mathbb{R}$ and $f : [t_0, a] \times G \to \mathbb{R}$ a given function such that $f(\cdot, y) \in L_{\Delta}[t_0, a)$ for all $y \in G$. If $y \in L_{\Delta}[t_0, a)$, then the problem (5.1), (5.2) is equivalent to the equation*

© Springer International Publishing AG, part of Springer Nature 2018
S. G. Georgiev, *Fractional Dynamic Calculus and Fractional Dynamic Equations on Time Scales*, https://doi.org/10.1007/978-3-319-73954-0_5

$$y(t) = \sum_{k=1}^{m} b_k h_{\alpha-k}(t, t_0) + I_{\Delta,t_0}^{\alpha} f\,(t, y(t))\,, \quad t \in [t_0, a). \tag{5.3}$$

Proof. 1. Let $y \in L_\Delta[a, b]$ be a solution of the problem (5.1), (5.2). Then we apply I_{Δ,t_0}^{α} to both sides of (5.1), and using (5.2), we get

$$I_{\Delta,t_0}^{\alpha} D_{\Delta,t_0}^{\alpha} y(t) = y(t) - \sum_{k=1}^{m} h_{\alpha-k}(t, t_0) D_{\Delta,t_0}^{\alpha-k} y(t_0)$$

$$= y(t) - \sum_{k=1}^{m} b_k h_{\alpha-k}(t, t_0)$$

$$= I_{\Delta,t_0}^{\alpha} f\,(t, y(t))\,, \qquad t \in [t_0, a),$$

i.e., y satisfies (5.3).

2. Let $y \in L_\Delta[a, b]$ satisfy equation (5.3). Then we apply D_{Δ,t_0}^{α} to both sides of (5.3), and we get

$$D_{\Delta,t_0}^{\alpha} y(t) = D_{\Delta,t_0}^{\alpha} \left(\sum_{k=1}^{m} b_k h_{\alpha-k}(t, t_0) + I_{\Delta,t_0}^{\alpha} f\,(t, y(t)) \right)$$

$$= D_{\Delta,t_0}^{\alpha} \left(\sum_{k=1}^{m} h_{\alpha-k}(t, t_0) b_k \right) + D_{\Delta,t_0}^{\alpha} I_{\Delta,t_0}^{\alpha} f\,(t, y(t))$$

$$= \sum_{k=1}^{m} D_{\Delta,t_0}^{\alpha} \left(h_{\alpha-k}(t, t_0) \right) b_k + f\,(t, y(t))$$

$$= f\,(t, y(t))\,, \quad t \in [t_0, a),$$

i.e., y satisfies (5.1). Also, for $j \in \{1, \ldots, m\}$, we have

$$D_{\Delta,t_0}^{\alpha-j} y(t) = D_{\Delta,t_0}^{\alpha-j} \left(\sum_{k=1}^{m} h_{\alpha-k}(t, t_0) b_k + I_{\Delta,t_0}^{\alpha} f\,(t, y(t)) \right)$$

$$= D_{\Delta,t_0}^{\alpha-j} \left(\sum_{k=1}^{m} h_{\alpha-k}(t, t_0) b_k \right) + D_{\Delta,t_0}^{\alpha-j} I_{\Delta,t_0}^{\alpha} f\,(t, y(t))$$

$$= \sum_{k=1}^{m} D_{\Delta,t_0}^{\alpha-j} \left(h_{\alpha-k}(t, t_0) \right) b_k + I_{\Delta,t_0}^{j} f\,(t, y(t))\,, \quad t \in [t_0, a).$$

Hence,

$$D_{\Delta,t_0}^{\alpha-j} y(t_0) = \left(\sum_{k=1}^{m} D_{\Delta,t_0}^{\alpha-j} \left(h_{\alpha-k}(t, t_0) \right) b_k + I_{\Delta,t_0}^j f(t, y(t)) \right) \Bigg|_{t=t_0}$$

$$= b_j, \quad j \in \{1, \ldots, m\}.$$

Therefore, y satisfies (5.2). This completes the proof.

Suppose

$$|f(t, y_1(t)) - f(t, y_2(t))| \leq A \, |y_1(t) - y_2(t)|, \quad t \in [t_0, a], \tag{5.4}$$

where $A > 0$ is a constant that does not depend on $t \in [t_0, a]$.

Theorem 5.2. *Let f be as in Theorem 5.1 and satisfy the Lipschitz condition (5.4), and suppose that $|f(t, y)| \leq M$ on $[t_0, a] \times G$ for some positive constant M. Let also $_\Delta F_{\alpha,1}(A, t, t_0)$ be defined on $[t_0, a]$. Then the Cauchy problem (5.1), (5.2) has a unique solution $y \in L_\Delta^\alpha[t_0, a)$.*

Proof. We define the sequence $\{y_l(t)\}_{l \in \mathbb{N}_0}$, $t \in [t_0, a)$, as follows:

$$y_0(t) = \sum_{k=1}^{m} h_{\alpha-k}(t, t_0) b_k,$$

$$y_l(t) = y_0(t) + I_{\Delta,t_0}^\alpha f(t, y_{l-1}(t)), \quad t \in [t_0, a), \quad l \in \mathbb{N}.$$

We have

$$|y_1(t) - y_0(t)| = \left| I_{\Delta,t_0}^\alpha f(t, y_0(t)) \right|$$

$$\leq I_{\Delta,t_0}^\alpha |f(t, y_0(t))|$$

$$\leq M I_{\Delta,t_0}^\alpha h_0(t, t_0)$$

$$= M h_\alpha(t, t_0), \quad t \in [t_0, a).$$

Assume that

$$|y_{l-1}(t) - y_{l-2}(t)| \leq M A^{l-2} h_{(l-1)\alpha}(t, t_0), \quad t \in [t_0, a), \tag{5.5}$$

for some $l \in \mathbb{N}$, $l \geq 2$. We will prove that

$$|y_l(t) - y_{l-1}(t)| \leq M A^{l-1} h_{l\alpha}(t, t_0), \quad t \in [t_0, a).$$

Indeed, we have

$$
\begin{aligned}
|y_l(t) - y_{l-1}(t)| &= \left| y_0 + I^\alpha_{\Delta,t_0} f\left(t, y_{l-1}(t)\right) - y_0 - I^\alpha_{\Delta,t_0} f\left(t, y_{l-2}(t)\right) \right| \\
&= \left| I^\alpha_{\Delta,t_0} \left(f\left(t, y_{l-1}(t)\right) - f\left(t, y_{l-2}(t)\right) \right) \right| \\
&\le I^\alpha_{\Delta,t_0} \left| f\left(t, y_{l-1}(t)\right) - f\left(t, y_{l-2}(t)\right) \right| \\
&\le A I^\alpha_{\Delta,t_0} |y_{l-1}(t) - y_{l-2}(t)| \\
&\le M A^{l-1} I^\alpha_{\Delta,t_0} h_{(l-1)\alpha}(t, t_0) \\
&= M A^{l-1} h_{l\alpha}(t, t_0), \quad t \in [t_0, a).
\end{aligned}
$$

Therefore, (5.5) holds for all $l \in \mathbb{N}$, $l \ge 2$. Note that

$$
\begin{aligned}
\left| \lim_{l\to\infty} (y_l(t) - y_0(t)) \right| &= \left| \sum_{l=1}^{\infty} (y_l(t) - y_{l-1}(t)) \right| \\
&\le \sum_{l=1}^{\infty} |y_l(t) - y_{l-1}(t)| \\
&\le M \sum_{l=1}^{\infty} A^{l-1} h_{l\alpha}(t, t_0) \\
&= \frac{M}{A} \sum_{l=1}^{\infty} A^l h_{l\alpha}(t, t_0) \\
&\le \frac{M}{A} \sum_{l=1}^{\infty} A^l h_{l\alpha}(a, t_0) \\
&= \frac{M}{A} \left({}_\Delta F_{\alpha,1}(A, a, t_0) - 1 \right) \\
&< \infty, \quad t \in [t_0, a).
\end{aligned}
$$

Consequently, the series

$$
\sum_{l=1}^{\infty} (y_l(t) - y_{l-1}(t))
$$

is uniformly convergent on $[t_0, a)$. Hence, there exists

$$
\lim_{l\to\infty} (y_l(t) - y_0(t)) = \sum_{l=1}^{\infty} (y_l(t) - y_{l-1}(t)), \quad t \in [t_0, a),
$$

and hence there exists

$$y(t) = \lim_{l \to \infty} y_l(t), \quad t \in [t_0, a),$$

and $y(t)$, $t \in [t_0, a)$, satisfies (5.3). From this and Theorem 5.1, it follows that $y(t)$, $t \in [t_0, a)$, is a solution of the Cauchy problem (5.1), (5.2). Assume that the Cauchy problem (5.1), (5.2) has another solution $z(t)$, $t \in [t_0, a)$. For this solution, we have

$$z(t) = y_0(t) + I^\alpha_{\Delta, t_0} f(t, z(t)), \quad t \in [t_0, a).$$

Note that

$$\begin{aligned}
|y_0(t) - z(t)| &= \left| I^\alpha_{\Delta, t_0} f(t, z(t)) \right| \\
&\leq I^\alpha_{\Delta, t_0} |f(t, z(t))| \\
&\leq M I^\alpha_{\Delta, t_0} h_0(t, t_0) \\
&= M h_\alpha(t, t_0), \quad t \in [t_0, a).
\end{aligned}$$

Assume that

$$|y_{l-1}(t) - z(t)| \leq M A^{l-1} h_{l\alpha}(t, t_0), \quad t \in [t_0, a), \tag{5.6}$$

for some $l \in \mathbb{N}$. We will prove that

$$|y_l(t) - z(t)| \leq M A^l h_{(l+1)\alpha}(t, t_0), \quad t \in [t_0, a).$$

In fact, we have

$$\begin{aligned}
|y_l(t) - z(t)| &= \left| I^\alpha_{\Delta, t_0} f(t, y_{l-1}(t)) - I^\alpha_{\Delta, t_0} f(t, z(t)) \right| \\
&= \left| I^\alpha_{\Delta, t_0} \left(f(t, y_{l-1}(t)) - f(t, z(t)) \right) \right| \\
&\leq I^\alpha_{\Delta, t_0} |f(t, y_{l-1}(t)) - f(t, z(t))| \\
&\leq A I^\alpha_{\Delta, t_0} |y_{l-1}(t) - z(t)| \\
&\leq M A^l I^\alpha_{\Delta, t_0} h_{l\alpha}(t, t_0) \\
&= M A^l h_{(l+1)\alpha}(t, t_0), \quad t \in [t_0, a).
\end{aligned}$$

Therefore, (5.6) holds for all $l \in \mathbb{N}$. Because

$$\lim_{l \to \infty} M A^l h_{(l+1)\alpha}(t, t_0) = 0, \quad t \in [t_0, a),$$

we conclude that

$$y(t) - z(t) = \lim_{l \to \infty} (y_l(t) - z(t))$$

$$= 0, \quad t \in [t_0, a).$$

This completes the proof.

5.2 The Dependence of the Solution on the Initial Data

Let $\alpha \in (0, 1)$. In this section we consider the Cauchy problems

$$\begin{aligned}
D^{\alpha}_{\Delta, t_0} y(t) &= f(t, y(t)), \quad t > t_0, \\
D^{\alpha}_{\Delta, t_0} y(t_0) &= \xi,
\end{aligned} \tag{5.7}$$

and

$$\begin{aligned}
D^{\alpha}_{\Delta, t_0} y(t) &= f(t, y(t)), \quad t > t_0, \\
D^{\alpha}_{\Delta, t_0} y(t_0) &= \eta,
\end{aligned} \tag{5.8}$$

where $f : \mathbb{T} \times \mathbb{R} \to \mathbb{R}$ is a given function, ξ and η are given constants. Let a and G be as in the previous section.

Theorem 5.3. *Suppose that f satisfies the Lipschitz condition (5.4), and $y(t)$ and $z(t)$ are the solutions of the Cauchy problems (5.7) and (5.8), respectively. Let also $_{\Delta}F_{\alpha,\alpha}(A, t, t_0)$ be defined on $[t_0, a)$. Then*

$$|y(t) - z(t)| \le |\xi - \eta|_{\Delta}F_{\alpha,\alpha}(A, a, t_0), \quad t \in [t_0, a).$$

Proof. Let

$$\begin{aligned}
y_0(t) &= \xi h_{\alpha-1}(t, t_0), \\
y_m(t) &= y_0(t) + I^{\alpha}_{\Delta, t_0} f(t, y_{m-1}(t)), \\
z_0(t) &= \eta h_{\alpha-1}(t, t_0), \\
z_m(t) &= z_0(t) + I^{\alpha}_{\Delta, t_0} f(t, z_{m-1}(t)), \quad t \in [t_0, a), \quad m \in \mathbb{N}.
\end{aligned}$$

By the proof of Theorem 5.2, we have

$$y(t) = \lim_{m \to \infty} y_m(t),$$

$$z(t) = \lim_{m \to \infty} z_m(t), \quad t \in [t_0, a).$$

We have that

$$
\begin{aligned}
|z_1(t) - y_1(t)| &= \left| z_0(t) + I^\alpha_{\Delta,t_0} f\left(t, z_0(t)\right) - y_0(t) - I^\alpha_{\Delta,t_0} f\left(t, y_0(t)\right) \right| \\
&= \left| z_0(t) - y_0(t) + I^\alpha_{\Delta,t_0} \left(f\left(t, z_0(t)\right) - f\left(t, y_0(t)\right) \right) \right| \\
&\leq |z_0(t) - y_0(t)| + \left| I^\alpha_{\Delta,t_0} \left(f\left(t, z_0(t)\right) - f\left(t, y_0(t)\right) \right) \right| \\
&\leq |z_0(t) - y_0(t)| + I^\alpha_{\Delta,t_0} |f\left(t, z_0(t)\right) - f\left(t, y_0(t)\right)| \\
&\leq |\xi - \eta| h_{\alpha-1}(t, t_0) + A I^\alpha_{\Delta,t_0} |z_0(t) - y_0(t)| \\
&\leq |\xi - \eta| h_{\alpha-1}(t, t_0) + A|\xi - \eta| I^\alpha_{\Delta,t_0} h_{\alpha-1}(t, t_0) \\
&= |\xi - \eta| \left(h_{\alpha-1}(t, t_0) + A h_{\alpha+\alpha-1}(t, t_0) \right), \quad t \in [t_0, a).
\end{aligned}
$$

Assume that

$$
|z_m(t) - y_m(t)| \leq |\xi - \eta| \sum_{j=0}^{m} A^j h_{j\alpha+\alpha-1}(t, t_0), \quad t \in [t_0, a), \tag{5.9}
$$

for some $m \in \mathbb{N}$. We will prove that

$$
|z_{m+1}(t) - y_{m+1}(t)| \leq |\xi - \eta| \sum_{j=0}^{m+1} A^j h_{j\alpha+\alpha-1}(t, t_0), \quad t \in [t_0, a).
$$

In fact, we have

$$
\begin{aligned}
|z_{m+1}(t) - y_{m+1}(t)| &= \left| z_0(t) + I^\alpha_{\Delta,t_0} f\left(t, z_m(t)\right) - y_0(t) - I^\alpha_{\Delta,t_0} f\left(t, y_m(t)\right) \right| \\
&= \left| z_0(t) - y_0(t) + I^\alpha_{\Delta,t_0} \left(f\left(t, z_m(t)\right) - f\left(t, y_m(t)\right) \right) \right| \\
&\leq |z_0(t) - y_0(t)| + \left| I^\alpha_{\Delta,t_0} \left(f\left(t, z_m(t)\right) - f\left(t, y_m(t)\right) \right) \right| \\
&\leq |z_0(t) - y_0(t)| + I^\alpha_{\Delta,t_0} |f\left(t, z_m(t)\right) - f\left(t, y_m(t)\right)| \\
&\leq |\xi - \eta| h_{\alpha-1}(t, t_0) + A I^\alpha_{\Delta,t_0} |z_m(t) - y_m(t)| \\
&\leq |\xi - \eta| h_{\alpha-1}(t, t_0) + A|\xi - \eta| I^\alpha_{\Delta,t_0} \sum_{j=0}^{m} A^j h_{j\alpha+\alpha-1}(t, t_0) \\
&= |\xi - \eta| \left(h_{\alpha-1}(t, t_0) + \sum_{j=0}^{m} A^{j+1} h_{(j+1)\alpha+\alpha-1}(t, t_0) \right) \\
&= |\xi - \eta| \sum_{j=0}^{m+1} A^j h_{(j+1)\alpha+\alpha-1}(t, t_0), \quad t \in [t_0, a).
\end{aligned}
$$

Therefore, (5.9) holds for all $m \in \mathbb{N}$. By (5.9), we get

$$
\begin{aligned}
|z(t) - y(t)| &= \lim_{m \to \infty} |z_m(t) - y_m(t)| \\
&\leq |\xi - \eta| \sum_{j=0}^{\infty} A^j h_{j\alpha+\alpha-1}(t, t_0) \\
&= |\xi - \eta|_{\Delta} F_{\alpha,\alpha}(A, t, t_0) \\
&\leq |\xi - \eta|_{\Delta} F_{\alpha,\alpha}(A, a, t_0), \quad t \in [t_0, a).
\end{aligned}
$$

This completes the proof.

Chapter 6
Riemann–Liouville Fractional Dynamic Equations with Constant Coefficients

In this chapter we suppose that $\mathbb{T}$ is a time scale with forward jump operator and delta differentiation operator σ and Δ, respectively, such that

$$\mathbb{T} = \{t_n : n \in \mathbb{N}_0\},$$

$$\lim_{n \to \infty} t_n = \infty,$$

$$\sigma(t_n) = t_{n+1}, \quad n \in \mathbb{N}_0,$$

$$w = \inf_{n \in \mathbb{N}_0} \mu(t_n) > 0.$$

6.1 Homogeneous Riemann–Liouville Fractional Dynamic Equations with Constant Coefficients

Let $l \in \mathbb{N}$, $\alpha \in (l-1, l]$, $n = -\overline{[-\alpha]}$, $a \in \mathbb{T}$, $t_0 < a$, $\lambda \in \mathbb{R}$. Consider the equation

$$D_{\Delta,t_0}^{\alpha} y(t) - \lambda y(t) = 0, \quad t \in [t_0, a), \tag{6.1}$$

subject to the initial conditions

$$D_{\Delta,t_0}^{\alpha-j} y(t_0) = d_j, \quad j \in \{1, \ldots, n\}, \tag{6.2}$$

where $d_j, j \in \{1, \ldots, n\}$, are given constants. We define the function

$$W_\alpha(t) = det\left(\left(D_{\Delta,t_0}^{\alpha-k} y_j\right)(t)\right)_{k,j=1}^{n}, \quad t \in [t_0, a),$$

where $y_1, \ldots, y_n : [t_0, a) \to \mathbb{R}$ are solutions of the equation (6.1).

© Springer International Publishing AG, part of Springer Nature 2018
S. G. Georgiev, *Fractional Dynamic Calculus and Fractional Dynamic Equations on Time Scales*, https://doi.org/10.1007/978-3-319-73954-0_6

Theorem 6.1. *Let $l > 1$. The solutions $y_1(t), y_2(t), \ldots, y_n(t)$, defined on $[t_0, a)$, of the equation (6.1) are linearly independent if and only if $W_\alpha(t^\star) \neq 0$ at some point $t^\star \in [t_0, a)$.*

Proof. 1. Let $W_\alpha(t^\star) \neq 0$ at some point $t^\star \in [t_0, a)$. Suppose that $y_j(t), t \in [t_0, a)$, $j \in \{1, \ldots, n\}$, are linearly dependent on $[t_0, a)$. Then there exist n constants c_1, $c_2, \ldots, c_n$, not all zero, such that

$$c_1 y_1(t) + c_2 y_2(t) + \cdots + c_n y_n(t) = 0, \quad t \in [t_0, a).$$

Hence,

$$c_1 D_{\Delta,t_0}^{\alpha-k} y_1(t) + c_2 D_{\Delta,t_0}^{\alpha-k} y_2(t) + \cdots + c_n D_{\Delta,t_0}^{\alpha-k} y_n(t) = 0, \quad t \in [t_0, a),$$

for $k \in \{1, \ldots, n\}$. Therefore,

$$\left(\left(D_{\Delta,t_0}^{\alpha-k} y_j \right)(t) \right)_{k,j=1}^n \begin{pmatrix} c_1 \\ c_2 \\ \vdots \\ c_n \end{pmatrix} = 0, \quad t \in [t_0, a).$$

Thus $W_\alpha(t) = 0$, $t \in [t_0, a)$. This is a contradiction. Therefore, $y_1(t), y_2(t), \ldots, y_n(t), t \in [t_0, a)$, are linearly independent on $[t_0, a)$.

2. Let $y_1(t), y_2(t), \ldots, y_n(t), t \in [t_0, a)$ be linearly independent on $[t_0, a)$. Assume that $W_\alpha(t) = 0$ for all $t \in [t_0, a)$. Consider

$$\left(\left(D_{\Delta,t_0}^{\alpha-k} y_j \right)(t^\star) \right)_{k,j=1}^n C = 0, \tag{6.3}$$

where $t^\star \in [t_0, a)$, $C = \begin{pmatrix} c_1 \\ c_2 \\ \vdots \\ c_n \end{pmatrix}$. Since $W_\alpha(t^\star) = 0$, the system (6.3) has a nontrivial solution $c_1, c_2, \ldots, c_n$. Let

$$y(t) = \sum_{j=1}^n c_j y_j(t), \quad t \in [t_0, a).$$

Note that y is a solution of the equation (6.1). By (6.3), we get that y satisfies the initial condition

$$D_{\Delta,t_0}^{\alpha-k} y(t^\star) = 0, \quad k \in \{1, \ldots, n\}. \tag{6.4}$$

Since $y(t) = 0, t \in [t_0, a)$, is also a solution of equation (6.1) satisfying the initial condition (6.4), we conclude that

$$\sum_{j=1}^{n} c_j y_j(t) = 0, \quad t \in [t_0, a).$$

Therefore, $y_1(t), y_2(t), \ldots, y_n(t), t \in [t_0, a)$, are linearly dependent on $[t_0, a)$. This is a contradiction. Consequently, there exists $t^* \in [t_0, a)$ such that $W_\alpha(t^*) \neq 0$. This completes the proof.

Theorem 6.2. *The system*

$$y_j(t) = {}_\Delta F_{\alpha, \alpha+1-j}(\lambda, t, t_0), \quad t \in [t_0, a), \quad j \in \{1, \ldots, l\},$$

yields the fundamental system of solutions of the equation (6.1). *Moreover,* $y_j(t)$, $j \in \{1, \ldots, l\}$, *satisfy*

$$D_{\Delta, t_0}^{\alpha-k} y_j(t_0) = 0, \quad k, j \in \{1, \ldots, l\}, \quad k \neq j,$$

$$D_{\Delta, t_0}^{\alpha-k} y_k(t_0) = 1, \quad k \in \{1, \ldots, n\},$$

and the solution of the problem (6.1), (6.2) *is given by the expression*

$$y(t) = \sum_{j=1}^{l} d_j y_j(t), \quad t \in [t_0, a).$$

Proof. We apply the Laplace transform to both sides of equation (6.1), and using (6.2), we get

$$0 = \mathscr{L}\left(D_{\Delta, t_0}^{\alpha} y(t)\right)(z, t_0) - \lambda \mathscr{L}(y(t))(z)$$

$$= z^\alpha \mathscr{L}(y(t))(z) - \sum_{j=1}^{l} d_j z^{j-1} - \lambda \mathscr{L}(y(t))(z)$$

$$= \left(z^\alpha - \lambda\right) \mathscr{L}(y(t))(z) - \sum_{j=1}^{l} d_j z^{j-1}.$$

Hence,

$$\left(z^\alpha - \lambda\right) \mathscr{L}(y(t))(z) = \sum_{j=1}^{l} d_j z^{j-1},$$

or

$$\mathscr{L}(y(t))(z) = \sum_{j=1}^{l} d_j \frac{z^{j-1}}{z^\alpha - \lambda}. \tag{6.5}$$

Now we apply Theorem 4.2 with $\beta = \alpha + 1 - j$, and we obtain

$$\mathscr{L}\left(_\Delta F_{\alpha, \alpha+1-j}(\lambda, t, t_0)\right)(z) = \frac{z^{j-1}}{z^\alpha - \lambda}.$$

From this and (6.5), we get

$$\mathscr{L}(y(t))(z) = \sum_{j=1}^{l} d_j \mathscr{L}\left(_\Delta F_{\alpha, \alpha+1-j}(\lambda, t, t_0)\right)(z)$$

$$= \mathscr{L}\left(\sum_{j=1}^{l} d_j {}_\Delta F_{\alpha, \alpha+1-j}(\lambda, t, t_0)\right)(z).$$

Therefore,

$$y(t) = \sum_{j=1}^{l} d_j {}_\Delta F_{\alpha, \alpha+1-j}(\lambda, t, t_0)$$

$$= \sum_{j=1}^{l} d_j y_j(t), \quad t \in [t_0, a).$$

Now we will check that the functions $y_j(t), j \in \{1, \ldots, l\}$, satisfy equation (6.1). In fact, we have

$$D^\alpha_{\Delta, t_0}(y_j(t)) = D^\alpha_{\Delta, t_0}\left(_\Delta F_{\alpha, \alpha+1-j}(\lambda, t, t_0)\right)$$

$$= D^\alpha_{\Delta, t_0}\left(\sum_{k=0}^{\infty} \lambda^k h_{k\alpha+\alpha-j}(t, t_0)\right)$$

$$= \sum_{k=0}^{\infty} \lambda^k D^\alpha_{\Delta, t_0}\left(h_{k\alpha+\alpha-j}(t, t_0)\right)$$

$$= \sum_{k=0}^{\infty} \lambda^k h_{k\alpha-j}(t, t_0)$$

$$= \sum_{k=1}^{\infty} \lambda^k h_{(k)\alpha-j}(t, t_0)$$

$$+ \lambda^0 h_{-j}(t, t_0)$$

$$= \sum_{k=0}^{\infty} \lambda^{k+1} h_{(k+1)\alpha-j}(t, t_0)$$

$$= \lambda \sum_{k=0}^{\infty} \lambda^k h_{k\alpha+\alpha-j}(t, t_0)$$

$$= \lambda_\Delta F_{\alpha,\alpha+1-j}(\lambda, t, t_0), \quad t \in [t_0, a), \quad j \in \{1, \ldots, l\}.$$

Moreover,

$$D_{\Delta,t_0}^{\alpha-k} y_j(t) = D_{\Delta,t_0}^{\alpha-k} \left({}_\Delta F_{\alpha,\alpha+1-j}(\lambda, t, t_0) \right)$$

$$= D_{\Delta,t_0}^{\alpha-k} \left(\sum_{s=0}^{\infty} \lambda^s h_{s\alpha+\alpha-j}(t, t_0) \right)$$

$$= \sum_{s=0}^{\infty} \lambda^s D_{\Delta,t_0}^{\alpha-k} h_{s\alpha+\alpha-j}(t, t_0)$$

$$= \sum_{s=0}^{\infty} \lambda^s h_{s\alpha+k-j}(t, t_0), \quad t \in [t_0, a), \quad j \in \{1, \ldots, l\}.$$

Hence, we get

$$\begin{aligned} D_{\Delta,t_0}^{\alpha-k} y_j(t_0) &= 0, \quad k, j \in \{1, \ldots, l\}, \quad k > j, \\ D_{\Delta,t_0}^{\alpha-k} y_k(t_0) &= 1, \quad k \in \{1, \ldots, l\}. \end{aligned} \tag{6.6}$$

If $k < j$, $k, j \in \{1, \ldots, l\}$, then

$$D_{\Delta,t_0}^{\alpha-k} y_j(t) = \sum_{s=1}^{\infty} \lambda^s h_{s\alpha+k-j}(t, t_0)$$

$$= \sum_{s=0}^{\infty} \lambda^{s+1} h_{s\alpha+\alpha+k-j}(t, t_0), \quad t \in [t_0, a),$$

and since

$$\alpha + k - j \geq \alpha + 1 - l > 0, \quad k, j \in \{1, \ldots, l\}, \quad k < j,$$

we obtain

$$D_{\Delta,t_0}^{\alpha-k} y_j(t_0) = 0, \quad k, j \in \{1, \ldots, l\}, \quad k < j. \tag{6.7}$$

By (6.6) and (6.7), we obtain

$$W_\alpha(t_0) = det \begin{pmatrix} 1 & 0 & \dots & 0 \\ 0 & 1 & \dots & 0 \\ \vdots & \vdots & \vdots & \vdots \\ 0 & 0 & \dots & 1 \end{pmatrix}$$

$$= 1.$$

By Theorem 6.1, it follows that $y_j(t)$, $t \in [t_0, a)$, $j \in \{1, \dots, l\}$, yield the fundamental system of solutions to (6.1). This completes the proof.

Example 6.1. Consider the equation

$$D^{\frac{1}{3}}_{\Delta, t_0} y(t) - 3y(t) = 0, \quad t \in [t_0, a),$$

subject to the initial condition

$$D^{-\frac{2}{3}}_{\Delta, t_0} y(t_0) = 1.$$

Here

$$\alpha = \frac{1}{3},$$

$$l = 1,$$

$$n = \left[\frac{1}{3} \right] + 1$$

$$= 0 + 1$$

$$= 1,$$

$$\lambda = 3.$$

Then

$$y(t) = {}_\Delta F_{\frac{1}{3}, \frac{1}{3} + 1 - 1}(3, t, t_0)$$

$$= {}_\Delta F_{\frac{1}{3}, \frac{1}{3}}(3, t, t_0), \quad t \in [t_0, a),$$

is the solution of the considered initial value problem (IVP).

Example 6.2. Consider the equation

$$D^{\frac{5}{3}}_{\Delta, t_0} y(t) - 2y(t) = 0, \quad t \in [t_0, a),$$

subject to the initial conditions

$$D_{\Delta,t_0}^{\frac{2}{3}} y(t_0) = 1, \quad D_{\Delta,t_0}^{-\frac{1}{3}} y(t_0) = 2.$$

Here

$$\alpha = \frac{5}{3},$$
$$l = 2,$$
$$n = \left[\frac{5}{3}\right] + 1$$
$$= 1 + 1$$
$$= 2,$$
$$\lambda = 2.$$

Then

$$y(t) = {}_\Delta F_{\frac{5}{3},\frac{5}{3}+1-1}(2, t, t_0) + 2 {}_\Delta F_{\frac{5}{3},\frac{5}{3}+1-2}(2, t, t_0)$$
$$= {}_\Delta F_{\frac{5}{3},\frac{5}{3}}(2, t, t_0) + 2 {}_\Delta F_{\frac{5}{3},\frac{2}{3}}(2, t, t_0), \quad t \in [t_0, a),$$

is the solution of the considered IVP.

Example 6.3. Consider the equation

$$D_{\Delta,t_0}^{\frac{11}{4}} y(t) - 7 y(t) = 0, \quad t \in [t_0, a),$$

subject to the initial conditions

$$D_{\Delta,t_0}^{\frac{7}{4}} y(t_0) = -2, \quad D_{\Delta,t_0}^{\frac{3}{4}} y(t_0) = 1, \quad D_{\Delta,t_0}^{-\frac{1}{4}} y(t_0) = 3.$$

Here

$$\alpha = \frac{11}{4},$$
$$l = 3,$$
$$n = \left[\frac{11}{4}\right] + 1$$
$$= 2 + 1$$
$$= 3,$$
$$\lambda = 7.$$

Then

$$y(t) = -2_\Delta F_{\frac{11}{4}, \frac{11}{4}+1-1}(7, t, t_0) + _\Delta F_{\frac{11}{4}, \frac{11}{4}+1-2}(t, t_0)$$

$$+3_\Delta F_{\frac{11}{4}, \frac{11}{4}+1-3}(7, t, t_0)$$

$$= -2_\Delta F_{\frac{11}{4}, \frac{11}{4}}(7, t, t_0) + _\Delta F_{\frac{11}{4}, \frac{7}{4}}(7, t, t_0)$$

$$+3_\Delta F_{\frac{11}{4}, \frac{3}{4}}(7, t, t_0), \quad t \in [t_0, a),$$

is the solution of the considered IVP.

Exercise 6.1. Find the solution of the following IVP:

$$D_{\Delta, t_0}^{\frac{4}{3}} y(t) - 5y(t) = 0, \quad t \in [t_0, a),$$

$$D_{\Delta, t_0}^{\frac{1}{3}} y(t_0) = D_{\Delta, t_0}^{-\frac{2}{3}} y(t_0) = 1.$$

Answer.

$$y(t) = _\Delta F_{\frac{4}{3}, \frac{4}{3}}(5, t, t_0) + _\Delta F_{\frac{4}{3}, \frac{1}{3}}(5, t, t_0), \quad t \in [t_0, a).$$

Exercise 6.2. Find α, λ, d_1, d_2, and d_3 such that the function

$$y(t) = -2_\Delta F_{\frac{5}{2}, \frac{1}{2}}(4, t, t_0), \quad t \in [t_0, a),$$

is the solution of the IVP (6.1), (6.2).

Answer.

$$\alpha = \frac{5}{2}, \quad \lambda = 4, \quad d_1 = d_2 = 0, \quad d_3 = -2.$$

Now we consider the equation

$$D_{\Delta, t_0}^{\alpha} y(t) - \lambda D_{\Delta, t_0}^{\beta} y(t) - \mu y(t) = 0, \quad t \in [t_0, a), \tag{6.8}$$

where $\alpha \in (l-1, l]$, $l \in \mathbb{N}$, $0 < \beta < \alpha$, $\beta \in (m-1, m]$, $\lambda, \mu \in \mathbb{R}$, subject to the initial conditions

$$d_{\alpha, j} = D_{\Delta, t_0}^{\alpha-j} y(t_0), \quad j \in \{1, \dots, l\},$$

$$d_{\beta, j} = D_{\Delta, t_0}^{\beta-j} y(t_0), \quad j \in \{1, \dots, m\}. \tag{6.9}$$

Theorem 6.3. *The functions*

$$y_j(t) = \sum_{k=0}^{\infty} \frac{\mu^k}{k!} \frac{\partial^k}{\partial \lambda^k} \Delta F_{\alpha-\beta,\alpha+k\beta+1-j}(\lambda, t, t_0), \quad t \in [t_0, a), \quad j \in \{1, \ldots, l\},$$

(6.10)

yield the fundamental system of (6.8), *provided that the series in* (6.10) *are convergent. Moreover, if* $\alpha + 1 - l > \beta > l - 1$, *then* $y_j(t), j \in \{1, \ldots, l\},$ *in* (6.10) *satisfy*

$$D_{\Delta,t_0}^{\alpha-k} y_j(t_0) = 0, \quad k, j \in \{1, \ldots, l\}, \quad k \neq j,$$

$$D_{\Delta,t_0}^{\alpha-k} y_k(t_0) = 1, \quad k \in \{1, \ldots, l\}.$$

Proof. Since $\beta < \alpha$, we have that $m \leq l$. We take the Laplace transform of both sides of (6.8), and we get

$$0 = \mathscr{L}\left(D_{\Delta,t_0}^{\alpha} y(t) - \lambda D_{\Delta,t_0}^{\beta} y(t) - \mu y(t)\right)(z, t_0)$$

$$= \mathscr{L}\left(D_{\Delta,t_0}^{\alpha} y(t)\right)(z, t_0) - \lambda \mathscr{L}\left(D_{\Delta,t_0}^{\beta} y(t)\right)(z, t_0)$$

$$- \mu \mathscr{L}(y(t))(z, t_0)$$

$$= z^{\alpha} \mathscr{L}(y(t))(z, t_0) - \sum_{j=1}^{l} d_{\alpha,j} z^{j-1}$$

$$- \lambda z^{\beta} \mathscr{L}(y(t))(z, t_0) + \lambda \sum_{j=1}^{m} d_{\beta,j} z^{j-1}$$

$$- \mu \mathscr{L}(y(t))(z, t_0)$$

$$= \left(z^{\alpha} - \lambda z^{\beta} - \mu\right) \mathscr{L}(y(t))(z, t_0) - \sum_{j=1}^{l} d_{\alpha,j} z^{j-1} + \lambda \sum_{j=1}^{m} d_{\beta,j} z^{j-1}.$$

Let

$$b_j = \begin{cases} d_{\alpha,j} - \lambda d_{\beta,j} & \text{if } j \in \{1, \ldots, m\}, \\ d_{\alpha,j} & \text{if } j \in \{m+1, \ldots, l\}, \end{cases} \quad \text{if } l > m.$$

Therefore,

$$0 = \left(z^{\alpha} - \lambda z^{\beta} - \mu\right) \mathscr{L}(y(t))(z, t_0) - \sum_{j=1}^{l} b_j z^{j-1},$$

or

$$\left(z^\alpha - \lambda z^\beta - \mu\right) \mathscr{L}(y(t))(z, t_0) = \sum_{j=1}^{l} b_j z^{j-1},$$

or

$$\mathscr{L}(y(t))(z, t_0) = \sum_{j=1}^{l} b_j \frac{z^{j-1}}{z^\alpha - \lambda z^\beta - \mu}. \tag{6.11}$$

Note that for $z \in \mathbb{C}$ and

$$\left| \frac{\mu z^{-\beta}}{z^{\alpha-\beta} - \lambda} \right| < 1, \tag{6.12}$$

we have

$$\frac{1}{z^\alpha - \lambda z^\beta - \mu} = \frac{z^{-\beta}}{z^{\alpha-\beta} - \lambda} \frac{1}{\frac{z^{-\beta}}{z^{\alpha-\beta}-\lambda}\left(z^\alpha - \lambda z^\beta - \mu\right)}$$

$$= \frac{z^{-\beta}}{z^{\alpha-\beta} - \lambda} \frac{1}{\frac{z^{\alpha-\beta}}{z^{\alpha-\beta}-\lambda} - \frac{\lambda}{z^{\alpha-\beta}-\lambda} - \frac{\mu z^{-\beta}}{z^{\alpha-\beta}-\lambda}}$$

$$= \frac{z^{-\beta}}{z^{\alpha-\beta} - \lambda} \frac{1}{1 - \frac{\mu z^{-\beta}}{z^{\alpha-\beta}-\lambda}}$$

$$= \frac{z^{-\beta}}{z^{\alpha-\beta} - \lambda} \sum_{k=0}^{\infty} \frac{\mu^k z^{-k\beta}}{\left(z^{\alpha-\beta} - \lambda\right)^k}$$

$$= \sum_{k=0}^{\infty} \frac{\mu^k z^{-\beta-k\beta}}{\left(z^{\alpha-\beta} - \lambda\right)^{k+1}}.$$

From this and (6.11), we get

$$\mathscr{L}(y(t))(z, t_0) = \sum_{j=1}^{l} b_j z^{j-1} \left(\sum_{k=0}^{\infty} \frac{\mu^k z^{-\beta-k\beta}}{\left(z^{\alpha-\beta} - \lambda\right)^{k+1}} \right)$$

$$= \sum_{j=1}^{l} b_j \left(\sum_{k=0}^{\infty} \frac{\mu^k z^{j-1-\beta-k\beta}}{\left(z^{\alpha-\beta} - \lambda\right)^{k+1}} \right).$$

For

$$\left| \lambda z^{\beta-\alpha} \right| < 1, \tag{6.13}$$

we have

$$\frac{z^{j-1-\beta-k\beta}}{\left(z^{\alpha-\beta}-\lambda\right)^{k+1}} = \frac{z^{\alpha-\beta-(\alpha+k\beta+1-j)}}{\left(z^{\alpha-\beta}-\lambda\right)^{k+1}}$$

$$= \frac{1}{k!}\mathscr{L}\left(\frac{\partial^k}{\partial\lambda^k}{}_{\Delta}F_{\alpha-\beta,\alpha+k\beta+1-j}(\lambda,t,t_0)\right)(z,t_0).$$

Therefore,

$$\mathscr{L}(y(t))(z,t_0) = \sum_{j=1}^{l} b_j \left(\sum_{k=0}^{\infty} \frac{\mu^k}{k!}\mathscr{L}\left(\frac{\partial^k}{\partial\lambda^k}{}_{\Delta}F_{\alpha-\beta,\alpha+k\beta+1-j}(\lambda,t,t_0)\right)(z,t_0)\right)$$

$$= \mathscr{L}\left(\sum_{j=1}^{l} b_j \sum_{k=0}^{\infty} \frac{\mu^k}{k!}\left(\frac{\partial^k}{\partial\lambda^k}{}_{\Delta}F_{\alpha-\beta,\alpha+k\beta+1-j}(\lambda,t,t_0)\right)\right)(z,t_0)$$

$$= \mathscr{L}\left(\sum_{j=1}^{l} b_j y_j(t)\right)(z,t_0),$$

provided (6.12), (6.13) hold. Hence,

$$y(t) = \sum_{j=1}^{l} b_j y_j(t), \quad t \in [t_0, a).$$

For $q, j \in \{1, \ldots, l\}$, we have

$$D_{\Delta,t_0}^{\alpha-q} y_j(t) = D_{\Delta,t_0}^{\alpha-q} \sum_{k=0}^{\infty} \frac{\mu^k}{k!}\frac{\partial^k}{\partial\lambda^k}{}_{\Delta}F_{\alpha-\beta,\alpha+k\beta+1-j}(\lambda,t,t_0)$$

$$= \sum_{k=0}^{\infty} \frac{\mu^k}{k!}\frac{\partial^k}{\partial\lambda^k}\left(D_{\Delta,t_0}^{\alpha-q}\left({}_{\Delta}F_{\alpha-\beta,\alpha+k\beta+1-j}(\lambda,t,t_0)\right)\right)$$

$$= \sum_{k=0}^{\infty} \frac{\mu^k}{k!}\frac{\partial^k}{\partial\lambda^k}\left(D_{\Delta,t_0}^{\alpha-q}\left(\sum_{s=0}^{\infty}\lambda^s h_{s(\alpha-\beta)+\alpha+k\beta-j}(t,t_0)\right)\right)$$

$$= \sum_{k=0}^{\infty} \frac{\mu^k}{k!}\frac{\partial^k}{\partial\lambda^k}\sum_{s=0}^{\infty}\lambda^s h_{s(\alpha-\beta)+k\beta+q-j}(t,t_0), \quad t \in [t_0, a).$$

For $q > j$, we have

$$D_{\Delta,t_0}^{\alpha-q} y_j(t_0) = 0;$$

for $q = j$, we have

$$D_{\Delta,t_0}^{\alpha-q} y_q(t_0) = 1.$$

Therefore,

$$W_\alpha(t_0) = 1.$$

From this and Theorem 6.1, we conclude that the functions $y_j(t)$, $j \in \{1, \ldots, n\}$, are linearly independent solutions, and then they yield the fundamental system of solutions of the equation (6.8). Furthermore, if $q < j$, then

$$
\begin{aligned}
D_{\Delta,t_0}^{\alpha-q} y_j(t) = {} & D_{\Delta,t_0}^{\alpha-q} h_{q-j}(t, t_0) \\
& + \sum_{s=1}^{\infty} \lambda^s h_{s(\alpha-\beta)+q-j}(t, t_0) \\
& + \sum_{k=1}^{\infty} \frac{\mu^k}{k!} \frac{\partial^k}{\partial \lambda^k} h_{k\beta+q-j}(t, t_0) \\
& + \sum_{k=1}^{\infty} \frac{\mu^k}{k!} \frac{\partial^k}{\partial \lambda^k} \sum_{s=1}^{\infty} \lambda^s h_{s(\alpha-\beta)+q-j}(t, t_0) \\
= {} & \sum_{s=1}^{\infty} \lambda^s h_{s(\alpha-\beta)+q-j}(t, t_0) \\
& + \sum_{k=1}^{\infty} \frac{\mu^k}{k!} \frac{\partial^k}{\partial \lambda^k} h_{k\beta+q-j}(t, t_0) \\
& + \sum_{k=1}^{\infty} \frac{\mu^k}{k!} \frac{\partial^k}{\partial \lambda^k} \sum_{s=1}^{\infty} \lambda^s h_{s(\alpha-\beta)+k\beta+q-j}(t, t_0), \quad t \in [t_0, a).
\end{aligned}
$$

If

$$\alpha + 1 - l > \beta > l - 1,$$

then

$$
\begin{aligned}
s(\alpha - \beta) + q - j &\geq \alpha - \beta + 1 - l \\
&> 0
\end{aligned}
$$

for $k = 0$, $q, j \in \{1, \ldots, l\}$, $s \in \mathbb{N}$, and

$$
\begin{aligned}
k\beta + q - j &\geq \beta + 1 - l \\
&> 0
\end{aligned}
$$

for $s = 0, q, j \in \{1, \ldots, l\}, k \in \mathbb{N}$. Also,

$$s(\alpha - \beta) + k\beta + q - j \geq \alpha - \beta + \beta + 1 - l$$
$$= \alpha + 1 - l$$
$$> 0$$

for $q, j \in \{1, \ldots, l\}, s, k \in \mathbb{N}$. Therefore,

$$D_{\Delta, t_0}^{\alpha - q} y_j(t_0) = 0, \quad k, j \in \{1, \ldots, l\}, \quad k \neq j,$$

and

$$D_{\Delta, t_0}^{\alpha - q} y_q(t_0) = 1.$$

This completes the proof.

Example 6.4. Consider the equation

$$D_{\Delta, t_0}^{\frac{1}{2}} y(t) - 2 D_{\Delta, t_0}^{\frac{1}{3}} y(t) - 3 y(t) = 0, \quad t \in [t_0, a]$$

subject to the initial conditions

$$D_{\Delta, t_0}^{-\frac{1}{2}} y(t_0) = 1, \quad D_{\Delta, t_0}^{-\frac{2}{3}} y(t_0) = 3.$$

Here

$$\alpha = \frac{1}{2},$$
$$\beta = \frac{1}{3},$$
$$\lambda = 2,$$
$$\mu = 3,$$
$$l = 1,$$
$$m = 1,$$
$$d_{\frac{1}{2}, 1} = 1,$$
$$d_{\frac{1}{3}, 1} = 3.$$

Then

$$b_1 = d_{\frac{1}{2},1} - 2d_{\frac{1}{3},1}$$
$$= 1 - 6$$
$$= -5,$$

$$y_1(t) = \sum_{k=0}^{\infty} \frac{3^k}{k!} \frac{\partial^k}{\partial \lambda^k} \Delta F_{\frac{1}{2}-\frac{1}{3},\frac{1}{2}+k\frac{1}{3}+1-1}(2, t, t_0)$$

$$= \sum_{k=0}^{\infty} \frac{3^k}{k!} \frac{\partial^k}{\partial \lambda^k} \Delta F_{\frac{1}{6},\frac{1}{2}+k\frac{1}{3}}(2, t, t_0)$$

$$= \sum_{k=0}^{\infty} \frac{3^k}{k!} \frac{\partial^k}{\partial \lambda^k} \left(\sum_{j=0}^{\infty} \lambda^j h_{\frac{1}{6}j+\frac{1}{2}+\frac{1}{3}k-1}(t, t_0) \right) \Big|_{\lambda=2}$$

$$= \sum_{k=0}^{\infty} \frac{3^k}{k!} \frac{\partial^k}{\partial \lambda^k} \left(\sum_{j=0}^{\infty} \lambda^j h_{\frac{1}{6}j-\frac{1}{2}+\frac{1}{3}k}(t, t_0) \right) \Big|_{\lambda=2},$$

and

$$y(t) = -5y_1(t), \quad t \in [t_0, a),$$

is the solution of the considered IVP, provided that the last series is convergent.

Example 6.5. Consider the equation

$$D_{\Delta,t_0}^{\frac{4}{3}} y(t) - D_{\Delta,t_0}^{\frac{1}{2}} y(t) - y(t) = 0, \quad t \in [t_0, a),$$

subject to the initial condition

$$D_{\Delta,t_0}^{\frac{1}{3}} y(t_0) = D_{\Delta,t_0}^{-\frac{2}{3}} y(t_0) = D_{\Delta,t_0}^{-\frac{1}{2}} y(t_0) = 1.$$

Here

$$\alpha = \frac{4}{3},$$
$$l = 2,$$
$$\beta = \frac{1}{2},$$
$$m = 1,$$
$$\lambda = 1,$$

$$\mu = 1,$$
$$d_{\frac{4}{3},1} = 1,$$
$$d_{\frac{4}{3},2} = 1,$$
$$d_{\frac{1}{2},1} = 1,$$
$$b_1 = d_{\frac{4}{3},1} - d_{\frac{1}{2},1}$$
$$= 1 - 1$$
$$= 0,$$
$$b_2 = d_{\frac{4}{3},2}$$
$$= 1.$$

Then

$$y_1(t) = \sum_{k=0}^{\infty} \frac{1}{k!} \frac{\partial^k}{\partial \lambda^k} \Delta F_{\frac{4}{3}-\frac{1}{2},\frac{4}{3}+\frac{1}{2}k+1-1}(1,t,t_0)$$

$$= \sum_{k=0}^{\infty} \frac{1}{k!} \frac{\partial^k}{\partial \lambda^k} \Delta F_{\frac{5}{6},\frac{4}{3}+\frac{1}{2}k}(1,t,t_0)$$

$$= \sum_{k=0}^{\infty} \frac{1}{k!} \frac{\partial^k}{\partial \lambda^k} \left(\sum_{j=0}^{\infty} \lambda^j h_{\frac{5}{6}j+\frac{4}{3}+\frac{1}{2}k-1}(t,t_0) \right)\Big|_{\lambda=1}$$

$$= \sum_{k=0}^{\infty} \frac{1}{k!} \frac{\partial^k}{\partial \lambda^k} \left(\sum_{j=0}^{\infty} \lambda^j h_{\frac{5}{6}j+\frac{1}{3}+\frac{1}{2}k}(t,t_0) \right)\Big|_{\lambda=1},$$

$$y_2(t) = \sum_{k=0}^{\infty} \frac{1}{k!} \frac{\partial^k}{\partial \lambda^k} \Delta F_{\frac{4}{3}-\frac{1}{2},\frac{4}{3}+\frac{1}{2}k+1-2}(1,t,t_0)$$

$$= \sum_{k=0}^{\infty} \frac{1}{k!} \frac{\partial^k}{\partial \lambda^k} \Delta F_{\frac{5}{6},\frac{1}{3}+\frac{1}{2}k}(1,t,t_0)$$

$$= \sum_{k=0}^{\infty} \frac{1}{k!} \frac{\partial^k}{\partial \lambda^k} \sum_{j=0}^{\infty} \lambda^j h_{\frac{5}{6}j+\frac{1}{3}+\frac{1}{2}k-1}(t,t_0)\Big|_{\lambda=1}$$

$$= \sum_{k=0}^{\infty} \frac{1}{k!} \frac{\partial^k}{\partial \lambda^k} \sum_{j=0}^{\infty} \lambda^j h_{\frac{5}{6}j-\frac{2}{3}+\frac{1}{2}k}(t,t_0)\Big|_{\lambda=1},$$

$$y(t) = b_1 y_1(t) + b_2 y_2(t)$$
$$= y_2(t), \quad t \in [t_0, a),$$

is the solution of the considered IVP, provided that the last series is convergent.

Example 6.6. Consider the equation

$$D_{\Delta,t_0}^{\frac{9}{4}} y(t) - D_{\Delta,t_0}^{\frac{3}{2}} y(t) - 4y(t) = 0, \quad t \in [t_0, a]$$

subject to the initial conditions

$$D_{\Delta,t_0}^{\frac{5}{4}} y(t_0) = 2, \quad D_{\Delta,t_0}^{\frac{1}{4}} y(t_0) = -3, \quad D_{\Delta,t_0}^{-\frac{3}{4}} y(t_0) = 1,$$

$$D_{\Delta,t_0}^{\frac{1}{2}} y(t_0) = 1, \quad D_{\Delta,t_0}^{-\frac{1}{2}} y(t_0) = 2.$$

Here

$$\alpha = \frac{9}{4},$$
$$l = 3,$$
$$\beta = \frac{3}{2},$$
$$m = 2,$$
$$\lambda = 1,$$
$$\mu = 4,$$
$$d_{\frac{5}{4},1} = 2,$$
$$d_{\frac{5}{4},2} = -3,$$
$$d_{\frac{5}{4},3} = 1,$$
$$d_{\frac{1}{2},1} = 1,$$
$$d_{\frac{1}{2},2} = 2,$$
$$b_1 = d_{\frac{5}{4},1} - d_{\frac{1}{2},1}$$
$$= 2 - 1$$
$$= 1,$$
$$b_2 = d_{\frac{5}{4},2} - d_{\frac{1}{2},2}$$
$$= -3 - 2$$
$$= -5,$$

$$b_3 = d_{\frac{5}{4},3}$$

$$= 1.$$

Then

$$y_1(t) = \sum_{k=0}^{\infty} \frac{4^k}{k!} \frac{\partial^k}{\partial \lambda^k} {}_\Delta F_{\frac{9}{4}-\frac{3}{2},\frac{9}{4}+\frac{3}{2}k+1-1}(1, t, t_0)$$

$$= \sum_{k=0}^{\infty} \frac{4^k}{k!} \frac{\partial^k}{\partial \lambda^k} {}_\Delta F_{\frac{3}{4},\frac{9}{4}+\frac{3}{2}k}(1, t, t_0)$$

$$= \sum_{k=0}^{\infty} \frac{4^k}{k!} \frac{\partial^k}{\partial \lambda^k} \sum_{j=0}^{\infty} \lambda^j h_{\frac{3}{4}j+\frac{9}{4}+\frac{3}{2}k-1}(t, t_0)\Big|_{\lambda=1}$$

$$= \sum_{k=0}^{\infty} \frac{4^k}{k!} \frac{\partial^k}{\partial \lambda^k} \sum_{j=0}^{\infty} \lambda^j h_{\frac{3}{4}j+\frac{5}{4}+\frac{3}{2}k}(t, t_0)\Big|_{\lambda=1},$$

$$y_2(t) = \sum_{k=0}^{\infty} \frac{4^k}{k!} \frac{\partial^k}{\partial \lambda^k} {}_\Delta F_{\frac{9}{4}-\frac{3}{2},\frac{9}{4}+\frac{3}{2}k+1-2}(1, t, t_0)$$

$$= \sum_{k=0}^{\infty} \frac{4^k}{k!} \frac{\partial^k}{\partial \lambda^k} {}_\Delta F_{\frac{3}{4},\frac{5}{4}+\frac{3}{2}k}(1, t, t_0)$$

$$= \sum_{k=0}^{\infty} \frac{4^k}{k!} \frac{\partial^k}{\partial \lambda^k} \left(\sum_{j=0}^{\infty} \lambda^j h_{\frac{3}{4}j+\frac{5}{4}+\frac{3}{2}k-1}(t, t_0) \right)\Big|_{\lambda=1}$$

$$= \sum_{k=0}^{\infty} \frac{4^k}{k!} \frac{\partial^k}{\partial \lambda^k} \left(\sum_{j=0}^{\infty} \lambda^j h_{\frac{3}{4}j+\frac{3}{2}k+\frac{1}{4}}(t, t_0) \right)\Big|_{\lambda=1},$$

$$y_3(t) = \sum_{k=0}^{\infty} \frac{4^k}{k!} \frac{\partial^k}{\partial \lambda^k} {}_\Delta F_{\frac{9}{4}-\frac{3}{2},\frac{9}{4}+\frac{3}{2}k+1-3}(t, t_0)$$

$$= \sum_{k=0}^{\infty} \frac{4^k}{k!} \frac{\partial^k}{\partial \lambda^k} {}_\Delta F_{\frac{3}{4},\frac{1}{4}+\frac{3}{2}k}(t, t_0)$$

$$= \sum_{k=0}^{\infty} \frac{4^k}{k!} \frac{\partial^k}{\partial \lambda^k} \left(\sum_{j=0}^{\infty} \lambda^j h_{\frac{3}{4}j+\frac{1}{4}+\frac{3}{2}k-1}(t, t_0) \right)\Big|_{\lambda=1}$$

$$= \sum_{k=0}^{\infty} \frac{4^k}{k!} \frac{\partial^k}{\partial \lambda^k} \left(\sum_{j=0}^{\infty} \lambda^j h_{\frac{3}{4}j-\frac{3}{4}+\frac{3}{2}k}(t, t_0) \right)\Big|_{\lambda=1},$$

$$y(t) = b_1 y_1(t) + b_2 y_2(t) + b_3 y_3(t)$$
$$= y_1(t) - 5y_2(t) + y_3(t), \quad t \in [t_0, a),$$

is the solution of the considered IVP, provided that the last series is convergent.

Exercise 6.3. Find the solution of the following IVP:

$$D^{\frac{7}{2}}_{\Delta, t_0} y(t) - 3D^{\frac{4}{3}}_{\Delta, t_0} y(t) - 5y(t) = 0, \quad t \in [t_0, a),$$

$$D^{\frac{5}{2}}_{\Delta, t_0} y(t_0) = 1, \quad D^{\frac{3}{2}}_{\Delta, t_0} y(t_0) = -3, \quad D^{\frac{1}{2}}_{\Delta, t_0} y(t_0) = -4,$$

$$D^{-\frac{1}{2}}_{\Delta, t_0} y(t_0) = 2, \quad D^{\frac{1}{3}}_{\Delta, t_0} y(t_0) = -1, \quad D^{-\frac{2}{3}}_{\Delta, t_0} y(t_0) = -2.$$

Answer.

$$y(t) = \sum_{k=0}^{\infty} \frac{4^k}{k!} \frac{\partial^k}{\partial \lambda^k} \sum_{j=0}^{\infty} \lambda^j$$

$$\times \Big(4h_{\frac{13}{6}j + \frac{5}{2} + \frac{4}{3}k}(t, t_0) + 3h_{\frac{13}{6}j + \frac{3}{2} + \frac{4}{3}k}(t, t_0)$$

$$-4h_{\frac{13}{6}j + \frac{1}{2} + \frac{4}{3}k}(t, t_0) + 2h_{\frac{13}{6}j - \frac{1}{2} + \frac{4}{3}k}(t, t_0) \Big) \Big|_{\lambda = 3}, \quad t \in [t_0, a),$$

provided that the last series is convergent.
Consider the equation

$$D^{\alpha}_{\Delta, t_0} y(t) - \lambda D^{\beta}_{\Delta, t_0} y(t) - \sum_{k=0}^{m-2} A_k D^{\alpha_k}_{\Delta, t_0} y(t) = 0, \quad t \in [t_0, a), \tag{6.14}$$

where

$$m > 2, \quad m \in \mathbb{N},$$
$$\alpha_0 = 0, \quad 0 < \alpha_1 < \ldots < \alpha_{m-2} < \beta < \alpha,$$
$$\lambda, A_0, \ldots, A_{m-2} \in \mathbb{R},$$
$$l_k - 1 < \alpha_k \le l_k, \quad l_k \in \mathbb{N}, \quad k \in \{1, \ldots, m-2\}$$
$$l_{m-1} - 1 < \beta \le l_{m-1},$$
$$l - 1 < \alpha \le l, \quad l \in \mathbb{N},$$

subject to the initial conditions

$$
\begin{aligned}
D_{\Delta,t_0}^{\alpha_j-r} y(t_0) &= d_{\alpha_j,r}, \quad d_{\alpha_j,r} \in \mathbb{R}, \quad r \in \{1,\ldots,l_j\}, \quad j \in \{1,\ldots,m-2\}, \\
D_{\Delta,t_0}^{\beta-r} y(t_0) &= d_{\beta,r}, \quad d_{\beta,r} \in \mathbb{R}, \quad r \in \{1,\ldots,l_{m-1}\}, \\
D_{\Delta,t_0}^{\alpha-r} y(t_0) &= d_{\alpha,r}, \quad d_{\alpha,r} \in \mathbb{R}, \quad r \in \{1,\ldots,l\}.
\end{aligned}
$$

$$(6.15)$$

Theorem 6.4. *The functions*

$$
\begin{aligned}
y_j(t) = \sum_{n=0}^{\infty} &\left(\sum_{k_0+\cdots+k_{m-2}=n} \right) \frac{1}{k_0!\ldots k_{m-2}!} \\
&\times \left(\prod_{\nu=0}^{m-2} (A_\nu)^{k_\nu} \right) \frac{\partial^n}{\partial \lambda^n} {}_{\Delta} F_{\alpha-\beta,\alpha+1-j+\sum\limits_{\nu=0}^{m-2}(\beta-\alpha_\nu)k_\nu} (\lambda,t,t_0),
\end{aligned}
$$

$$(6.16)$$

$t \in [t_0,a)$, $j \in \{1,\ldots,l\}$, *yield the fundamental system of solutions of the equation* (6.14), *provided that the series in* (6.16) *are convergent. Moreover, if*

$$
\alpha + 1 - l > \beta > \alpha_m + l - 1,
$$

then

$$
\begin{aligned}
D_{\Delta,t_0}^{\alpha-k} y_j(t_0) &= 0, \quad k,j \in \{1,\ldots,l\}, \quad k \neq j, \\
D_{\Delta,t_0}^{\alpha-k} y_k(t_0) &= 1, \quad k \in \{1,\ldots,l\}.
\end{aligned}
$$

Proof. We take the Laplace transform of both sides of (6.14), and we get

$$
\begin{aligned}
0 &= \mathscr{L} \left(D_{\Delta,t_0}^{\alpha} y(t) - \lambda D_{\Delta,t_0}^{\beta} y(t) - \sum_{k=0}^{m-2} A_k D_{\Delta,t_0}^{\alpha_k} y(t) \right)(z,t_0) \\
&= \mathscr{L} \left(D_{\Delta,t_0}^{\alpha} y(t) \right)(z,t_0) - \lambda \mathscr{L} \left(D_{\Delta,t_0}^{\beta} y(t) \right)(z,t_0) \\
&\quad - \sum_{k=0}^{m-2} A_k \mathscr{L} \left(D_{\Delta,t_0}^{\alpha_k} y(t) \right)(z,t_0) \\
&= z^{\alpha} \mathscr{L}(y(t))(z,t_0) - \sum_{j=1}^{l} d_{\alpha,j} z^{j-1} \\
&\quad - \lambda z^{\beta} \mathscr{L}(y(t))(z,t_0) + \lambda \sum_{j=1}^{l_{m-1}} d_{\beta,j} z^{j-1}
\end{aligned}
$$

$$-\sum_{k=0}^{m-2} A_k z^{\alpha_k} \mathscr{L}(y(t))(z, t_0) + \sum_{k=1}^{m-2} A_k \left(\sum_{j=1}^{l_k} d_{\alpha_k, j} z^{j-1} \right)$$

$$= \left(z^\alpha - \lambda z^\beta - \sum_{k=0}^{m-2} A_k z^{\alpha_k} \right) \mathscr{L}(y(t))(z, t_0)$$

$$- \sum_{j=1}^{l} d_{\alpha, j} z^{j-1} + \lambda \sum_{j=1}^{l_{m-2}} d_{\beta, j} z^{j-1}$$

$$+ \sum_{k=1}^{m-2} \left(\sum_{j=1}^{l_k} A_k d_{\alpha_k, j} z^{j-1} \right).$$

Let

$$d_j = d_{\alpha, j} - \lambda d_{\beta, j} - \sum_{k=1}^{m-2} A_k d_{\alpha_k, j}, \quad j \in \{1, \ldots, l_1\},$$

$$d_j = d_{\alpha, j} - \lambda d_{\beta, j} - \sum_{k=1}^{m-2} A_k d_{\alpha_k, j}, \quad j \in \{l_1 + 1, \ldots, l_2\}, \quad l_2 > l_1, \quad m \geq 3,$$

$$\vdots$$

$$d_j = d_{\alpha, j} - \lambda d_{\beta, j} - A_{m-2} d_{\alpha_{m-2}, j}, \quad j \in \{l_{m-3} + 1, \ldots, l_{m-2}\}, \quad l_{m-2} > l_{m-3},$$

$$d_j = d_{\alpha, j} - \lambda d_{\beta, j}, \quad j \in \{l_{m-2} + 1, \ldots, l_{m-1}\}, \quad l_{m-1} > l_{m-2},$$

$$d_j = d_{\alpha, j}, \quad j \in \{l_{m-1} + 1, \ldots, l\}, \quad l > l_{m-1}.$$

Then

$$0 = \left(z^\alpha - \lambda z^\beta - \sum_{k=0}^{m-2} A_k z^{\alpha_k} \right) \mathscr{L}(y(t))(z, t_0)$$

$$- \sum_{j=1}^{l} d_j z^{j-1},$$

whereupon

$$\left(z^\alpha - \lambda z^\beta - \sum_{k=0}^{m-2} A_k z^{\alpha_k} \right) \mathscr{L}(y(t))(z, t_0) = \sum_{j=1}^{l} d_j z^{j-1},$$

or

$$\mathscr{L}(y(t))(z, t_0) = \sum_{j=1}^{l} d_j \frac{z^{j-1}}{z^\alpha - \lambda z^\beta - \sum_{k=0}^{m-2} A_k z^{\alpha_k}}.$$

Now, using that

$$(x_0 + \cdots + x_{m-2})^n = \left(\sum_{k_0 + \cdots + k_{m-2} = n} \right) \frac{n!}{k_0! \ldots k_{m-2}!} \prod_{\nu=0}^{m-2} x_\nu^{k_\nu},$$

we get

$$\frac{1}{z^\alpha - \lambda z^\beta - \sum_{k=0}^{m-2} A_k z^{\alpha_k}} = \frac{z^{-\beta}}{z^{\alpha-\beta} - \lambda} \frac{1}{\frac{z^{-\beta}}{z^{\alpha-\beta}-\lambda} \left(z^\alpha - \lambda z^\beta - \sum_{k=0}^{m-2} A_k z^{\alpha_k} \right)}$$

$$= \frac{z^{-\beta}}{z^{\alpha-\beta} - \lambda} \frac{1}{\frac{1}{z^{\alpha-\beta}-\lambda} \left(z^{\alpha-\beta} - \lambda - \sum_{k=0}^{m-2} A_k z^{\alpha_k - \beta} \right)}$$

$$= \frac{z^{-\beta}}{z^{\alpha-\beta} - \lambda} \frac{1}{1 - \sum_{k=0}^{m-2} A_k \frac{z^{\alpha_k - \beta}}{z^{\alpha-\beta}-\lambda}}$$

$$= \sum_{n=0}^{\infty} \frac{z^{-\beta}}{(z^{\alpha-\beta} - \lambda)^{n+1}} \left(\sum_{k=0}^{m-2} A_k z^{\alpha_k - \beta} \right)^n$$

$$= \sum_{n=0}^{\infty} \left(\sum_{k_0 + \cdots + k_{m-2} = n} \right) \frac{n!}{k_0! \ldots k_{m-2}!}$$

$$\times \left(\prod_{\nu=0}^{m-2} (A_\nu)^{k_\nu} \right) \frac{z^{-\beta - \sum_{\nu=0}^{m-2} (\beta-\alpha_\nu)k_\nu}}{(z^{\alpha-\beta} - \lambda)^{n+1}},$$

if

$$\left| \sum_{k=0}^{m-2} A_k \frac{z^{\alpha_k - \beta}}{z^{\alpha-\beta} - \lambda} \right| < 1. \tag{6.17}$$

From this, using that

$$
\frac{z^{j-1-\beta-\sum\limits_{v=0}^{m-2}(\beta-\alpha_v)k_v}}{\left(z^{\alpha-\beta}-\lambda\right)^{n+1}}=\frac{z^{\alpha-\beta-\left(\alpha+1-j+\sum\limits_{v=0}^{m-2}(\beta-\alpha_v)k_v\right)}}{\left(z^{\alpha-\beta}-\lambda\right)^{n+1}}
$$

$$
=\frac{1}{n!}\mathscr{L}\left(\frac{\partial^n}{\partial\lambda^n}{}_\Delta F_{\alpha-\beta,\alpha+1-j+\sum\limits_{v=0}^{m-2}(\beta-\alpha_v)k_v}(\lambda,t,t_0)\right)(z,t_0),
$$

if (6.17) holds, we obtain

$$
\mathscr{L}(y(t))(z,t_0)=\sum_{j=1}^{l}d_j\sum_{n=0}^{\infty}\left(\sum_{k_0+\cdots+k_{m-2}=n}\right)\frac{n!}{k_0!\ldots k_{m-2}!}
$$

$$
\times\left(\prod_{v=0}^{m-2}(A_v)^{k_v}\right)\mathscr{L}\left(\frac{\partial^n}{\partial\lambda^n}{}_\Delta F_{\alpha-\beta,\alpha+1-j+\sum\limits_{v=0}^{m-2}(\beta-\alpha_v)k_v}(\lambda,t,t_0)\right)(z,t_0),
$$

whereupon

$$
y(t)=\sum_{j=1}^{l}d_j\sum_{n=0}^{\infty}\left(\sum_{k_0+\cdots+k_{m-2}=n}\right)\frac{n!}{k_0!\ldots k_{m-2}!}
$$

$$
\times\left(\prod_{v=0}^{m-2}(A_v)^{k_v}\right)\frac{\partial^n}{\partial\lambda^n}{}_\Delta F_{\alpha-\beta,\alpha+1-j+\sum\limits_{v=0}^{m-2}(\beta-\alpha_v)k_v}(\lambda,t,t_0),\quad t\in[t_0,a).
$$

For $q,j\in\{1,\ldots,l\}$, we have

$$
D^{\alpha-q}_{\Delta,t_0}y_j(t)=D^{\alpha-q}_{\Delta,t_0}\left(\sum_{n=0}^{\infty}\left(\sum_{k_0+\cdots+k_{m-2}=n}\right)\frac{n!}{k_0!\ldots k_{m-2}!}\right.
$$

$$
\times\left(\prod_{v=0}^{m-2}(A_v)^{k_v}\right)\frac{\partial^n}{\partial\lambda^n}{}_\Delta F_{\alpha-\beta,\alpha+1-j+\sum\limits_{v=0}^{m-2}(\beta-\alpha_v)k_v}(\lambda,t,t_0)\right)
$$

$$
=\sum_{n=0}^{\infty}\left(\sum_{k_0+\cdots+k_{m-2}=n}\right)\frac{n!}{k_0!\ldots k_{m-2}!}
$$

$$
\times\left(\prod_{v=0}^{m-2}(A_v)^{k_v}\right)\frac{\partial^n}{\partial\lambda^n}D^{\alpha-q}_{\Delta,t_0}{}_\Delta F_{\alpha-\beta,\alpha+1-j+\sum\limits_{v=0}^{m-2}(\beta-\alpha_v)k_v}(\lambda,t,t_0)
$$

$$= \sum_{n=0}^{\infty} \left(\sum_{k_0 + \cdots + k_{m-2} = n} \right) \frac{n!}{k_0! \ldots k_{m-2}!}$$

$$\times \left(\prod_{\nu=0}^{m-2} (A_\nu)^{k_\nu} \right) \frac{\partial^n}{\partial \lambda^n} D_{\Delta, t_0}^{\alpha-q} \sum_{s=0}^{\infty} \lambda^s h_{s(\alpha-\beta)+\alpha-j+1+\sum_{\nu=0}^{m-2}(\beta-\alpha_\nu)k_\nu - 1}(t, t_0)$$

$$= \sum_{n=0}^{\infty} \left(\sum_{k_0 + \cdots + k_{m-2} = n} \right) \frac{n!}{k_0! \ldots k_{m-2}!}$$

$$\times \left(\prod_{\nu=0}^{m-2} (A_\nu)^{k_\nu} \right) \frac{\partial^n}{\partial \lambda^n} \sum_{s=0}^{\infty} \lambda^s h_{s(\alpha-\beta)+q-j+\sum_{\nu=0}^{m-2}(\beta-\alpha_\nu)k_\nu}(t, t_0), \quad t > t_0.$$

If $q < j, j \in \{1, \ldots, l\}$, then

$$D_{\Delta, t_0}^{\alpha-q} y(t) = D_{\Delta, t_0}^{\alpha-q} h_{\alpha-j}(t, t_0)$$

$$+ \sum_{s=1}^{\infty} \lambda^s h_{s(\alpha-\beta)+q-j}(t, t_0)$$

$$+ \sum_{n=1}^{\infty} \left(\sum_{k_0 + \cdots + k_{m-2} = n} \right) \frac{n!}{k_0! \cdots k_{m-2}!}$$

$$\times \left(\prod_{\nu=0}^{m-2} (A_\nu)^{k_\nu} \right) \frac{\partial^n}{\partial \lambda^n} h_{m-2}_{\sum_{\nu=0}^{m-2}(\beta-\alpha_\nu)k_\nu+q-j}(t, t_0)$$

$$+ \sum_{n=1}^{\infty} \left(\sum_{k_0 + \cdots + k_{m-2} = n} \right) \frac{n!}{k_0! \ldots k_{m-2}!}$$

$$\times \left(\prod_{\nu=0}^{m-2} (A_\nu)^{k_\nu} \right) \frac{\partial^n}{\partial \lambda^n} \sum_{s=1}^{\infty} \lambda^s h_{s(\alpha-\beta)+q-j+\sum_{\nu=0}^{m-2}(\beta-\alpha_\nu)k_\nu}(t, t_0)$$

$$= \sum_{s=1}^{\infty} \lambda^s h_{s(\alpha-\beta)+q-j}(t, t_0)$$

$$+ \sum_{n=1}^{\infty} \left(\sum_{k_0 + \cdots + k_{m-2} = n} \right) \frac{1}{k_0! \ldots k_{m-2}!}$$

$$\times \left(\prod_{v=0}^{m-2} (A_v)^{k_v} \right) \frac{\partial^n}{\partial \lambda^n} h_{m-2 \atop \sum_{v=0} (\beta - \alpha_v)k_v + q - j} (t, t_0)$$

$$+ \sum_{n=1}^{\infty} \left(\sum_{k_0 + \cdots + k_{m-2} = n} \right) \frac{n!}{k_0! \ldots k_{m-2}!}$$

$$\times \left(\prod_{v=0}^{m-2} (A_v)^{k_v} \right) \frac{\partial^n}{\partial \lambda^n} \sum_{s=1}^{\infty} \lambda^s h_{s(\alpha - \beta) + q - j + \sum_{v=0}^{m-2} (\beta - \alpha_v)k_v} (t, t_0), \quad t \in [t_0, a).$$

If

$$\alpha + 1 - l > \beta$$

$$> \alpha_{m-2} + l - 1,$$

then

$$s(\alpha - \beta) + q - j \geq \alpha - \beta + 1 - l$$

$$> 0$$

for $n = 0$, $q, j \in \{1, \ldots, l\}$, $s \in \mathbb{N}$, and

$$\sum_{v=0}^{m-2} (\beta - \alpha_v) k_v + q - j \geq \beta - \alpha_1 + 1 - l$$

$$\geq \beta - \alpha_{m-2} + 1 - l$$

$$> 0$$

for $s = 0$, $q, j \in \{1, \ldots, l\}$, $n \in \mathbb{N}$. Also,

$$s(\alpha - \beta) + \sum_{v=0}^{m-2} (\beta - \alpha_v) k_v + q - j \geq \alpha - \beta + \beta - \alpha_1 + 1 - l$$

$$\geq \alpha - \alpha_{m-2} + 1 - l$$

$$> 0$$

for $q, j \in \{1, \ldots, l\}$, $s, n \in \mathbb{N}$. Therefore,

$$D_{\Delta, t_0}^{\alpha - q} y_j(t_0) = 0, \quad q, j \in \{1, \ldots, l\}, \quad q \neq j,$$

$$D_{\Delta, t_0}^{\alpha - q} y_q(t_0) = 1, \quad q \in \{1, \ldots, l\}.$$

This completes the proof.

Example 6.7. Consider the equation

$$D_{\Delta,t_0}^{\frac{5}{2}}y(t) - 3D_{\Delta,t_0}^{\frac{4}{3}}y(t) - 2D_{\Delta,t_0}^{\frac{3}{4}}y(t) - 4y(t) = 0, \quad t \in [t_0, a),$$

subject to the initial conditions

$$D_{\Delta,t_0}^{\frac{3}{2}}y(t_0) = 1, \quad D_{\Delta,t_0}^{\frac{1}{2}}y(t_0) = 2, \quad D_{\Delta,t_0}^{-\frac{1}{2}}y(t_0) = 1,$$

$$D_{\Delta,t_0}^{\frac{1}{3}}y(t_0) = 2, \quad D_{\Delta,t_0}^{-\frac{2}{3}}y(t_0) = 1,$$

$$D_{\Delta,t_0}^{-\frac{1}{4}}y(t_0) = -3.$$

Here

$$\alpha = \frac{5}{2},$$

$$\beta = \frac{4}{3},$$

$$\alpha_1 = \frac{3}{4},$$

$$\alpha_0 = 0,$$

$$\lambda = 3,$$

$$A_0 = 4,$$

$$A_1 = 2,$$

$$m = 3,$$

$$d_1 = 1 - 3(2) - 2(-3)$$

$$= 1 - 6 + 6$$

$$= 1,$$

$$d_2 = 2 - 3(1)$$

$$= 2 - 3$$

$$= -1,$$

$$d_3 = 1.$$

Then

$$
y_1(t) = \sum_{n=0}^{\infty} \left(\sum_{k_0+k_1=n} \right) \frac{n!}{k_0!k_1!} \left(\prod_{\nu=0}^{1} (A_\nu)^{k_\nu} \right)
$$

$$
\times \frac{\partial^n}{\partial \lambda^n} {}_\Delta F_{\frac{5}{2}-\frac{4}{3},\,\frac{5}{2}+1-1+\sum_{\nu=0}^{1}\left(\frac{4}{3}-\alpha_\nu\right)k_\nu} (3, t, t_0)
$$

$$
= \sum_{n=0}^{\infty} \left(\sum_{k_0+k_1=n} \right) \frac{n!}{k_0!k_1!} \left(\prod_{\nu=0}^{1} (A_\nu)^{k_\nu} \right)
$$

$$
\times \frac{\partial^n}{\partial \lambda^n} {}_\Delta F_{\frac{7}{6},\,\frac{5}{2}+\frac{4}{3}k_0+\frac{7}{12}k_1} (3, t, t_0)
$$

$$
= \sum_{n=0}^{\infty} \left(\sum_{k_0+k_1=n} \right) \frac{n!}{k_0!k_1!} \left(4^{k_0} 2^{k_1} \right)
$$

$$
\times \frac{\partial^n}{\partial \lambda^n} {}_\Delta F_{\frac{7}{6},\,\frac{5}{2}+\frac{4}{3}k_0+\frac{7}{12}k_1} (3, t, t_0)
$$

$$
= \sum_{n=0}^{\infty} \left(\sum_{k_0+k_1=n} \right) \frac{n!}{k_0!k_1!} 2^{2k_0+k_1} \frac{\partial^n}{\partial \lambda^n} \left(\sum_{j=0}^{\infty} \lambda^j h_{\frac{7}{6}j+\frac{5}{2}+\frac{4}{3}k_0+\frac{7}{12}k_1-1} (t, t_0) \right) \Bigg|_{\lambda=3}
$$

$$
= \sum_{j=0}^{\infty} \lambda^j h_{\frac{7}{6}j+\frac{3}{2}+\frac{4}{3}k_0+\frac{7}{12}k_1} (t, t_0) \Big|_{\lambda=3}
$$

$$
+ \sum_{n=1}^{\infty} \left(\sum_{k_0+k_1=n} \right) \frac{n!}{k_0!k_1!} 2^{2k_0+k_1} \sum_{j=n}^{\infty} \lambda \ldots (\lambda-n+1) h_{\frac{7}{6}j+\frac{3}{2}+\frac{4}{3}k_0+\frac{7}{12}k_1} (t, t_0) \Big|_{\lambda=3}
$$

$$
= \sum_{j=0}^{\infty} 3^j h_{\frac{7}{6}j+\frac{3}{2}+\frac{4}{3}k_0+\frac{7}{12}k_1} (t, t_0)
$$

$$
+ \sum_{n=1}^{\infty} \left(\sum_{k_0+k_1=n} \right) \frac{n!}{k_0!k_1!} 2^{2k_0+k_1} \sum_{j=n}^{\infty} 3 \ldots (4-n) h_{\frac{7}{6}j+\frac{3}{2}+\frac{4}{3}k_0+\frac{7}{12}k_1} (t, t_0)
$$

$$
= \sum_{j=0}^{\infty} 3^j h_{\frac{7}{6}j+\frac{3}{2}+\frac{4}{3}k_0+\frac{7}{12}k_1} (t, t_0)
$$

$$
+ \sum_{n=1}^{3} \left(\sum_{k_0+k_1=n} \right) \frac{n!}{k_0!k_1!} 2^{2k_0+k_1} \sum_{j=n}^{\infty} 3 \ldots (4-n) h_{\frac{7}{6}j+\frac{3}{2}+\frac{4}{3}k_0+\frac{7}{12}k_1} (t, t_0),
$$

$$y_2(t) = \sum_{n=0}^{\infty} \left(\sum_{k_0+k_1=n} \right) \frac{n!}{k_0!k_1!} \left(\prod_{\nu=0}^{1} (A_\nu)^{k_\nu} \right)$$

$$\times \frac{\partial^n}{\partial \lambda^n} {}_\Delta F_{\frac{5}{2}-\frac{4}{3},\frac{5}{2}+1-2+\sum_{\nu=0}^{1}\left(\frac{4}{3}-\alpha_\nu\right)k_\nu}(3,t,t_0)$$

$$= \sum_{n=0}^{\infty} \left(\sum_{k_0+k_1=n} \right) \frac{n!}{k_0!k_1!} \left(\prod_{\nu=0}^{1} (A_\nu)^{k_\nu} \right)$$

$$\times \frac{\partial^n}{\partial \lambda^n} {}_\Delta F_{\frac{7}{6},\frac{3}{2}+\frac{4}{3}k_0+\frac{7}{12}k_1}(3,t,t_0)$$

$$= \sum_{n=0}^{\infty} \left(\sum_{k_0+k_1=n} \right) \frac{n!}{k_0!k_1!} \left(4^{k_0} 2^{k_1} \right)$$

$$\times \frac{\partial^n}{\partial \lambda^n} {}_\Delta F_{\frac{7}{6},\frac{3}{2}+\frac{4}{3}k_0+\frac{7}{12}k_1}(3,t,t_0)$$

$$= \sum_{n=0}^{\infty} \left(\sum_{k_0+k_1=n} \right) \frac{n!}{k_0!k_1!} 2^{2k_0+k_1} \frac{\partial^n}{\partial \lambda^n} \left(\sum_{j=0}^{\infty} \lambda^j h_{\frac{7}{6}j+\frac{3}{2}+\frac{4}{3}k_0+\frac{7}{12}k_1-1}(t,t_0) \right) \Bigg|_{\lambda=3}$$

$$= \sum_{j=0}^{\infty} \lambda^j h_{\frac{7}{6}j+\frac{1}{2}+\frac{4}{3}k_0+\frac{7}{12}k_1}(t,t_0) \Bigg|_{\lambda=3}$$

$$+ \sum_{n=1}^{\infty} \left(\sum_{k_0+k_1=n} \right) \frac{n!}{k_0!k_1!} 2^{2k_0+k_1} \sum_{j=n}^{\infty} \lambda \ldots (\lambda-n+1) h_{\frac{7}{6}j+\frac{1}{2}+\frac{4}{3}k_0+\frac{7}{12}k_1}(t,t_0) \Bigg|_{\lambda=3}$$

$$= \sum_{j=0}^{\infty} 3^j h_{\frac{7}{6}j+\frac{1}{2}+\frac{4}{3}k_0+\frac{7}{12}k_1}(t,t_0)$$

$$+ \sum_{n=1}^{\infty} \left(\sum_{k_0+k_1=n} \right) \frac{n!}{k_0!k_1!} 2^{2k_0+k_1} \sum_{j=n}^{\infty} 3 \ldots (4-n) h_{\frac{7}{6}j+\frac{1}{2}+\frac{4}{3}k_0+\frac{7}{12}k_1}(t,t_0)$$

$$= \sum_{j=0}^{\infty} 3^j h_{\frac{7}{6}j+\frac{1}{2}+\frac{4}{3}k_0+\frac{7}{12}k_1}(t,t_0)$$

$$+ \sum_{n=1}^{3} \left(\sum_{k_0+k_1=n} \right) \frac{n!}{k_0!k_1!} 2^{2k_0+k_1} \sum_{j=n}^{\infty} 3 \ldots (4-n) h_{\frac{7}{6}j+\frac{1}{2}+\frac{4}{3}k_0+\frac{7}{12}k_1}(t,t_0),$$

$$y_3(t) = \sum_{n=0}^{\infty} \left(\sum_{k_0+k_1=n} \right) \frac{n!}{k_0!k_1!} \left(\prod_{\nu=0}^{1} (A_\nu)^{k_\nu} \right)$$

$$\times \frac{\partial^n}{\partial \lambda^n} {}_{\Delta}F_{\frac{5}{2}-\frac{4}{3},\frac{5}{2}+1-3+\sum_{\nu=0}^{1}\left(\frac{4}{3}-\alpha_\nu\right)k_\nu} (3,t,t_0)$$

$$= \sum_{n=0}^{\infty} \left(\sum_{k_0+k_1=n} \right) \frac{n!}{k_0!k_1!} \left(\prod_{\nu=0}^{1} (A_\nu)^{k_\nu} \right)$$

$$\times \frac{\partial^n}{\partial \lambda^n} {}_{\Delta}F_{\frac{7}{6},\frac{1}{2}+\frac{4}{3}k_0+\frac{7}{12}k_1} (3,t,t_0)$$

$$= \sum_{n=0}^{\infty} \left(\sum_{k_0+k_1=n} \right) \frac{n!}{k_0!k_1!} \left(4^{k_0}2^{k_1} \right)$$

$$\times \frac{\partial^n}{\partial \lambda^n} {}_{\Delta}F_{\frac{7}{6},\frac{1}{2}+\frac{4}{3}k_0+\frac{7}{12}k_1} (3,t,t_0)$$

$$= \sum_{n=0}^{\infty} \left(\sum_{k_0+k_1=n} \right) \frac{n!}{k_0!k_1!} 2^{2k_0+k_1} \frac{\partial^n}{\partial \lambda^n} \left(\sum_{j=0}^{\infty} \lambda^j h_{\frac{7}{6}j+\frac{1}{2}+\frac{4}{3}k_0+\frac{7}{12}k_1-1}(t,t_0) \right) \Bigg|_{\lambda=3}$$

$$= \sum_{j=0}^{\infty} \lambda^j h_{\frac{7}{6}j-\frac{1}{2}+\frac{4}{3}k_0+\frac{7}{12}k_1}(t,t_0) \Bigg|_{\lambda=3}$$

$$+ \sum_{n=1}^{\infty} \left(\sum_{k_0+k_1=n} \right) \frac{n!}{k_0!k_1!} 2^{2k_0+k_1} \sum_{j=n}^{\infty} \lambda \dots (\lambda-n+1) h_{\frac{7}{6}j-\frac{1}{2}+\frac{4}{3}k_0+\frac{7}{12}k_1}(t,t_0) \Bigg|_{\lambda=3}$$

$$= \sum_{j=0}^{\infty} 3^j h_{\frac{7}{6}j-\frac{1}{2}+\frac{4}{3}k_0+\frac{7}{12}k_1}(t,t_0)$$

$$+ \sum_{n=1}^{\infty} \left(\sum_{k_0+k_1=n} \right) \frac{n!}{k_0!k_1!} 2^{2k_0+k_1} \sum_{j=n}^{\infty} 3 \dots (4-n) h_{\frac{7}{6}j-\frac{1}{2}+\frac{4}{3}k_0+\frac{7}{12}k_1}(t,t_0)$$

$$= \sum_{j=0}^{\infty} 3^j h_{\frac{7}{6}j-\frac{1}{2}+\frac{4}{3}k_0+\frac{7}{12}k_1}(t,t_0)$$

$$+ \sum_{n=1}^{3} \left(\sum_{k_0+k_1=n} \right) \frac{n!}{k_0!k_1!} 2^{2k_0+k_1} \sum_{j=n}^{\infty} 3 \dots (4-n) h_{\frac{7}{6}j-\frac{1}{2}+\frac{4}{3}k_0+\frac{7}{12}k_1}(t,t_0), \quad t > t_0.$$

Consequently,

$$y(t) = \sum_{j=1}^{3} d_j y_j(t)$$

$$= y_1(t) - y_2(t) + y_3(t)$$

$$= \sum_{j=0}^{\infty} 3^j h_{\frac{7}{6}j+\frac{3}{2}+\frac{4}{3}k_0+\frac{7}{12}k_1}(t, t_0)$$

$$+ \sum_{n=1}^{3} \left(\sum_{k_0+k_1=n} \right) \frac{n!}{k_0! k_1!} 2^{2k_0+k_1} \sum_{j=n}^{\infty} 3 \ldots (4-n) h_{\frac{7}{6}j+\frac{3}{2}+\frac{4}{3}k_0+\frac{7}{12}k_1}(t, t_0)$$

$$- \sum_{j=0}^{\infty} 3^j h_{\frac{7}{6}j+\frac{1}{2}+\frac{4}{3}k_0+\frac{7}{12}k_1}(t, t_0)$$

$$- \sum_{n=1}^{3} \left(\sum_{k_0+k_1=n} \right) \frac{n!}{k_0! k_1!} 2^{2k_0+k_1} \sum_{j=n}^{\infty} 3 \ldots (4-n) h_{\frac{7}{6}j+\frac{1}{2}+\frac{4}{3}k_0+\frac{7}{12}k_1}(t, t_0)$$

$$+ \sum_{j=0}^{\infty} 3^j h_{\frac{7}{6}j-\frac{1}{2}+\frac{4}{3}k_0+\frac{7}{12}k_1}(t, t_0)$$

$$+ \sum_{n=1}^{3} \left(\sum_{k_0+k_1=n} \right) \frac{n!}{k_0! k_1!} 2^{2k_0+k_1} \sum_{j=n}^{\infty} 3 \ldots (4-n) h_{\frac{7}{6}j-\frac{1}{2}+\frac{4}{3}k_0+\frac{7}{12}k_1}(t, t_0)$$

$$= \sum_{j=0}^{\infty} 3^j \left(h_{\frac{7}{6}j+\frac{3}{2}+\frac{4}{3}k_0+\frac{7}{12}k_1}(t, t_0) \right.$$

$$- h_{\frac{7}{6}j+\frac{1}{2}+\frac{4}{3}k_0+\frac{7}{12}k_1}(t, t_0)$$

$$\left. + h_{\frac{7}{6}j-\frac{1}{2}+\frac{4}{3}k_0+\frac{7}{12}k_1}(t, t_0) \right)$$

$$+ \sum_{n=1}^{3} \left(\sum_{k_0+k_1=n} \right) \frac{n!}{k_0! k_1!} 2^{2k_0+k_1} \sum_{j=n}^{\infty} 3 \ldots (4-n) \left(h_{\frac{7}{6}j+\frac{3}{2}+\frac{4}{3}k_0+\frac{7}{12}k_1}(t, t_0) \right.$$

$$- h_{\frac{7}{6}j+\frac{1}{2}+\frac{4}{3}k_0+\frac{7}{12}k_1}(t, t_0)$$

$$\left. + h_{\frac{7}{6}j-\frac{1}{2}+\frac{4}{3}k_0+\frac{7}{12}k_1}(t, t_0) \right), \quad t \in [t_0, a),$$

is the solution of the considered IVP, provided that the last series are convergent.

Exercise 6.4. Find the solution of the equation

$$D_{\Delta,t_0}^{\frac{7}{3}}y(t) - 2D_{\Delta,t_0}^{\frac{3}{2}}y(t) - 2D_{\Delta,t_0}^{\frac{1}{4}}y(t) - y(t) = 0, \quad t \in [t_0, a),$$

subject to the initial conditions

$$D_{\Delta,t_0}^{\frac{4}{3}}y(t_0) = 2, \quad D_{\Delta,t_0}^{\frac{1}{3}}y(t_0) = -3, \quad D_{\Delta,t_0}^{-\frac{2}{3}}y(t_0) = -1,$$

$$D_{\Delta,t_0}^{\frac{1}{2}}y(t_0) = -12, \quad D_{\Delta,t_0}^{-\frac{1}{2}}y(t_0) = -1,$$

$$D_{\Delta,t_0}^{-\frac{3}{4}}y(t_0) = 1.$$

6.2 Inhomogeneous Riemann–Liouville Fractional Dynamic Equations with Constant Coefficients

Consider the equation

$$\sum_{k=1}^{m} A_k D_{\Delta,t_0}^{\alpha_k} y(t) + A_0 y(t) = f(t), \quad t \in [t_0, a), \tag{6.18}$$

where

$$m \in \mathbb{N},$$
$$0 < \alpha_1 < \ldots < \alpha_m,$$
$$\alpha_i \in (l_i - 1, l_i], \quad l_i \in \mathbb{N}, \quad i \in \{1, \ldots, m\},$$
$$A_0, A_k \in \mathbb{R}, \quad k \in \{1, \ldots, m\}, \quad A_m \neq 0,$$

$f : \mathbb{T} \to \mathbb{R}$ is a given function, subject to the initial conditions

$$D_{\Delta,t_0}^{\alpha_k - j} y(t_0) = d_{\alpha_k, j}, \quad j \in \{1, \ldots, m_k\}, \quad k \in \{1, \ldots, m\}, \tag{6.19}$$

$m_k = [\alpha_k] + 1, k \in \{1, \ldots, m\}$. We will search for a particular solution y_p of the equation (6.18) subject to the initial conditions

$$D_{\Delta,t_0}^{\alpha_k - j} y_p(t_0) = 0, \quad j \in \{1, \ldots, m_k\}, \quad k \in \{1, \ldots, m\}. \tag{6.20}$$

We take the Laplace transform of both sides of (6.18), and we get

$$\mathscr{L}(f(t))(z) = \mathscr{L}\left(\sum_{k=1}^{m} A_k D_{\Delta,t_0}^{\alpha_k} y_p(t) + A_0 y_p(t)\right)(z, t_0)$$

$$= \mathscr{L}\left(\sum_{k=1}^{m} A_k D_{\Delta,t_0}^{\alpha_k} y_p(t)\right)(z, t_0) + \mathscr{L}\left(A_0 y_p(t)\right)(z, t_0)$$

$$= \sum_{k=1}^{m} A_k \mathscr{L}\left(D_{\Delta,t_0}^{\alpha_k} y_p(t)\right)(z, t_0) + A_0 \mathscr{L}\left(y_p(t)\right)(z, t_0)$$

$$= \sum_{k=1}^{m} A_k z^{\alpha_k} \mathscr{L}(y_p(t))(z, t_0) + A_0 \mathscr{L}(y_p(t))(z, t_0)$$

$$= \left(A_0 + \sum_{k=1}^{m} A_k z^{\alpha_k}\right) \mathscr{L}(y_p(t))(z, t_0),$$

whereupon

$$\mathscr{L}\left(y_p(t)\right)(z, t_0) = \frac{\mathscr{L}\left(f(t)\right)(z, t_0)}{A_0 + \sum\limits_{k=1}^{m} A_k z^{\alpha_k}}.$$

Let

$$P_{\alpha_1,\ldots,\alpha_m}(z) = A_0 + \sum_{k=1}^{m} A_k z^{\alpha_k}.$$

Then

$$\mathscr{L}\left(y_p(t)\right)(z, t_0) = \frac{\mathscr{L}\left(f(t)\right)(z, t_0)}{P_{\alpha_1,\ldots,\alpha_m}(z)}.$$

From this, we obtain

$$y_p(t) = \mathscr{L}^{-1}\left(\frac{\mathscr{L}\left(f(t)\right)(z)}{P_{\alpha_1,\ldots,\alpha_m}(z)}\right)(t), \quad t \in [t_0, a).$$

We introduce the Laplace fractional analogue of the Green function as follows:

$$G_{\alpha_1,\ldots,\alpha_m}(t) = \mathscr{L}^{-1}\left(\frac{1}{P_{\alpha_1,\ldots,\alpha_m}(z)}\right)(t), \quad t \in [t_0, a).$$

Then

$$y_p(t) = \mathscr{L}^{-1}\left(\mathscr{L}\left(f(t)\right)(z)\mathscr{L}\left(G_{\alpha_1,\ldots,\alpha_m}(t)\right)(z)\right)(t)$$

$$= \mathscr{L}^{-1}\left(\mathscr{L}\left(f \star G_{\alpha_1,\ldots,\alpha_m}\right)(t, t_0)\right)(z)$$

$$= \left(f \star G_{\alpha_1,\ldots,\alpha_m}\right)(t), \quad t \in [t_0, a).$$

Let y_h be the solution of the equation

$$\sum_{k=1}^{m} A_k D_{\Delta,t_0}^{\alpha_k} y(t) + A_0 y(t) = 0, \quad t \in [t_0, a),$$

subject to the initial conditions (6.19). Then

$$y(t) = y_p(t) + y_h(t), \quad t \in [t_0, a),$$

is the solution of the problem (6.18), (6.19).

Example 6.8. Consider the equation

$$D_{\Delta,t_0}^{\frac{1}{2}} y(t) - y(t) = f(t), \quad t \in [t_0, a), \tag{6.21}$$

subject to the initial condition

$$D_{\Delta,t_0}^{-\frac{1}{2}} y(t_0) = 1. \tag{6.22}$$

Here $f : \mathbb{T} \to \mathbb{R}$ is a given function. First, we will find a particular solution y_p of the equation (6.21) subject to the initial condition

$$D_{\Delta,t_0}^{\frac{1}{2}} y(t_0) = 0. \tag{6.23}$$

We take the Laplace transform of both sides of equation (6.21), and using (6.23), we obtain

$$\mathscr{L}\left(f(t)\right)(z) = \mathscr{L}\left(D_{\Delta,t_0}^{\frac{1}{2}} y(t) - y(t)\right)(z)$$

$$= \mathscr{L}\left(D_{\Delta,t_0}^{\frac{1}{2}} y(t)\right)(z) - \mathscr{L}\left(y(t)\right)(z)$$

$$= z^{\frac{1}{2}} \mathscr{L}\left(y(t)\right)(z) - \mathscr{L}\left(y(t)\right)(z)$$

$$= \left(\sqrt{z} - 1\right) \mathscr{L}\left(y(t)\right)(z).$$

Hence,

$$\mathscr{L}\left(y(t)\right)(z) = \frac{\mathscr{L}\left(f(t)\right)(z)}{\sqrt{z} - 1}.$$

We will find

$$\mathscr{L}^{-1}\left(\frac{1}{\sqrt{z}-1}\right)(t).$$

Let

$$\gamma = \left\{z \in \mathbb{C} : |z| = \frac{1}{w} + 1, \quad \sqrt{1} = 1\right\}.$$

Then

$$-\frac{1}{\mu(t_k)}, 1 \in \left\{z \in \mathbb{C} : |z| \leq \frac{1}{w} + 1, \quad \sqrt{1} = 1\right\}$$

for all $k \in \mathbb{N}$. Therefore,

$$\mathscr{L}^{-1}\left(\frac{1}{\sqrt{z}-1}\right)(t_n) = \frac{1}{2\pi i}\int_\gamma \frac{1}{\sqrt{z}-1}\prod_{k=0}^{n-1}(1+\mu(t_k)z)\,dz$$

$$= \frac{1}{2\pi i}\prod_{k=0}^{n-1}\mu(t_k)\int_\gamma \frac{1}{\sqrt{z}-1}\prod_{k=0}^{n-1}\left(z+\frac{1}{\mu(t_k)}\right)\,dz$$

$$= \frac{1}{2\pi i}\prod_{k=0}^{n-1}\mu(t_k)\int_\gamma \frac{1}{\sqrt{z}-1}\sum_{l=0}^{n}a_{n-l}z^{n-l}\,dz$$

$$= \frac{1}{2\pi i}\prod_{k=0}^{n-1}\mu(t_k)\left(\sum_{l=0}^{n}a_{n-l}\int_\gamma \frac{z^{n-l}}{\sqrt{z}-1}\,dz\right)$$

$$= \frac{1}{2\pi i}\prod_{k=0}^{n-1}\mu(t_k)\left(\sum_{l=0}^{n}(2\pi i)2a_{n-l}\right)$$

$$= 2\prod_{k=0}^{n-1}\mu(t_k)\sum_{l=0}^{n}a_{n-l},$$

where $a_l, l \in \{0, \ldots, n\}$, are given by (4.7). Let

$$g(t_n) = 2\prod_{k=0}^{n-1}\mu(t_k)\sum_{l=0}^{n}a_{n-l}, \quad n \in \mathbb{N}.$$

Then

$$g(t_n) = \mathscr{L}^{-1}\left(\frac{1}{\sqrt{z}-1}\right)(t_n)$$

and

$$\frac{1}{\sqrt{z}-1} = \mathscr{L}\left(g(t_n)\right)(z), \quad n \in \mathbb{N}.$$

Consequently,

$$\mathscr{L}\left(y(t_n)\right) = \mathscr{L}\left(f(t_n)\right)(z)\mathscr{L}\left(g(t_n)\right)(z)$$
$$= \mathscr{L}\left((f \star g)(t_n)\right)(z)$$

and

$$y_p(t_n) = (f \star g)(t_n), \quad n \in \mathbb{N},$$

is the solution of the problem (6.21), (6.23). Now we will find the solution of the equation

$$D^{\frac{1}{2}}_{\Delta,t_0} y(t) - y(t) = 0, \quad t > t_0, \tag{6.24}$$

subject to the initial condition (6.22). Here

$$\alpha = \frac{1}{2},$$

$$m = \left[\frac{1}{2}\right] + 1$$
$$= 0 + 1$$
$$= 1,$$

$$d_{\frac{1}{2},1} = 1,$$

$$l = 1,$$

$$\lambda = 1.$$

Then

$$y(t) = {}_{\Delta}F_{\frac{1}{2},\frac{1}{2}+1-1}(\lambda, t, t_0)\Big|_{\lambda=1}$$

$$= {}_{\Delta}F_{\frac{1}{2},\frac{1}{2}}(\lambda, t, t_0)\Big|_{\lambda=1}$$

$$= \sum_{j=0}^{\infty} \lambda^j h_{\frac{1}{2}j+\frac{1}{2}-1}(t, t_0)\Big|_{\lambda=1}$$

$$= \sum_{j=0}^{\infty} h_{\frac{1}{2}j-\frac{1}{2}}(t, t_0), \quad t \in \mathbb{T}, \quad t > t_0,$$

provided that the last series is convergent. Therefore,

$$y_h(t_n) = \sum_{j=0}^{\infty} h_{\frac{1}{2}j-\frac{1}{2}}(t_n, t_0), \quad n \in \mathbb{N},$$

is the solution of the IVP (6.24), (6.22), provided that the series is convergent. From this, we have that

$$y(t_n) = y_h(t_n) + y_p(t_n)$$

$$= \sum_{j=0}^{\infty} h_{\frac{1}{2}j-\frac{1}{2}}(t_n, t_0) + (f \star g)(t_n), \quad n \in \mathbb{N},$$

is the solution of the problem (6.21), (6.22), provided that the series is convergent.

Exercise 6.5. Find the solution of the equation

$$D_{\Delta,t_0}^{\frac{9}{4}} y(t) - 4D_{\Delta,t_0}^{\frac{4}{3}} y(t) - 8D_{\Delta,t_0}^{\frac{1}{4}} y(t) - 3y(t) = f(t), \quad t > t_0,$$

subject to the initial conditions

$$D_{\Delta,t_0}^{\frac{5}{4}} y(t_0) = -2, \quad D_{\Delta,t_0}^{\frac{1}{4}} y(t_0) = -1, \quad D_{\Delta,t_0}^{-\frac{3}{4}} y(t_0) = -2,$$

$$D_{\Delta,t_0}^{\frac{1}{3}} y(t_0) = 1, \quad D_{\Delta,t_0}^{-\frac{2}{3}} y(t_0) = 2.$$

Here $f : \mathbb{T} \to \mathbb{R}$ is a given function.

6.3   Advanced Practical Problems

Problem 6.1. Find the solution of the following IVP:

$$D_{\Delta,t_0}^{\frac{5}{4}} y(t) - 3y(t) = 0, \quad t > t_0,$$

$$D_{\Delta,t_0}^{\frac{1}{4}} y(t_0) = 1, \quad D_{\Delta,t_0}^{-\frac{3}{4}} y(t_0) = -4.$$

Answer.

$$y(t) = {}_\Delta F_{\frac{5}{4},\frac{5}{4}}(3, t, t_0) - 4{}_\Delta F_{\frac{5}{4},\frac{1}{4}}(3, t, t_0), \quad t > t_0.$$

Problem 6.2. Find α, λ, d_1, d_2, d_3, and d_4 such that the function

$$y(t) = 4F_{\frac{10}{3},\frac{10}{3}}(12, t, t_0) - 3F_{\frac{10}{3},\frac{4}{3}}(12, t, t_0), \quad t > t_0,$$

is the solution of the IVP (6.1), (6.2).

Answer.

$$\alpha = \frac{10}{3}, \quad \lambda = 12, \quad d_1 = 4, \quad d_2 = 0, \quad d_3 = -3, \quad d_4 = 0.$$

Problem 6.3. Find the solution of the following IVP:

$$D_{\Delta,t_0}^{\frac{13}{4}}y(t) - 7D_{\Delta,t_0}^{\frac{7}{3}}y(t) - 4y(t) = 0, \quad t > t_0,$$

$$D_{\Delta,t_0}^{\frac{9}{4}}y(t_0) = 1, \quad D_{\Delta,t_0}^{\frac{5}{4}}y(t_0) = -1, \quad D_{\Delta,t_0}^{\frac{1}{4}}y(t_0) = -2,$$

$$D_{\Delta,t_0}^{-\frac{3}{4}}y(t_0) = -3, \quad D_{\Delta,t_0}^{\frac{4}{3}}y(t_0) = 1, \quad D_{\Delta,t_0}^{\frac{1}{3}}y(t_0) = -1,$$

$$D_{\Delta,t_0}^{-\frac{2}{3}}y(t_0) = 2.$$

Answer.

$$y(t) = \sum_{k=0}^{\infty} \frac{4^k}{k!} \frac{\partial^k}{\partial \lambda^k} \sum_{j=0}^{\infty} \lambda^j$$

$$\times \left(-6h_{\frac{11}{12}j+\frac{9}{4}+\frac{7}{3}k}(t, t_0) + 6h_{\frac{11}{12}j+\frac{5}{4}+\frac{7}{3}k}(t, t_0) \right.$$

$$\left. -16h_{\frac{11}{12}j+\frac{1}{4}+\frac{7}{3}k}(t, t_0) - 3h_{\frac{11}{12}j-\frac{3}{4}+\frac{7}{3}k}(t, t_0) \right)\Big|_{\lambda=7}, \quad t > t_0,$$

provided that the last series is convergent.

Problem 6.4. Find the solution of the equation

$$D_{\Delta,t_0}^{\frac{13}{6}}y(t) + 3D_{\Delta,t_0}^{\frac{3}{2}}y(t) - D_{\Delta,t_0}^{\frac{1}{4}}y(t) - 7y(t) = 0, \quad t > t_0,$$

subject to the initial conditions

$$D_{\Delta,t_0}^{\frac{7}{6}}y(t_0) = 1, \quad D_{\Delta,t_0}^{\frac{1}{6}}y(t_0) = -1, \quad D_{\Delta,t_0}^{-\frac{5}{6}}y(t_0) = 10,$$

$$D_{\Delta,t_0}^{\frac{1}{2}}y(t_0) = -1, \quad D_{\Delta,t_0}^{-\frac{1}{2}}y(t_0) = 1,$$

$$D_{\Delta,t_0}^{-\frac{3}{4}}y(t_0) = -1.$$

Problem 6.5. Find the solution of the equation

$$D_{\Delta,t_0}^{\frac{5}{2}} y(t) - D_{\Delta,t_0}^{\frac{4}{3}} y(t) - D_{\Delta,t_0}^{\frac{1}{4}} y(t) + 2y(t) = f(t), \quad t > t_0,$$

subject to the initial conditions

$$D_{\Delta,t_0}^{\frac{3}{2}} y(t_0) = 2, \quad D_{\Delta,t_0}^{\frac{1}{2}} y(t_0) = -2, \quad D_{\Delta,t_0}^{-\frac{1}{2}} y(t_0) = 1,$$

$$D_{\Delta,t_0}^{\frac{1}{3}} y(t_0) = -1, \quad D_{\Delta,t_0}^{-\frac{2}{3}} y(t_0) = -2,$$

$$D_{\Delta,t_0}^{-\frac{3}{4}} y(t_0) = -3.$$

Here $f : \mathbb{T} \to \mathbb{R}$ is a given function.

Chapter 7
The Caputo Fractional Δ-Derivative on Time Scales

In this chapter we suppose that $\mathbb{T}$ is a time scale with forward jump operator and delta differentiation operator σ and Δ, respectively, such that

$$\mathbb{T} = \{t_n : n \in \mathbb{N}_0\},$$

$$\lim_{n \to \infty} t_n = \infty,$$

$$\sigma(t_n) = t_{n+1}, \quad n \in \mathbb{N}_0,$$

$$w = \inf_{n \in \mathbb{N}_0} \mu(t_n) > 0.$$

7.1 Definition of the Caputo Fractional Δ-Derivative and Examples

Definition 7.1. Let $t \in \mathbb{T}$. The Caputo fractional Δ-derivative of order $\alpha \geq 0$ is defined via the Riemann–Liouville fractional Δ-derivative as follows:

$$^{C}D^{\alpha}_{\Delta,t_0} f(t) = D^{\alpha}_{\Delta,t_0} \left(f(t) - \sum_{k=0}^{m-1} h_k(t,t_0) f^{\Delta^k}(t_0) \right), \quad t > t_0, \tag{7.1}$$

where $m = \overline{[\alpha]} + 1$ if $\alpha \notin \mathbb{N}$, $m = \overline{[\alpha]}$ if $\alpha \in \mathbb{N}$.

Example 7.1. Let $\alpha \in (0,1)$. Then $m = 1$ and

$$^{C}D^{\alpha}_{\Delta,t_0} f(t) = D^{\alpha}_{\Delta,t_0} \left(f(t) - \sum_{k=0}^{0} h_k(t,t_0) f^{\Delta^k}(t_0) \right)$$

$$= D^{\alpha}_{\Delta,t_0} \left(f(t) - h_0(t,t_0) f(t_0) \right)$$

$$= D^{\alpha}_{\Delta, t_0} \left(f(t) - f(t_0) \right), \quad t \in \mathbb{T}, \quad t > t_0.$$

If $f(t_0) = 0$, then the Caputo fractional Δ-derivative coincides with the Riemann–Liouville fractional Δ-derivative.

Example 7.2. Let $\alpha = m \in \mathbb{N}$. Then

$$^C D^m_{\Delta, t_0} f(t) = D^m_{\Delta, t_0} \left(f(t) - \sum_{k=0}^{m-1} h_k(t, t_0) f^{\Delta^k}(t_0) \right)$$

$$= D^m_{\Delta} \left(f(t) - \sum_{k=0}^{m-1} h_k(t, t_0) f^{\Delta^k}(t_0) \right)$$

$$= D^m_{\Delta} f(t) - \sum_{k=0}^{m-1} D^m_{\Delta} h_k(t, t_0) f^{\Delta^k}(t_0)$$

$$= f^{\Delta^m}(t), \quad t \in \mathbb{T}, \quad t > t_0,$$

i.e., when $\alpha = m \in \mathbb{N}$, then the Caputo fractional Δ-derivative coincides with the delta derivative.

We can rewrite (7.1) in the following way:

$$^C D^{\alpha}_{\Delta, t_0} f(t) = D^m_{\Delta} I^{m-\alpha}_{\Delta, t_0} \left(f(t) - \sum_{k=0}^{m-1} h_k(t, t_0) f^{\Delta^k}(t_0) \right)$$

$$= D^m_{\Delta} \left(\int_{t_0}^{t} h_{m-\alpha-1}(t, \sigma(u)) \left(f(u) - \sum_{k=0}^{m-1} h_k(u, t_0) f^{\Delta^k}(t_0) \right) \Delta u \right)$$

$$= D^m_{\Delta} \left(\int_{t_0}^{t} h_{m-\alpha-1}(t, \sigma(u)) f(u) \Delta u \right)$$

$$- \sum_{k=0}^{m-1} f^{\Delta^k}(t_0) D^m_{\Delta} \left(\int_{t_0}^{t} h_{m-\alpha-1}(t, \sigma(u)) h_k(u, t_0) \Delta u \right)$$

$$= D^m_{\Delta} \left(\int_{t_0}^{t} h_{m-\alpha-1}(t, \sigma(u)) f(u) \Delta u \right) \qquad (7.2)$$

$$- \sum_{k=0}^{m-1} f^{\Delta^k}(t_0) D^m_{\Delta} h_{m+k-\alpha}(t, t_0)$$

$$= D^m_{\Delta} \left(h_{m-\alpha-1} \star f \right)(t, t_0)$$

$$- \sum_{k=0}^{m-1} f^{\Delta^k}(t_0) h_{k-\alpha}(t, t_0), \quad t \in \mathbb{T}, \quad t > t_0.$$

Example 7.3. We will find

$$^C D^{\alpha}_{\Delta, t_0} h_{\alpha}(t, t_0), \qquad t \in \mathbb{T}, \quad t > t_0.$$

We have, using (7.2),

$$
\begin{aligned}
^C D^{\alpha}_{\Delta, t_0} h_{\alpha}(t, t_0) &= D^m_{\Delta} (h_{m-\alpha-1} \star h_{\alpha}) (t, t_0) \\
&\quad - \sum_{k=0}^{m-1} h_{\alpha}^{\Delta^k}(t_0, t_0) h_{k-\alpha}(t, t_0) \\
&= D^m_{\Delta} h_m(t, t_0) \\
&= 1, \quad t \in \mathbb{T}, \quad t > t_0.
\end{aligned}
$$

Example 7.4. Let $\beta > \alpha$. We will find

$$^C D^{\alpha}_{\Delta, t_0} h_{\beta}(t, t_0), \quad t \in \mathbb{T}, \quad t > t_0.$$

Using (7.2), we obtain

$$
\begin{aligned}
^C D^{\alpha}_{\Delta, t_0} h_{\beta}(t, t_0) &= D^m_{\Delta} \left(h_{m-\alpha-1} \star h_{\beta} \right) (t, t_0) \\
&\quad - \sum_{k=0}^{m-1} h_{\beta}^{\Delta^k}(t_0, t_0) h_{k-\alpha}(t, t_0) \\
&= D^m_{\Delta} h_{m+\beta-\alpha}(t, t_0) \\
&= h_{\beta-\alpha}(t, t_0), \quad t \in \mathbb{T}, \quad t > t_0.
\end{aligned}
$$

Exercise 7.1. Find

$$^C D^{\frac{1}{2}}_{\Delta, t_0} \left(h_{\frac{4}{3}} \star h_{\frac{1}{4}} \right)(t, t_0), \quad t \in \mathbb{T}, \quad t > t_0.$$

Answer. $h_{\frac{25}{12}}(t, t_0), t \in \mathbb{T}, t > t_0.$

7.2 Properties of the Caputo Fractional Δ-Derivative

Theorem 7.1. *Let* $\alpha \geq 0$, $m = \overline{\lceil \alpha \rceil} + 1$ *if* $\alpha \notin \mathbb{N}$, $m = \alpha$ *if* $\alpha \in \mathbb{N}$.
1. If $\alpha \notin \mathbb{N}$, *then*

$$
\begin{aligned}
^C D^{\alpha}_{\Delta, t_0} f(t) &= \left(h_{m-\alpha-1}(\cdot, t_0) \star f^{\Delta^m} \right)(t) \\
&= I^{m-\alpha}_{\Delta, t_0} D^m_{\Delta} f(t), \quad t \in \mathbb{T}, \quad t > t_0.
\end{aligned}
$$

2. *If* $\alpha = m \in \mathbb{N}$, *then*

$$^C D_{\Delta,t_0}^\alpha f(t) = f^{\Delta^m}(t), \quad t \in \mathbb{T}, \quad t > t_0.$$

Proof. 1. By Taylor's formula, we have

$$f(t) = \sum_{k=0}^{m-1} h_k(t, t_0) f^{\Delta^k}(t_0) + \int_{t_0}^t h_{m-1}(t, \sigma(\tau)) f^{\Delta^m}(\tau) \Delta \tau$$

$$= \sum_{k=0}^{m-1} h_k(t, t_0) f^{\Delta^k}(t_0) + I_{\Delta,t_0}^m f^{\Delta^m}(t), \quad t \in \mathbb{T}, \quad t > t_0.$$

Hence,

$$^C D_{\Delta,t_0}^\alpha f(t) = D_{\Delta,t_0}^\alpha \left(f(t) - \sum_{k=0}^{m-1} h_k(t, t_0) f^{\Delta^k}(t_0) \right)$$

$$= D_{\Delta,t_0}^\alpha I_{\Delta,t_0}^m f^{\Delta^m}(t)$$

$$= I_{\Delta,t_0}^{m-\alpha} f^{\Delta^m}(t), \quad t \in \mathbb{T}, \quad t > t_0.$$

2. Let $\alpha = m \in \mathbb{N}$. Then we have

$$^C D_{\Delta,t_0}^m f(t) = D_\Delta^m \left(f(t) - \sum_{k=0}^{m-1} h_k(t, t_0) f^{\Delta^k}(t_0) \right)$$

$$= D_\Delta^m f(t)$$

$$= f^{\Delta^m}(t), \quad t \in \mathbb{T}, \quad t > t_0.$$

This completes the proof.

Theorem 7.2. *Let* $\alpha > 0$. *Then*

$$^C D_{\Delta,t_0}^\alpha I_{\Delta,t_0}^\alpha f(t) = f(t), \quad t \in \mathbb{T}, \quad t > t_0.$$

Proof. Let $\alpha \notin \mathbb{N}$. Then

$$^C D_{\Delta,t_0}^\alpha I_{\Delta,t_0}^\alpha f(t) = D_{\Delta,t_0}^\alpha \left(I_{\Delta,t_0}^\alpha f(t) - \sum_{k=0}^{m-1} h_k(t, t_0) \left(I_{\Delta,t_0}^\alpha f \right)^{\Delta^k}(t_0) \right)$$

$$= D_{\Delta,t_0}^\alpha \left(I_{\Delta,t_0}^\alpha f(t) \right)$$

$$= f(t), \quad t \in \mathbb{T}, \quad t > t_0.$$

Let $\alpha = m \in \mathbb{N}$. Then

$$^C D^m_{\Delta, t_0} I^m_{\Delta, t_0} f(t) = D^m_{\Delta} I^m_{\Delta, t_0} f(t)$$

$$= f(t), \quad t \in \mathbb{T}, \quad t > t_0.$$

This completes the proof.

Theorem 7.3. *Let $\alpha > 0$, $m = -\overline{[-\alpha]}$. Then*

$$I^\alpha_{\Delta, t_0} {}^C D^\alpha_{\Delta, t_0} f(t) = f(t) - \sum_{k=0}^{m-1} h_k(t, t_0) D^k_{\Delta, t_0} f(t_0), \quad t \in \mathbb{T}, \quad t > t_0.$$

Proof. Let $\alpha \notin \mathbb{N}$. Then, using that

$$^C D^\alpha_{\Delta, t_0} f(t) = I^{m-\alpha}_{\Delta, t_0} D^m_{\Delta} f(t), \quad t \in \mathbb{T}, \quad t > t_0,$$

we get

$$I^\alpha_{\Delta, t_0} {}^C D^\alpha_{\Delta, t_0} f(t) = I^\alpha_{\Delta, t_0} I^{m-\alpha}_{\Delta, t_0} D^m_{\Delta} f(t)$$

$$= I^m_{\Delta, t_0} D^m_{\Delta} f(t)$$

$$= f(t) - \sum_{k=1}^{m} h_{m-k}(t, t_0) D^{m-k}_{\Delta} f(t_0)$$

$$= f(t) - \sum_{k=0}^{m-1} h_k(t, t_0) D^k_{\Delta} f(t_0), \quad t \in \mathbb{T}, \quad t > t_0.$$

Let $\alpha = m \in \mathbb{N}$. Then, using that

$$^C D^m_{\Delta, t_0} f(t) = f^{\Delta^m}(t), \quad t \in \mathbb{T}, \quad t > t_0,$$

we get

$$I^m_{\Delta, t_0} {}^C D^m_{\Delta, t_0} f(t) = I^m_{\Delta, t_0} f^{\Delta^m}(t)$$

$$= f(t) - \sum_{k=1}^{m} h_{m-k}(t, t_0) D^{m-k}_{\Delta} f(t_0)$$

$$= f(t) - \sum_{k=0}^{m-1} h_k(t, t_0) D^k_{\Delta} f(t_0), \quad t \in \mathbb{T}, \quad t > t_0.$$

This completes the proof.

Remark 7.1. When $\alpha \in (0, 1)$, then

$$I_{\Delta,t_0}^{\alpha} {}^{C}D_{\Delta,t_0}^{\alpha} f(t) = f(t) - f(t_0), \quad t \in \mathbb{T}, \quad t > t_0.$$

Example 7.5. We have

$$I_{\Delta,t_0}^{\frac{1}{2}} {}^{C}D_{\Delta,t_0}^{\frac{1}{2}} e_1(t, t_0) = e_1(t, t_0) - e_1(t_0, t_0)$$
$$= e_1(t, t_0) - 1, \quad t \in \mathbb{T}, \quad t > t_0.$$

Exercise 7.2. Find

$$I_{\Delta,t_0}^{\frac{1}{3}} {}^{C}D_{\Delta,t_0}^{\frac{1}{3}} \left(-3e_2(t, t_0) + \sin_1(t, t_0) + h_2(t, t_0)\right), \quad t \in \mathbb{T}, \quad t > t_0.$$

Answer.

$$-3e_2(t, t_0) + \sin_1(t, t_0) + h_2(t, t_0) + 3, \quad t \in \mathbb{T}, \quad t > t_0.$$

Theorem 7.4. *Let* $m - 1 < \beta < \alpha < m$, $m \in \mathbb{N}$. *Then for all* $k \in \{1, \ldots, m - 1\}$, *we have*

$$^{C}D_{\Delta,t_0}^{\alpha-m+k} D_{\Delta}^{m-k} f(t) = {}^{C}D_{\Delta,t_0}^{\alpha} f(t), \quad t \in \mathbb{T}, \quad t > t_0, \tag{7.3}$$

$$^{C}D_{\Delta,t_0}^{\alpha-\beta} {}^{C}D_{\Delta,t_0}^{\beta} f(t) = {}^{C}D_{\Delta,t_0}^{\alpha} f(t), \quad t \in \mathbb{T}, \quad t > t_0. \tag{7.4}$$

Proof. For each $k \in \{1, \ldots, m - 1\}$, we have

$$^{C}D_{\Delta,t_0}^{\alpha} f(t) = I_{\Delta,t_0}^{m-\alpha} D_{\Delta}^{m} f(t)$$

$$= \int_{t_0}^{t} h_{m-\alpha-1}(t, \sigma(s)) D_{\Delta}^{m} f(s) \Delta s$$

$$= \int_{t_0}^{t} h_{k-(\alpha-(m-k))-1}(t, \sigma(s)) D_{\Delta}^{k} D_{\Delta}^{m-k} f(s) \Delta s, \quad t \in \mathbb{T}, \quad t > t_0.$$

Note that

$$\alpha - (m - k) > m - 1 - (m - k)$$
$$= k - 1,$$
$$\alpha - (m - k) < m - (m - k)$$
$$= k,$$

i.e.,

$$\alpha - (m - k) \in (k - 1, k), \quad k \in \{1, \ldots, m - 1\}.$$

Therefore,

$$
\begin{aligned}
{}^{C}D^{\alpha}_{\Delta, t_0} f(t) &= I^{k-(\alpha-(m-k))}_{\Delta, t_0} D^{k}_{\Delta} D^{m-k}_{\Delta} f(t) \\
&= {}^{C}D^{\alpha-m+k}_{\Delta, t_0} D^{m-k}_{\Delta} f(t), \quad t \in \mathbb{T}, \quad t > t_0.
\end{aligned}
$$

Thus (7.3) holds. Let $\alpha_0, \beta_0 \in (0, 1)$ and $\alpha_0 + \beta_0 < 1$. Then ${}^{C}D^{\beta_0}_{\Delta, t_0} f(t_0) = 0$ and

$$
\begin{aligned}
{}^{C}D^{\alpha_0}_{\Delta, t_0} {}^{C}D^{\beta_0}_{\Delta, t_0} f(t) &= {}^{C}D^{\alpha_0}_{\Delta, t_0} \left(I^{1-\beta_0}_{\Delta, t_0} D^{1}_{\Delta} f(t) \right) \\
&= I^{1-\alpha_0}_{\Delta, t_0} D^{1}_{\Delta} I^{1-\beta_0}_{\Delta, t_0} D^{1}_{\Delta} f(t) \\
&= I^{1-\alpha_0}_{\Delta, t_0} I^{-\beta_0}_{\Delta, t_0} D^{1}_{\Delta} f(t) \\
&= I^{1-\alpha_0-\beta_0}_{\Delta, t_0} D^{1}_{\Delta} f(t) \\
&= {}^{C}D^{\alpha_0+\beta_0}_{\Delta, t_0} f(t), \quad t \in \mathbb{T}, \quad t > t_0,
\end{aligned}
$$

and

$$
\begin{aligned}
{}^{C}D^{\beta_0}_{\Delta, t_0} {}^{C}D^{\alpha_0}_{\Delta, t_0} f(t) &= {}^{C}D^{\beta_0}_{\Delta, t_0} I^{1-\alpha_0}_{\Delta, t_0} D^{1}_{\Delta} f(t) \\
&= I^{1-\beta_0}_{\Delta, t_0} D^{1}_{\Delta} I^{1-\alpha_0}_{\Delta, t_0} D^{1}_{\Delta} f(t) \\
&= I^{1-\beta_0}_{\Delta, t_0} I^{-\alpha_0}_{\Delta, t_0} D^{1}_{\Delta} f(t) \\
&= I^{1-\alpha_0-\beta_0}_{\Delta, t_0} D^{1}_{\Delta} f(t) \\
&= {}^{C}D^{\alpha_0+\beta_0}_{\Delta, t_0} f(t), \quad t \in \mathbb{T}, \quad t > t_0,
\end{aligned}
$$

i.e.,

$$
\begin{aligned}
{}^{C}D^{\alpha_0}_{\Delta, t_0} {}^{C}D^{\beta_0}_{\Delta, t_0} f(t) &= {}^{C}D^{\beta_0}_{\Delta, t_0} {}^{C}D^{\alpha_0}_{\Delta, t_0} f(t) \\
&= {}^{C}D^{\alpha_0+\beta_0}_{\Delta, t_0} f(t), \quad t \in \mathbb{T}, \quad t > t_0.
\end{aligned}
$$

Hence, using (7.3), we obtain

$$
\begin{aligned}
{}^{C}D^{\alpha}_{\Delta, t_0} f(t) &= {}^{C}D^{\alpha-(m-1)+(m-1)}_{\Delta, t_0} f(t) \\
&= {}^{C}D^{\alpha-m+1}_{\Delta, t_0} D^{m-1}_{\Delta} f(t)
\end{aligned}
$$

$$= {}^{C}D_{\Delta,t_0}^{(\alpha-\beta)+(\beta-(m-1))'}D_{\Delta}^{m-1}f(t)$$

$$= {}^{C}D_{\Delta,t_0}^{\alpha-\beta}{}^{C}D_{\Delta,t_0}^{\beta-(m-1)}D_{\Delta}^{m-1}f(t)$$

$$= {}^{C}D_{\Delta,t_0}^{\alpha-\beta}{}^{C}D_{\Delta,t_0}^{\beta}f(t), \quad t \in \mathbb{T}, \quad t > t_0.$$

Thus (7.4) is established. This completes the proof.

Exercise 7.3. Let $m \in \mathbb{N}$ and $m - 1 < \beta < \alpha < m$. Prove that

$$I_{\Delta,t_0}^{\alpha}{}^{C}D_{\Delta,t_0}^{\alpha-m+k}D_{\Delta,t_0}^{m-k}f(t) = f(t) - \sum_{k=0}^{m-1} h_k(t,t_0)D_{\Delta}^{k}f(t_0), \quad t \in \mathbb{T}, \quad t > t_0.$$

Theorem 7.5. *Let $\alpha > 0$, $m - 1 < \alpha \leq m$, $m \in \mathbb{N}$. Then*

$$\mathscr{L}\left({}^{C}D_{\Delta,t_0}^{\alpha}f(t)\right)(z) = z^{\alpha}\mathscr{L}\left(f(t)\right)(z) - \sum_{k=0}^{m-1} z^{\alpha-k-1}f^{\Delta^k}(t_0)$$

for all $z \in \mathbb{C}$ for which

$$\lim_{t\to\infty}\left(f^{\Delta^k}(t)e_{\ominus z}(t,t_0)\right) = 0, \quad k \in \{0,\ldots,m-1\}. \tag{7.5}$$

Proof. Suppose that $z \in \mathbb{C}$ satisfies (7.5). Then

$$\mathscr{L}\left({}^{C}D_{\Delta,t_0}^{\alpha}f(t)\right)(z) = \mathscr{L}\left(I_{\Delta,t_0}^{m-\alpha}D_{\Delta}^{m}f(t)\right)(z)$$

$$= \frac{1}{z^{m-\alpha}}\mathscr{L}\left(D_{\Delta}^{m}f(t)\right)(z)$$

$$= \frac{1}{z^{m-\alpha}}\left(z^{m}\mathscr{L}\left(f(t)\right)(z) - \sum_{k=0}^{m-1} z^{m-k-1}f^{\Delta^k}(t_0)\right)$$

$$= z^{\alpha}\mathscr{L}\left(f(t)\right)(z) - \sum_{k=0}^{m-1} z^{\alpha-k-1}f^{\Delta^k}(t_0).$$

This completes the proof.

Example 7.6. Let $\alpha > 0$, $m - 1 < \alpha \leq m$, $m \in \mathbb{N}$, $\beta \in \mathbb{C}$. Then

$$\mathscr{L}\left({}^{C}D_{\Delta,t_0}^{\alpha}e_{\beta}(t,t_0)\right)(z) = z^{\alpha}\mathscr{L}\left(e_{\beta}(t,t_0)\right)(z) - \sum_{k=0}^{m-1} z^{\alpha-k-1}e_{\beta}^{\Delta^k}(t_0,t_0)$$

$$= \frac{z^{\alpha}}{z-\beta} - \sum_{k=0}^{m-1}\beta^{k}z^{\alpha-k-1},$$

provided that

$$\lim_{t\to\infty} \left(\beta^k e_\beta(t, t_0) e_{\ominus z}(t, t_0) \right) = 0, \quad k \in \{0, \ldots, m-1\}.$$

Example 7.7. Let $\alpha > 0$, $m - 1 < \alpha \leq m$, $m \in \mathbb{N}$, $l \in \mathbb{N}$. Then

$$\mathscr{L}\left({}^C D^\alpha_{\Delta, t_0} h_l(t, t_0) \right)(z) = z^\alpha \mathscr{L}\left(h_l(t, t_0) \right)(z)$$

$$- \sum_{k=0}^{m-1} z^{\alpha-k-1} h_l^{\Delta^k}(t_0, t_0)$$

$$= \begin{cases} \frac{z^\alpha}{z^{l+1}} & \text{if} \quad l > m-1 \\ \frac{z^\alpha}{z^{l+1}} - z^{\alpha-l-1} & \text{if} \quad l \leq m-1 \end{cases}$$

$$= \begin{cases} z^{\alpha-l-1} & \text{if} \quad l > m-1 \\ 0 & \text{if} \quad l \leq m-1, \end{cases}$$

provided that

$$\lim_{t\to\infty} \left(h_l^{\Delta^k}(t, t_0) e_{\ominus z}(t, t_0) \right) = 0, \quad k \in \{0, \ldots, m-1\}.$$

Example 7.8. We have

$$\mathscr{L}\left({}^C D^{\frac{1}{2}}_{\Delta, t_0} \sin_1(t, t_0) \right) = z^{\frac{1}{2}} \mathscr{L}\left(\sin_1(t, t_0) \right)(z) - z^{-\frac{1}{2}} \sin_1(t_0, t_0)$$

$$= \frac{z^{\frac{1}{2}}}{z^2 + 1},$$

provided that

$$\lim_{t\to\infty} \left(\sin_1(t, t_0) e_{\ominus z}(t, t_0) \right) = 0.$$

Exercise 7.4. Find

$$\mathscr{L}\left({}^C D^{\frac{4}{3}}_{\Delta, t_0} e_2(t, t_0) - 2\, {}^C D^{\frac{1}{2}}_{\Delta, t_0} \cos_1(t, t_0) \right)(z).$$

Answer.

$$\frac{z^{\frac{4}{3}}}{z - 2} - z^{\frac{1}{3}} - 2z^{-\frac{2}{3}} - 2\frac{z^{\frac{3}{2}}}{z^2 + 1} + 2z^{-\frac{1}{2}}.$$

7.3   Advanced Practical Problems

Problem 7.1. Find

1. $^C D_{\Delta,t_0}^{\frac{3}{2}} \left(h_1 \star h_{\frac{1}{2}} \right) (t, t_0),   t \in \mathbb{T},   t > t_0,$

2. $^C D_{\Delta,t_0}^{\frac{4}{3}} \left(h_3 \star h_{\frac{5}{6}} \right) (t, t_0)   t \in \mathbb{T},   t > t_0,$

3. $^C D_{\Delta,t_0}^{\frac{4}{3}} \left(-4h_2 - 7 \left(h_1 \star h_{\frac{5}{6}} \right) \right) (t, t_0),   t \in \mathbb{T},   t > t_0.$

Answer. 1. $h_1(t, t_0),   t \in \mathbb{T},   t > t_0,$
2. $h_{\frac{7}{2}}(t, t_0),   t \in \mathbb{T},   t > t_0,$
3. $-4h_{\frac{2}{3}}(t, t_0) - 7h_{\frac{3}{2}}(t, t_0),   t \in \mathbb{T},   t > t_0.$

Problem 7.2. Let $\alpha \in (0, 1)$. Prove that

$$I_{\Delta,t_0}^{\alpha}{}^C D_{\Delta,t_0}^{\alpha} a = 0,$$

where a is a constant.

Problem 7.3. Find

$$I_{\Delta,t_0}^{\frac{1}{3}}{}^C D_{\Delta,t_0}^{\frac{1}{3}} \left(\cos_2(t, t_0) - 4h_{10}(t, t_0) + 5 \right),   t \in \mathbb{T},   t > t_0.$$

Answer.

$$\cos_2(t, t_0) - 4h_{10}(t, t_0) - 1,   t \in \mathbb{T},   t > t_0.$$

Problem 7.4. Let $m \in \mathbb{N}$ and $m - 1 < \beta < \alpha < m$. Prove that

$$I_{\Delta,t_0}^{\alpha}{}^C D_{\Delta,t_0}^{\alpha-\beta}{}^C D_{\Delta,t_0}^{\beta} f(t) = f(t) - \sum_{k=0}^{m-1} h_k(t, t_0) D_\Delta^k f(t_0),   t \in \mathbb{T},   t > t_0.$$

Problem 7.5. Find

$$\mathscr{L} \left({}^C D_{\Delta,t_0}^{\frac{3}{2}} e_1(t, t_0) + {}^C D_{\Delta,t_0}^{\frac{1}{3}} \sin_1(t, t_0) \right) (z).$$

Answer.

$$\frac{z^{\frac{3}{2}}}{z - 1} - z^{\frac{1}{2}} - z^{-\frac{1}{2}} + \frac{z^{\frac{1}{3}}}{z^2 + 1}.$$

Chapter 8
Cauchy-Type Problems with the Caputo Fractional Δ-Derivative

In this chapter we suppose that $\mathbb{T}$ is a time scale with forward jump operator and delta differentiation operator σ and Δ, respectively, such that

$$\mathbb{T} = \{t_n : n \in \mathbb{N}_0\},$$

$$\lim_{n \to \infty} t_n = \infty,$$

$$\sigma(t_n) = t_{n+1}, \quad n \in \mathbb{N}_0,$$

$$w = \inf_{n \in \mathbb{N}_0} \mu(t_n) > 0.$$

8.1 Existence and Uniqueness of the Solution to the Cauchy-Type Problem

Suppose that $\alpha > 0$, $m = -\overline{[-\alpha]}$. Consider the equation

$$^C D^\alpha_{\Delta, t_0} y(t) = f(t, y(t)), \quad t \in \Omega, \tag{8.1}$$

subject to the initial conditions

$$D^k_\Delta y(t_0) = b_k, \quad k \in \{0, \dots, m-1\}, \tag{8.2}$$

where $f : \mathbb{T} \times \mathbb{R} \to \mathbb{R}$ is a given function; $b_k \in \mathbb{R}$, $k \in \{0, \dots, m-1\}$, are given constants; and Ω is a time scale interval that contains t_0.

© Springer International Publishing AG, part of Springer Nature 2018
S. G. Georgiev, *Fractional Dynamic Calculus and Fractional Dynamic Equations on Time Scales*, https://doi.org/10.1007/978-3-319-73954-0_8

Definition 8.1. Let I be a time scale interval. A function $g : I \to \mathbb{R}$ is said to be right-dense absolutely continuous if for every $\epsilon > 0$ there exists $\delta > 0$ such that

$$\sum_{k=1}^{n} |g(b_k) - g(a_k)| < \epsilon$$

whenever a disjoint collection of time scale intervals $[a_k, b_k) \subset I, k \in \{1, \ldots, n\}$, satisfies

$$\sum_{k=1}^{n} (b_k - a_k) < \delta.$$

We will now define $g \in AC_\Delta(I)$. If $g^{\Delta^k} \in AC_\Delta(I)$ for every $k \in \{0, \ldots, l\}$, for some $l \in \mathbb{N}_0$, then we will write $g \in AC_\Delta^l(I)$.

Theorem 8.1. *Let G be an open set in $\mathbb{R}$ and let $f : \Omega \times G \to \mathbb{R}$ be a function such that for all $y \in G, f(\cdot, y) \in AC_\Delta(\Omega)$. Then $y \in AC_\Delta^m(\Omega)$ satisfies the Cauchy problem* (8.1), (8.2) *if and only if y satisfies the Volterra integral equation*

$$y(t) = \sum_{j=0}^{m-1} h_j(t, t_0)b_j + \int_{t_0}^{t} h_{\alpha-1}(t, \sigma(\tau))f(\tau, y(\tau))\Delta\tau, \quad t \in \Omega. \tag{8.3}$$

Proof. 1. Let $y \in AC_\Delta^m(\Omega)$ satisfy the Volterra integral equation (8.3) for $f(\cdot, y) \in AC_\Delta^m(\Omega)$ for all $y \in G$. Then we apply ${}^C D_{\Delta,t_0}^\alpha$ to both sides of (8.3), and we get

$$^C D_{\Delta,t_0}^\alpha y(t) = \sum_{j=0}^{m-1} b_j {}^C D_{\Delta,t_0}^\alpha h_j(t, t_0)$$

$$+ {}^C D_{\Delta,t_0}^\alpha I_{\Delta,t_0}^\alpha f(t, y(t))$$

$$= f(t, y(t)), \quad t \in \Omega.$$

Also, for all $k \in \{0, \ldots, m - 1\}$, we get

$$D_\Delta^k y(t) = \sum_{j=0}^{m-1} b_j D_\Delta^k h_j(t, t_0) + D_\Delta^k I_{\Delta,t_0}^\alpha f(t, y(t))$$

$$= \sum_{j=0}^{m-1} b_j h_{j-k}(t, t_0) + \int_{t_0}^{t} h_{\alpha-k-1}(t, \sigma(\tau))f(\tau, y(\tau))\Delta\tau, \quad t \in \Omega,$$

whereupon

$$D_\Delta^k y(t_0) = \sum_{j=0}^{m-1} b_j h_{j-k}(t_0, t_0)$$

$$= b_k.$$

2. Let $y \in AC_\Delta^m(\Omega)$ be a solution of the Cauchy problem (8.1), (8.2). Then we apply the operator I_{Δ,t_0}^α to both sides of (8.1), and using (8.2), we get

$$I_{\Delta,t_0}^\alpha {}^C D_{\Delta,t_0}^\alpha y(t) = y(t) - \sum_{j=0}^{m-1} h_j(t, t_0) D_\Delta^j y(t_0)$$

$$= y(t) - \sum_{j=0}^{m-1} h_j(t, t_0) y^{\Delta^j}(t_0)$$

$$= y(t) - \sum_{j=0}^{m-1} b_j h_j(t, t_0)$$

$$= I_{\Delta,t_0}^\alpha f(t, y(t)), \quad t \in \Omega.$$

This completes the proof.

Below we assume the Lipschitz condition

$$|f(t, y_1(t)) - f(t, y_2(t))| \le A |y_1(t) - y_2(t)|, \quad t \in \Omega, \tag{8.4}$$

for a constant $A > 0$ that does not depend on $t \in \Omega$.

Theorem 8.2. *Let G be an open set in $\mathbb{R}$ and $f : \Omega \times G \to \mathbb{R}$ a function such that for all $y \in G$, $f(\cdot, y) \in AC_\Delta(\Omega)$, $y(t) \in AC_\Delta^m(\Omega)$. Let also f satisfy the Lipschitz condition (8.4) and*

$$\max_{y \in G, t, s \in \Omega} \{|f(t, y)|, |h_{\alpha-1}(t, s)|\} \le M.$$

Then there exists a unique solution of the Cauchy problem (8.1), (8.2).

Proof. Define the sequence $\{y_l\}_{l \in \mathbb{N}_0}$ as follows:

$$y_0(t) = \sum_{j=0}^{m-1} b_j h_j(t, t_0),$$

$$y_l(t) = y_0(t) + \int_{t_0}^t h_{\alpha-1}(t, \sigma(\tau)) f(\tau, y_{l-1}(\tau)) \, \Delta\tau, \quad l \in \mathbb{N}, \quad t \in \Omega.$$

We have

$$
\begin{aligned}
|y_1(t) - y_0(t)| &= \left| \int_{t_0}^t h_{\alpha-1}(t, \sigma(\tau)) f(\tau, y_{l-1}(\tau)) \, \Delta\tau \right| \\
&\le \int_{t_0}^t |h_{\alpha-1}(t, \sigma(\tau))| \, |f(\tau, y_{l-1}(\tau))| \, \Delta\tau \\
&\le M^2 (t - t_0) \\
&= M^2 h_1(t, t_0), \quad t \in \Omega.
\end{aligned}
$$

Assume that

$$
|y_l(t) - y_{l-1}(t)| \le M^{l+1} A^{l-1} h_l(t, t_0), \quad t \in \Omega, \tag{8.5}
$$

for some $l \in \mathbb{N}$. We will prove that

$$
|y_{l+1}(t) - y_l(t)| \le M^{l+2} A^l h_{l+1}(t, t_0), \quad t \in \Omega.
$$

In fact, we have

$$
\begin{aligned}
|y_{l+1}(t) - y_l(t)| &= \left| y_0(t) + \int_{t_0}^t h_{\alpha-1}(t, \sigma(\tau)) f(\tau, y_l(\tau)) \, \Delta\tau \right. \\
&\quad \left. - y_0(t) - \int_{t_0}^t h_{\alpha-1}(t, \sigma(\tau)) f(\tau, y_{l-1}(\tau)) \, \Delta\tau \right| \\
&= \left| \int_{t_0}^t h_{\alpha-1}(t, \sigma(\tau)) \left(f(\tau, y_l(\tau)) - f(\tau, y_{l-1}(\tau)) \right) \, \Delta\tau \right| \\
&\le \int_{t_0}^t |h_{\alpha-1}(t, \sigma(\tau))| \, |f(\tau, y_l(\tau)) - f(\tau, y_{l-1}(\tau))| \, \Delta\tau \\
&\le MA \int_{t_0}^t |y_l(\tau) - y_{l-1}(\tau)| \, \Delta\tau \\
&\le M^{l+2} A^l \int_{t_0}^t h_l(\tau, t_0) \, \Delta\tau \\
&= M^{l+2} A^l h_{l+1}(t, t_0), \quad t \in \Omega.
\end{aligned}
$$

Therefore, (8.5) holds for all $l \in \mathbb{N}$. Note that

$$
\lim_{l \to \infty} |y_l(t) - y_0(t)| = \left| \sum_{r=1}^{\infty} (y_r(t) - y_{r-1}(t)) \right|
$$

$$\leq \sum_{r=1}^{\infty} |y_r(t) - y_{r-1}(t)|$$

$$\leq \sum_{r=1}^{\infty} M^{r+1} A^{r-1} h_r(t, t_0)$$

$$= \frac{M}{A} \sum_{r=1}^{\infty} (MA)^r h_r(t, t_0)$$

$$\leq \frac{M}{A} \sum_{r=1}^{\infty} (MA)^r \frac{(\sigma(t) - t_0)^r}{r!}, \quad t \in \Omega.$$

Therefore, $y_l(t)$ converges uniformly on Ω to a solution of the Cauchy problem (8.1), (8.2). Suppose that the Cauchy problem (8.1), (8.2) has another solution $z(t)$. For $z(t)$ we have the representation

$$z(t) = y_0(t) + \int_{t_0}^{t} h_{\alpha-1}(t, \sigma(\tau)) f(\tau, z(\tau)) \Delta\tau, \quad t \in \Omega.$$

We have

$$|y_0(t) - z(t)| = \left| \int_{t_0}^{t} h_{\alpha-1}(t, \sigma(\tau)) f(\tau, z(\tau)) \Delta\tau \right|$$

$$\leq \int_{t_0}^{t} |h_{\alpha-1}(t, \sigma(\tau))| |f(\tau, z(\tau))| \Delta\tau$$

$$\leq M^2 (t - t_0)$$

$$= M^2 h_1(t, t_0), \quad t \in \Omega.$$

Suppose that

$$|y_l(t) - z(t)| \leq A^l M^{l+2} h_{l+1}(t, t_0), \quad t \in \Omega, \tag{8.6}$$

for some $l \in \mathbb{N}$. We will prove that

$$|y_{l+1}(t) - z(t)| \leq A^{l+1} M^{l+3} h_{l+2}(t, t_0), \quad t \in \Omega.$$

In fact, we have

$$|y_{l+1}(t) - z(t)| = \left| y_0(t) + \int_{t_0}^{t} h_{\alpha-1}(t, \sigma(\tau)) f(\tau, y_l(\tau)) \Delta\tau \right.$$

$$-y_0(t) - \int_{t_0}^t h_{\alpha-1}(t, \sigma(\tau)) f(\tau, z(\tau)) \Delta\tau \Bigg|$$

$$= \left| \int_{t_0}^t h_{\alpha-1}(t, \sigma(\tau)) \left(f(\tau, y_l(\tau)) - f(\tau, z(\tau)) \right) \Delta\tau \right|$$

$$\leq \int_{t_0}^t |h_{\alpha-1}(t, \sigma(\tau))| \, |f(\tau, y_l(\tau)) - f(\tau, z(\tau))| \, \Delta\tau$$

$$\leq MA \int_{t_0}^t |y_l(\tau) - z(\tau)| \, \Delta\tau$$

$$\leq M^{l+3} A^{l+1} \int_{t_0}^t h_{l+1}(\tau, t_0) \Delta\tau$$

$$= M^{l+3} A^{l+1} h_{l+2}(t, t_0), \quad t \in \Omega.$$

Therefore, (8.6) holds for all $l \in \mathbb{N}$. By (8.6), we get

$$|y_l(t) - z(t)| \leq M^{l+2} A^l h_{l+1}(t, t_0)$$

$$\leq M^{l+2} A^l \frac{(\sigma(t) - t_0)^{l+1}}{(l+1)!}$$

$$\to 0, \quad as \quad l \to \infty.$$

Consequently, $y(t) = z(t), t \in \Omega$. This completes the proof.

8.2   The Dependence of the Solution on the Initial Value

Let $\alpha > 0$, $m = -\overline{[-\alpha]}$, and let Ω be a finite time scale interval that contains t_0. Consider the Cauchy problems

$$\begin{aligned} {}^C D^\alpha_{\Delta, t_0} y(t) &= f(t, y(t)), \quad t \in \Omega, \\ D^k_\Delta y(t_0) &= b_k, \quad k \in \{0, \ldots, m-1\}, \end{aligned} \tag{8.7}$$

$$\begin{aligned} {}^C D^\alpha_{\Delta, t_0} y(t) &= f(t, y(t)), \quad t \in \Omega, \\ D^k_\Delta y(t_0) &= c_k, \quad k \in \{0, \ldots, m-1\}, \end{aligned} \tag{8.8}$$

where $f : \mathbb{T} \times \mathbb{R} \to \mathbb{R}$ is a given function, $b_k, c_k, k \in \{0, \ldots, m-1\}$, are given constants. Let

$$y_0(t) = \sum_{j=0}^{m-1} h_j(t, t_0) b_j,$$

$$z_0(t) = \sum_{j=0}^{m-1} h_j(t, t_0)c_j, \quad t \in \Omega.$$

We define $\|y\| = \sup_{t \in \Omega} |y(t)|$.

Theorem 8.3. *Let $y(t)$ and $z(t)$ be the solutions of the Cauchy problems (8.7) and (8.8), respectively. Let also $|h_{\alpha-1}(t, s)| \leq M$ for all $t, s \in \Omega$, and let the function f satisfy the Lipschitz condition (8.4). Then*

$$|z(t) - y(t)| \leq \|z_0 - y_0\| \sum_{j=0}^{\infty} (AM)^j \frac{(\sigma(t) - t_0)^j}{j!}, \quad t \in \Omega.$$

Proof. By the proof of Theorem 8.2, we have that

$$y(t) = \lim_{l \to \infty} y_l(t),$$

$$z(t) = \lim_{l \to \infty} z_l(t), \quad t \in \Omega,$$

where

$$y_l(t) = y_0(t) + \int_{t_0}^{t} h_{\alpha-1}(t, \sigma(\tau)) f(\tau, y_{l-1}(\tau)) \, \Delta\tau,$$

$$z_l(t) = z_0(t) + \int_{t_0}^{t} h_{\alpha-1}(t, \sigma(\tau)) f(\tau, z_{l-1}(\tau)) \, \Delta\tau, \quad l \in \mathbb{N}, \quad t \in \Omega.$$

We have

$$|z_1(t) - y_1(t)| = \left| z_0(t) + \int_{t_0}^{t} h_{\alpha-1}(t, \sigma(\tau)) f(\tau, z_0(\tau)) \, \Delta\tau \right.$$

$$\left. -y_0(t) - \int_{t_0}^{t} h_{\alpha-1}(t, \sigma(\tau)) f(\tau, y_0(\tau)) \, \Delta\tau \right|$$

$$= \left| z_0(t) - y_0(t) \right.$$

$$\left. + \int_{t_0}^{t} h_{\alpha-1}(t, \sigma(\tau)) \left(f(\tau, z_0(\tau)) - f(\tau, y_0(\tau)) \right) \Delta\tau \right|$$

$$\leq |z_0(t) - y_0(t)|$$

$$+ \left| \int_{t_0}^{t} h_{\alpha-1}(t, \sigma(\tau)) \left(f(\tau, z_0(\tau)) - f(\tau, y_0(\tau)) \right) \Delta\tau \right|$$

$$\leq \|z_0 - y_0\| + \int_{t_0}^{t} |h_{\alpha-1}(t, \sigma(\tau))| \, |f(\tau, z_0(\tau)) - f(\tau, y_0(\tau))| \, \Delta\tau$$

$$\leq \|z_0 - y_0\| + MA \int_{t_0}^{t} |z_0(\tau) - y_0(\tau)| \, \Delta\tau$$

$$\leq \|z_0 - y_0\| + MA\|z_0 - y_0\|(t - t_0)$$

$$= \|z_0 - y_0\| + MA\|z_0 - y_0\|h_1(t, t_0)$$

$$= \|z_0 - y_0\| \left(1 + MAh_1(t, t_0)\right), \quad t \in \Omega.$$

Suppose that

$$|z_l(t) - y_l(t)| \leq \|z_0 - y_0\| \sum_{j=0}^{l}(AM)^j h_j(t, t_0), \quad t \in \Omega. \tag{8.9}$$

We will prove that

$$|z_{l+1}(t) - y_{l+1}| \leq \|z_0 - y_0\| \sum_{j=0}^{l+1}(AM)^j h_j(t, t_0), \quad t \in \Omega.$$

In fact, we have

$$|z_{l+1}(t) - y_{l+1}(t)| = \left| z_0(t) + \int_{t_0}^{t} h_{\alpha-1}(t, \sigma(\tau)) f(\tau, z_l(\tau)) \, \Delta\tau \right.$$

$$\left. -y_0(t) - \int_{t_0}^{t} h_{\alpha-1}(t, \sigma(\tau)) f(\tau, y_l(\tau)) \, \Delta\tau \right|$$

$$= \left| z_0(t) - y_0(t) \right.$$

$$\left. + \int_{t_0}^{t} h_{\alpha-1}(t, \sigma(\tau)) \left(f(\tau, z_l(\tau)) - f(\tau, y_l(\tau)) \right) \Delta\tau \right|$$

$$\leq |z_0(t) - y_0(t)|$$

$$+ \left| \int_{t_0}^{t} h_{\alpha-1}(t, \sigma(\tau)) \left(f(\tau, z_l(\tau)) - f(\tau, y_l(\tau)) \right) \Delta\tau \right|$$

$$\leq \|z_0 - y_0\| + \int_{t_0}^{t} |h_{\alpha-1}(t, \sigma(\tau))| \, |f(\tau, z_l(\tau)) - f(\tau, y_l(\tau))| \, \Delta\tau$$

$$\leq \|z_0 - y_0\| + MA \int_{t_0}^{t} |z_l(\tau) - y_l(\tau)| \, \Delta\tau$$

$$\leq \|z_0 - y_0\| + \sum_{j=0}^{l}(MA)^{j+1}\|z_0 - y_0\| \int_{t_0}^{t} h_j(\tau, t_0)\Delta\tau$$

$$= \|z_0 - y_0\| + \sum_{j=0}^{l} (MA)^{j+1} \|z_0 - y_0\| h_{j+1}(t, t_0)$$

$$= \|z_0 - y_0\| \left(1 + \sum_{j=0}^{l} (MA)^{j+1} h_{j+1}(t, t_0) \right)$$

$$= \|z_0 - y_0\| \sum_{j=0}^{l+1} (MA)^j h_j(t, t_0), \quad t \in \Omega.$$

Consequently, (8.9) holds for all $l \in \mathbb{N}$. By (8.9), we get

$$|z(t) - y(t)| = \lim_{l \to \infty} |z_l(t) - y_l(t)|$$

$$\leq \|z_0 - y_0\| \sum_{j=0}^{\infty} (AM)^j h_j(t, t_0)$$

$$\leq \|z_0 - y_0\| \sum_{j=0}^{\infty} \frac{(AM)^j (\sigma(t) - t_0)^j}{j!}, \quad t \in \Omega.$$

This completes the proof.

Chapter 9
Caputo Fractional Dynamic Equations with Constant Coefficients

In this chapter we suppose that $\mathbb{T}$ is a time scale with forward jump operator and delta differentiation operator σ and Δ, respectively, such that

$$\mathbb{T} = \{t_n : n \in \mathbb{N}_0\},$$

$$\lim_{n \to \infty} t_n = \infty,$$

$$\sigma(t_n) = t_{n+1}, \quad n \in \mathbb{N}_0,$$

$$w = \inf_{n \in \mathbb{N}_0} \mu(t_n) > 0.$$

9.1 Homogeneous Caputo Fractional Dynamic Equations with Constant Coefficients

Consider the equation

$$^C D^\alpha_{\Delta, t_0} y(t) - \lambda y(t) = 0, \quad t \in \Omega, \tag{9.1}$$

subject to the initial conditions

$$D^k_\Delta y(t_0) = d_k, \quad k \in \{0, \ldots, m-1\}, \tag{9.2}$$

where $\alpha > 0$, $m = -\overline{[-\alpha]}$, $\lambda \in \mathbb{C}$, $d_k \in \mathbb{R}$, $k \in \{0, \ldots, m-1\}$, are given constants and Ω is a finite time scale interval that contains t_0. We define the function

$$W(t) = det\left(\left(D^k_\Delta y_j\right)(t)\right)^n_{k,j=1}, \quad t \in \Omega.$$

© Springer International Publishing AG, part of Springer Nature 2018
S. G. Georgiev, *Fractional Dynamic Calculus and Fractional Dynamic
Equations on Time Scales*, https://doi.org/10.1007/978-3-319-73954-0_9

Theorem 9.1. *Let $m > 1$. The solutions $y_1(t)$, $y_2(t)$, ..., $y_n(t)$ of equation (9.1) are linearly independent if and only if $W(t^\star) \neq 0$ at some point $t^\star \in \Omega$.*

Proof. 1. Let $W(t^\star) \neq 0$ at some point $t^\star \in \Omega$. Suppose that $y_j(t)$, $j \in \{1, \ldots, n\}$, are linearly dependent on Ω. Then there exist n constants $c_1, c_2, \ldots, c_n$, not all zero, such that

$$c_1 y_1(t) + c_2 y_2(t) + \cdots + c_n y_n(t) = 0, \quad t \in \Omega.$$

Hence,

$$c_1 D_\Delta^k y_1(t) + c_2 D_\Delta^k y_2(t) + \cdots + c_n D_\Delta^k y_n(t) = 0, \quad t \in \Omega,$$

for $k \in \{1, \ldots, n\}$. Therefore,

$$\left(\left(D_\Delta^k y_j \right)(t) \right)_{k,j=1}^n \begin{pmatrix} c_1 \\ c_2 \\ \vdots \\ c_n \end{pmatrix} = 0, \quad t \in \Omega.$$

Thus $W(t) = 0$, $t \in \Omega$. This is a contradiction. Therefore, $y_1(t)$, $y_2(t)$, ..., $y_n(t)$ are linearly independent on Ω.

2. Let $y_1(t)$, $y_2(t)$, ..., $y_n(t)$ be linearly independent on Ω. Assume that $W(t) = 0$ for all $t \in \Omega$. Consider

$$\left(\left(D_\Delta^k y_j \right)(t^\star) \right)_{k,j=1}^n C = 0, \tag{9.3}$$

where $t^\star \in \Omega$, $C = \begin{pmatrix} c_1 \\ c_2 \\ \vdots \\ c_n \end{pmatrix}$. Since $W(t^\star) = 0$, the system (9.3) has a nontrivial solution $c_1, c_2, \ldots, c_n$. Let

$$y(t) = \sum_{j=1}^n c_j y_j(t), \quad t \in \Omega.$$

Note that y is a solution of the equation (9.1). By (9.3), we get that y satisfies the initial condition

$$D_\Delta^k y(t^\star) = 0, \quad k \in \{1, \ldots, n\}. \tag{9.4}$$

Since $y(t) = 0$ is also a solution of equation (9.1) satisfying the initial condition (9.4), we conclude that

$$\sum_{j=1}^{n} c_j y_j(t) = 0, \quad t \in \Omega.$$

Therefore, $y_1(t)$, $y_2(t)$, ..., $y_n(t)$ are linearly dependent on Ω. This is a contradiction. Consequently, there exists $t^* \in \Omega$ such that $W(t^*) \neq 0$. This completes the proof.

Theorem 9.2. *The functions*

$$y_j(t) = {}_\Delta F_{\alpha, j+1}(\lambda, t, t_0), \quad j \in \{0, \ldots, m-1\},$$

yield the fundamental system of solutions of (9.1).

Proof. We take the Laplace transform of both sides of (9.1), and using the initial conditions (9.2), we get

$$0 = \mathscr{L}\left({}^C D^\alpha_{\Delta, t_0} y(t) - \lambda y(t)\right)(z)$$

$$= \mathscr{L}\left({}^C D^\alpha_{\Delta, t_0} y(t)\right)(z) - \lambda \mathscr{L}(y(t))(z)$$

$$= z^\alpha \mathscr{L}(y(t))(z) - \sum_{k=0}^{m-1} z^{\alpha-k-1} D^k_\Delta y(t_0)$$

$$- \lambda \mathscr{L}(y(t))(z)$$

$$= z^\alpha \mathscr{L}(y(t))(z) - \sum_{k=0}^{m-1} z^{\alpha-k-1} d_k - \lambda \mathscr{L}(y(t))(z)$$

$$= \left(z^\alpha - \lambda\right) \mathscr{L}(y(t))(z) - \sum_{k=0}^{m-1} z^{\alpha-k-1} d_k,$$

whereupon

$$\left(z^\alpha - \lambda\right) \mathscr{L}(y(t))(z) = \sum_{k=0}^{m-1} z^{\alpha-k-1} d_k,$$

or

$$\mathscr{L}(y(t))(z) = \frac{1}{z^\alpha - \lambda} \sum_{k=0}^{m-1} z^{\alpha-k-1} d_k$$

$$= \sum_{k=0}^{m-1} d_k \frac{z^{\alpha-k-1}}{z^\alpha - \lambda}$$

$$= \sum_{k=0}^{m-1} d_k \mathscr{L}\left({}_\Delta F_{\alpha,k+1}(\lambda, t, t_0)\right)(z)$$

$$= \mathscr{L}\left(\sum_{k=0}^{m-1} d_k {}_\Delta F_{\alpha,k+1}(\lambda, t, t_0)\right)(z), \quad t \in \Omega.$$

Therefore,

$$y(t) = \sum_{k=0}^{m-1} d_k {}_\Delta F_{\alpha,k+1}(\lambda, t, t_0), \quad t \in \mathbb{T}, \quad t > t_0.$$

Now we will check that the functions $y_j(t)$, $j \in \{0, \ldots, m-1\}$, satisfy equation (9.1). We have

$${}^C D^\alpha_{\Delta,t_0} y_j(t) = {}^C D^\alpha_{\Delta,t_0} {}_\Delta F_{\alpha,j+1}(\lambda, t, t_0)$$

$$= {}^C D^\alpha_{\Delta,t_0} \sum_{k=0}^{\infty} \lambda^k h_{k\alpha+j}(t, t_0)$$

$$= \sum_{k=1}^{\infty} \lambda^k {}^C D^\alpha_{\Delta,t_0} h_{k\alpha+j}(t, t_0)$$

$$= \sum_{k=1}^{\infty} \lambda^k h_{(k-1)\alpha+j}(t, t_0)$$

$$= \sum_{k=0}^{\infty} \lambda^{k+1} h_{k\alpha+j}(t, t_0)$$

$$= \lambda \sum_{k=0}^{\infty} \lambda^k h_{k\alpha+j}(t, t_0)$$

$$= \lambda {}_\Delta F_{\alpha,j+1}(\lambda, t, t_0)$$

$$= \lambda y_j(t), \quad t \in \Omega,$$

or

$${}^C D^\alpha_{\Delta,t_0} y_j(t) - \lambda y_j(t) = 0, \quad t \in \Omega.$$

Also, for $k \in \{0, \ldots, m-1\}$, we have

$$D_\Delta^k y_j(t) = D_\Delta^k \left(\sum_{r=0}^{\infty} \lambda^r h_{r\alpha+j}(t, t_0) \right)$$

$$= \sum_{r=0}^{\infty} \lambda^r D_\Delta^k h_{r\alpha+j}(t, t_0)$$

$$= \sum_{r=0}^{\infty} \lambda^r h_{r\alpha+j-k}(t, t_0), \quad t \in \Omega.$$

Hence,

$$D_\Delta^k y_j(t_0) = \sum_{r=0}^{\infty} \lambda^r h_{r\alpha+j-k}(t_0, t_0)$$

$$= \begin{cases} 1 & \text{if } j = k, \\ 0 & \text{if } j \neq k, \end{cases}$$

and

$$W(t) = det \left(D_\Delta^k y_j(t) \right)_{k,j=0}^{m-1}$$

$$= det \begin{pmatrix} 1 & 0 & \ldots & 0 \\ 0 & 1 & \ldots & 0 \\ \vdots & \vdots & \vdots & \vdots \\ 0 & 0 & \ldots & 1 \end{pmatrix}$$

$$= 1, \quad t \in \Omega.$$

This completes the proof.

Example 9.1. Consider the Cauchy problem

$$^C D_{\Delta, t_0}^{\frac{3}{2}} y(t) - 2y(t) = 0, \quad t \in \Omega,$$

$$y^\Delta(t_0) = y(t_0) = 1.$$

Here

$$\alpha = \frac{3}{2},$$

$$m = -\left\lceil -\frac{3}{2} \right\rceil$$

$$= 2,$$

$$\lambda = 2,$$

$$d_0 = 1,$$

$$d_1 = 1.$$

The fundamental system of solutions of the considered equation is

$$y_0(t) = {}_{\Delta}F_{\frac{3}{2},1}(2, t, t_0),$$

$$y_1(t) = {}_{\Delta}F_{\frac{3}{2},2}(2, t, t_0), \quad t \in \Omega.$$

Therefore,

$$y(t) = d_0 y_0(t) + d_1 y_1(t)$$

$$= {}_{\Delta}F_{\frac{3}{2},1}(2, t, t_0) + {}_{\Delta}F_{\frac{3}{2},2}(2, t, t_0), \quad t \in \Omega,$$

is the solution of the considered Cauchy problem.

Example 9.2. Consider the Cauchy problem

$$^{C}D_{\Delta,t_0}^{\frac{11}{4}} y(t) - 3y(t) = 0, \quad t \in \Omega,$$

$$y(t_0) = 1, \quad y^{\Delta}(t_0) = -2, \quad y^{\Delta^2}(t_0) = 3.$$

Here

$$\alpha = \frac{11}{4}.$$

$$m = -\left\lceil -\frac{11}{4} \right\rceil$$

$$= 3,$$

$$\lambda = 3,$$

$$d_0 = 1,$$

$$d_1 = -2,$$

$$d_2 = 3.$$

The fundamental system of solutions of the considered equation is

$$y_0(t) = {}_\Delta F_{\frac{11}{4},1}(3, t, t_0),$$

$$y_1(t) = {}_\Delta F_{\frac{11}{4},2}(3, t, t_0),$$

$$y_2(t) = {}_\Delta F_{\frac{11}{4},3}(3, t, t_0), \quad t \in \Omega.$$

Therefore,

$$
\begin{aligned}
y(t) &= d_0 y_0(t) + d_1 y_1(t) + d_2 y_2(t) \\
&= y_0(t) - 2y_1(t) + 3y_2(t) \\
&= {}_\Delta F_{\frac{11}{4},1}(3, t, t_0) - 2\,{}_\Delta F_{\frac{11}{4},2}(3, t, t_0) \\
&\quad + 3\,{}_\Delta F_{\frac{11}{4},3}(3, t, t_0), \quad t \in \Omega,
\end{aligned}
$$

is the solution of the considered Cauchy problem.

Example 9.3. Consider the Cauchy problem

$$^C D_{\Delta,t_0}^{\frac{10}{3}} y(t) - 4y(t) = 0, \quad t \in \Omega,$$

$$y(t_0) = 1, \quad y^\Delta(t_0) = -3,$$

$$y^{\Delta^2}(t_0) = 3, \quad y^{\Delta^3}(t_0) = -4.$$

Here

$$\alpha = \frac{10}{3},$$

$$m = -\left[-\frac{10}{3}\right]$$

$$= 4,$$

$$\lambda = 4,$$

$$d_0 = 1,$$

$$d_1 = -3,$$

$$d_2 = 3,$$

$$d_3 = -4.$$

The fundamental system of solutions of the considered equation is

$$y_0(t) = {}_\Delta F_{\frac{10}{3},1}(4, t, t_0),$$

$$y_1(t) = {}_\Delta F_{\frac{10}{3},2}(4, t, t_0),$$

$$y_2(t) = {}_\Delta F_{\frac{10}{3},3}(4, t, t_0),$$

$$y_3(t) = {}_\Delta F_{\frac{10}{3},4}(4, t, t_0), \quad t \in \Omega.$$

Therefore,

$$y(t) = d_0 y_0(t) + d_1 y_1(t) + d_2 y_2(t) + d_3 y_3(t)$$
$$= y_0(t) - 3y_1(t) + 3y_2(t) - 4y_3(t)$$
$$= {}_\Delta F_{\frac{10}{3},1}(4, t, t_0) - 3 {}_\Delta F_{\frac{10}{3},2}(4, t, t_0)$$
$$+ 3 {}_\Delta F_{\frac{10}{3},3}(4, t, t_0) - 4 {}_\Delta F_{\frac{10}{3},4}(4, t, t_0), \quad t \in \Omega,$$

is the solution of the considered Cauchy problem.

Exercise 9.1. Find the solution of the Cauchy problem

$$^C D_{\Delta,t_0}^{\frac{16}{3}} y(t) - y(t) = 0, \quad t \in \Omega,$$

$$y(t_0) = -1, \quad y^\Delta(t_0) = 2,$$
$$y^{\Delta^2}(t_0) = -3, \quad y^{\Delta^3}(t_0) = 5.$$

Consider the equation

$$^C D_{\Delta,t_0}^{\alpha} y(t) - \lambda\, ^C D_{\Delta,t_0}^{\beta} y(t) - \sum_{k=0}^{r-2} A_k\, ^C D_{\Delta,t_0}^{\alpha_k} y(t) = 0, \quad t \in \Omega, \tag{9.5}$$

where $r \in \mathbb{N}\backslash\{1, 2\}$, λ, A_k, $k \in \{0, \ldots, r-2\}$, are given constants,

$$0 = \alpha_0 < \alpha_1 < \ldots < \alpha_{r-2} < \beta < \alpha,$$
$$0 = l_0 < l_1 \leq \ldots \leq l_{r-1} \leq l,$$
$$l - 1 < \alpha \leq l,$$
$$l_{r-1} - 1 < \beta \leq l_{r-1},$$
$$l_j - 1 < \alpha_j \leq l_j, \quad j \in \{0, \ldots, r-2\},$$

subject to the initial conditions

$$D_\Delta^k y(t_0) = b_k, \quad k \in \{0, \ldots, l-1\}, \tag{9.6}$$

where b_k, $k \in \{0, \ldots, l-1\}$, are given constants.

Theorem 9.3. *The fundamental system of solutions of equation* (9.5) *is given by*

$$
y_j(t) = \sum_{n=0}^{\infty} \left(\sum_{k_0+\cdots+k_{r-2}=n} \right) \frac{1}{k_0! \ldots k_{r-2}!} \left(\prod_{\nu=0}^{r-2} (A_\nu)^{k_\nu} \right)
$$

$$
\times \left(\frac{\partial^n}{\partial \lambda^n} \Delta F_{\alpha-\beta, j+1+\sum_{\nu=0}^{r-2}(\beta-\alpha_\nu)k_\nu}(\lambda, t, t_0) \right.
$$

$$
- \lambda \frac{\partial^n}{\partial \lambda^n} \Delta F_{\alpha-\beta, \alpha-\beta+j+1+\sum_{\nu=0}^{r-2}(\beta-\alpha_\nu)k_\nu}(\lambda, t, t_0) \tag{9.7}
$$

$$
\left. - \sum_{k=0}^{r-2} A_k \frac{\partial^n}{\partial \lambda^n} \Delta F_{\alpha-\beta, \alpha-\alpha_k+j+1+\sum_{\nu=0}^{r-2}(\beta-\alpha_\nu)k_\nu}(\lambda, t, t_0) \right),
$$

for $j \in \{0, \ldots, l_{r-2}-1\}$,

$$
y_j(t) = \sum_{n=0}^{\infty} \left(\sum_{k_0+\cdots+k_{r-2}=n} \right) \frac{1}{k_0! \ldots k_{r-2}!} \left(\prod_{\nu=0}^{r-2} (A_\nu)^{k_\nu} \right)
$$

$$
\times \left(\frac{\partial^n}{\partial \lambda^n} \Delta F_{\alpha-\beta, j+1+\sum_{\nu=0}^{r-2}(\beta-\alpha_\nu)k_\nu}(\lambda, t, t_0) \right. \tag{9.8}
$$

$$
\left. - \lambda \frac{\partial^n}{\partial \lambda^n} \Delta F_{\alpha-\beta, \alpha-\beta+j+1+\sum_{\nu=0}^{r-2}(\beta-\alpha_\nu)k_\nu}(\lambda, t, t_0) \right),
$$

for $j \in \{l_{r-2}, \ldots, l_{r-1}-1\}$, $l_{r-1} > l_{r-2}$,

$$
y_j(t) = \sum_{n=0}^{\infty} \left(\sum_{k_0+\cdots+k_{r-2}=n} \right) \frac{1}{k_0! \ldots k_{r-2}!} \left(\prod_{\nu=0}^{r-2} (A_\nu)^{k_\nu} \right)
$$

$$
\times \left(\frac{\partial^n}{\partial \lambda^n} \Delta F_{\alpha-\beta, j+1+\sum_{\nu=0}^{r-2}(\beta-\alpha_\nu)k_\nu}(\lambda, t, t_0) \right), \quad t \in \Omega, \tag{9.9}
$$

for $j \in \{l_{r-1}, \ldots, l-1\}$, $l > l_{r-1}$.

Proof. We take the Laplace transform of both sides of equation (9.5), and using the initial conditions (9.6), we get

$$
0 = \mathscr{L} \left({}^C D^\alpha_{\Delta,t_0} y(t) - \lambda {}^C D^\beta_{\Delta,t_0} y(t) - \sum_{k=0}^{m-2} A_k {}^C D^{\alpha_k}_{\Delta,t_0} y(t) \right)(z)
$$

$$
= \mathscr{L} \left({}^C D^\alpha_{\Delta,t_0} y(t) \right)(z) - \lambda \mathscr{L} \left({}^C D^\beta_{\Delta,t_0} y(t) \right)(z)
$$

$$
- \sum_{k=0}^{m-2} A_k \mathscr{L} \left({}^C D^{\alpha_k}_{\Delta,t_0} y(t) \right)(z)
$$

$$= z^\alpha \mathscr{L}(y(t))(z) - \sum_{j=0}^{l-1} z^{\alpha-j-1} y^{\Delta^j}(t_0)$$

$$-\lambda z^\beta \mathscr{L}(y(t))(z) + \lambda \sum_{j=0}^{l_{r-1}-1} z^{\beta-j-1} y^{\Delta^j}(t_0)$$

$$-\sum_{k=0}^{r-2} A_k z^{\alpha_k} + \sum_{k=0}^{r-2} A_k \sum_{j=0}^{l_k-1} z^{\alpha_k-j-1} y^{\Delta^j}(t_0)$$

$$= \left(z^\alpha - \lambda z^\beta - \sum_{k=0}^{r-2} A_k z^{\alpha_k} \right) \mathscr{L}(y(t))(z)$$

$$-\sum_{j=0}^{l-1} b_j z^{\alpha-j-1} + \lambda \sum_{j=0}^{l_{r-1}-1} b_j z^{\beta-j-1}$$

$$+\sum_{k=0}^{r-2} A_k \sum_{j=0}^{l_k-1} b_j z^{\alpha_k-j-1},$$

whereupon

$$\left(z^\alpha - \lambda z^\beta - \sum_{k=0}^{r-2} A_k z^{\alpha_k} \right) \mathscr{L}(y(t))(z) = \sum_{j=0}^{l-1} b_j z^{\alpha-j-1}$$

$$-\lambda \sum_{j=0}^{l_{r-1}-1} b_j z^{\beta-j-1}$$

$$-\sum_{k=0}^{r-2} A_k \sum_{j=0}^{l_k-1} b_j z^{\alpha_k-j-1},$$

or

$$\mathscr{L}(y(t))(z) = \sum_{j=0}^{l-1} b_j \frac{z^{\alpha-j-1}}{z^\alpha - \lambda z^\beta - \sum_{k=0}^{r-2} A_k z^{\alpha_k}}$$

$$-\lambda \sum_{j=0}^{l_{r-1}-1} b_j \frac{z^{\beta-j-1}}{z^\alpha - \lambda z^\beta - \sum_{k=0}^{r-2} A_k z^{\alpha_k}} \qquad (9.10)$$

$$-\sum_{k=0}^{r-2} A_k \sum_{j=0}^{l_k-1} b_j \frac{z^{\alpha_k-j-1}}{z^\alpha - \lambda z^\beta - \sum_{k=0}^{r-2} A_k z^{\alpha_k}}.$$

Note that for $z \in \mathbb{C}$ and

$$\left| \sum_{k=0}^{r-2} A_k \frac{z^{\alpha_k - \beta}}{z^{\alpha - \beta} - \lambda} \right| < 1,$$

we have

$$
\frac{1}{z^{\alpha} - \lambda z^{\beta} - \sum_{k=0}^{r-2} A_k z^{\alpha_k}} = \frac{z^{-\beta}}{z^{\alpha-\beta} - \lambda} \frac{1}{1 - \left(\sum_{k=0}^{r-2} A_k \frac{z^{\alpha_k - \beta}}{z^{\alpha-\beta} - \lambda} \right)}
$$

$$
= \sum_{n=0}^{\infty} \frac{z^{-\beta}}{(z^{\alpha-\beta} - \lambda)^{n+1}} \left(\sum_{k=0}^{r-2} A_k z^{\alpha_k - \beta} \right)^n
$$

$$
= \sum_{n=0}^{\infty} \left(\sum_{k_0 + \cdots + k_{r-2} = n} \right) \frac{n!}{k_0! \ldots k_{r-2}!}
$$

$$
\times \left(\prod_{v=0}^{r-2} (A_v)^{k_v} \right) \frac{z^{-\beta - \sum_{k=0}^{r-2} (\beta - \alpha_v) k_v}}{(z^{\alpha-\beta} - \lambda)^{n+1}}.
$$

Also, for $z \in \mathbb{C}$ and

$$\left| \lambda z^{\beta - \alpha} \right| < 1,$$

we have

$$
\frac{z^{\alpha - j - 1 - \beta - \sum_{v=0}^{r-2}(\beta - \alpha_v)k_v}}{(z^{\alpha-\beta} - \lambda)^{n+1}} = \frac{1}{n!} \mathscr{L} \left(\frac{\partial^n}{\partial \lambda^n} \Delta F_{\alpha - \beta, j+1 + \sum_{v=0}^{r-2}(\beta - \alpha_v)k_v} (\lambda, t, t_0) \right)(z),
$$

$$\tag{9.11}$$

$$
\frac{z^{\beta - 1 - j - \beta - \sum_{v=0}^{r-2}(\beta - \alpha_v)k_v}}{(z^{\alpha-\beta} - \lambda)^{n+1}} = \frac{1}{n!} \mathscr{L} \left(\frac{\partial^n}{\partial \lambda^n} \Delta F_{\alpha - \beta, \alpha - \beta + j+1 + \sum_{v=0}^{r-2}(\beta - \alpha_v)k_v} (\lambda, t, t_0) \right)(z),
$$

$$\tag{9.12}$$

$$
\frac{z^{\alpha_k - j - 1 - \beta - \sum_{v=0}^{r-2}(\beta - \alpha_v)k_v}}{(z^{\alpha-\beta} - \lambda)^{n+1}} = \frac{1}{n!} \mathscr{L} \left(\frac{\partial^n}{\partial \lambda^n} \Delta F_{\alpha - \beta, \alpha - \alpha_k + j+1 + \sum_{v=0}^{r-2}(\beta - \alpha_v)k_v} (\lambda, t, t_0) \right)(z).
$$

$$\tag{9.13}$$

By (9.10), (9.11), (9.12), (9.13), we get the formulas (9.7), (9.8), (9.9), respectively. For $k \in \{0, \ldots, l-1\}$, we obtain

$$
D_{\Delta}^k y_j(t) = \sum_{n=0}^{\infty} \left(\sum_{k_0 + \cdots + k_{r-2} = n} \right) \frac{1}{k_0! \ldots k_{r-2}!} \left(\prod_{v=0}^{r-2} (A_v)^{k_v} \right)
$$

$$
\times \left(\frac{\partial^n}{\partial \lambda^n} \sum_{s=0}^{\infty} \lambda^s h_{s(\alpha - \beta) + \sum_{v=0}^{r-2}(\beta - \alpha_v)k_v + j - k}(t, t_0) \right.
$$

$$-\lambda\frac{\partial^n}{\partial\lambda^n}\sum_{s=0}^{\infty}\lambda^s h_{s(\alpha-\beta)+\sum_{\nu=0}^{r-2}(\beta-\alpha_\nu)k_\nu+j-k+\alpha-\beta}(t,t_0)$$

$$-\sum_{k=0}^{r-2}A_k\frac{\partial^n}{\partial\lambda^n}\sum_{s=0}^{\infty}\lambda^s h_{s(\alpha-\beta)+\sum_{\nu=0}^{r-2}(\beta-\alpha_\nu)k_\nu+j-k+\alpha-\alpha_k}(t,t_0)\Bigg)$$

for $j \in \{0, \ldots, l_{r-2} - 1\}$, $l_{r-2} > 1$,

$$D_\Delta^k y_j(t) = \sum_{n=0}^{\infty}\left(\sum_{k_0+\cdots+k_{r-2}=n}\right)\frac{1}{k_0!\ldots k_{r-2}!}\left(\prod_{\nu=0}^{r-2}(A_\nu)^{k_\nu}\right)$$

$$\times\Bigg(\frac{\partial^n}{\partial\lambda^n}\sum_{s=0}^{\infty}\lambda^s h_{s(\alpha-\beta)+\sum_{\nu=0}^{r-2}(\beta-\alpha_\nu)k_\nu+j-k}(t,t_0)$$

$$-\lambda\frac{\partial^n}{\partial\lambda^n}\sum_{s=0}^{\infty}\lambda^s h_{s(\alpha-\beta)+\sum_{\nu=0}^{r-2}(\beta-\alpha_\nu)k_\nu+j-k+\alpha-\beta}(t,t_0)\Bigg)$$

for $j \in \{l_{r-2}, \ldots, l_{r-1} - 1\}$, $l_{r-1} > l_{r-2}$,

$$D_\Delta^k y_j(t) = \sum_{n=0}^{\infty}\left(\sum_{k_0+\cdots+k_{r-2}=n}\right)\frac{1}{k_0!\ldots k_{r-2}!}\left(\prod_{\nu=0}^{r-2}(A_\nu)^{k_\nu}\right)$$

$$\times\frac{\partial^n}{\partial\lambda^n}\sum_{s=0}^{\infty}\lambda^s h_{s(\alpha-\beta)+\sum_{\nu=0}^{r-2}(\beta-\alpha_\nu)k_\nu+j-k}(t,t_0), \quad t \in \Omega,$$

for $j \in \{l_{r-1}, \ldots, l - 1\}$, $l > l_{r-1}$. Hence, for $j > k$, we have that $D_\Delta^k y_j(t_0) = 0$, and for $j = k$ we have $D_\Delta^k y_k(t_0) = 1$. Therefore, $W(t_0) = 1$. This completes the proof.

Example 9.4. Let $A_k = 0$, $k \in \{0, \ldots, r - 2\}$. Then the fundamental system of solutions for equation (9.5) is given by

$$y_j(t) = {}_\Delta F_{\alpha-\beta,j+1}(\lambda, t, t_0)$$

$$-\lambda {}_\Delta F_{\alpha-\beta,\alpha-\beta+j+1}(\lambda, t, t_0), \quad t \in \Omega,$$

for $j \in \{0, \ldots, l_{r-1} - 1\}$, and

$$y_j(t) = {}_\Delta F_{\alpha-\beta,j+1}(\lambda, t, t_0), \quad t \in \Omega,$$

for $j \in \{l_{r-1}, \ldots, l - 1\}$.

Exercise 9.2. Find the solution of the Cauchy problem

$${}^C D_{\Delta,t_0}^{\frac{13}{3}} y(t) + {}^C D_{\Delta,t_0}^{\frac{3}{2}} y(t) + {}^C D_{\Delta,t_0}^{\frac{1}{4}} y(t) - 8y(t) = 0, \quad t \in \Omega,$$

$$y(t_0) = 1, \quad y^{\Delta}(t_0) = -2,$$
$$y^{\Delta^2}(t_0) = 4, \quad y^{\Delta^3}(t_0) = 1, \quad y^{\Delta^4}(t_0) = 3.$$

9.2 Inhomogeneous Caputo Fractional Dynamic Equations with Constant Coefficients

Consider the equation

$$\sum_{k=1}^{m} A_k \, ^C D_{\Delta,t_0}^{\alpha_k} y(t) + A_0 y(t) = f(t), \quad t \in \Omega, \tag{9.14}$$

where

$$0 < \alpha_1 < \ldots < \alpha_m,$$
$$\alpha_i \in (l_i - 1, l_i], \quad l_i \in \mathbb{N}, \quad i \in \{1, \ldots, m\},$$
$$A_0, A_k \in \mathbb{R}, \quad k \in \{1, \ldots, m\}, \quad A_m \neq 0,$$

$f : \mathbb{T} \to \mathbb{R}$ is a given function, subject to the initial conditions

$$D_{\Delta}^j y(t_0) = d_j, \quad j \in \{0, \ldots, l_m - 1\}, \tag{9.15}$$

where $d_j, j \in \{0, \ldots, l_m - 1\}$, are given constants. We will search for a particular solution y_p of equation (9.14) subject to the initial conditions

$$D_{\Delta}^j y_p(t_0) = 0, \quad j \in \{0, \ldots, l_m - 1\}.$$

We take the Laplace transform of both sides of (9.14), and we get

$$\mathscr{L}(f(t))(z) = \mathscr{L}\left(\sum_{k=1}^{m} A_k \, ^C D_{\Delta,t_0}^{\alpha_k} y_p(t) + A_0 y_p(t)\right)(z, t_0)$$

$$= \mathscr{L}\left(\sum_{k=1}^{m} A_k \, ^C D_{\Delta,t_0}^{\alpha_k} y_p(t)\right)(z, t_0) + \mathscr{L}\left(A_0 y_p(t)\right)(z, t_0)$$

$$= \sum_{k=1}^{m} A_k \mathscr{L}\left(^C D_{\Delta,t_0}^{\alpha_k} y_p(t)\right)(z, t_0) + A_0 \mathscr{L}\left(y_p(t)\right)(z, t_0)$$

$$= \sum_{k=1}^{m} A_k z^{\alpha_k} \mathscr{L}(y_p(t))(z, t_0) + A_0 \mathscr{L}(y_p(t))(z, t_0)$$

$$= \left(A_0 + \sum_{k=1}^{m} A_k z^{\alpha_k}\right) \mathscr{L}(y_p(t))(z, t_0),$$

whereupon

$$\mathscr{L}\left(y_p(t)\right)(z, t_0) = \frac{\mathscr{L}\left(f(t)\right)(z)}{A_0 + \sum_{k=1}^{m} A_k z^{\alpha_k}}.$$

Let

$$P_{\alpha_1,\dots,\alpha_m}(z) = A_0 + \sum_{k=1}^{m} A_k z^{\alpha_k}.$$

Then

$$\mathscr{L}\left(y_p(t)\right)(z, t_0) = \frac{\mathscr{L}\left(f(t)\right)(z)}{P_{\alpha_1,\dots,\alpha_m}(z)}.$$

From this, we obtain

$$y_p(t) = \mathscr{L}^{-1}\left(\frac{\mathscr{L}\left(f(t)\right)(z)}{P_{\alpha_1,\dots,\alpha_m}(z)}\right)(t), \quad t \in \Omega.$$

We introduce the Laplace fractional analogue of the Green function as follows:

$$G_{\alpha_1,\dots,\alpha_m}(t) = \mathscr{L}^{-1}\left(\frac{1}{P_{\alpha_1,\dots,\alpha_m}(z)}\right)(t), \quad t \in \Omega.$$

Then

$$\begin{aligned} y_p(t) &= \mathscr{L}^{-1}\left(\mathscr{L}\left(f(t)\right)(z)\mathscr{L}\left(G_{\alpha_1,\dots,\alpha_m}(t)\right)(z)\right)(t) \\ &= \mathscr{L}^{-1}\left(\mathscr{L}\left(f \star G_{\alpha_1,\dots,\alpha_m}\right)(z)\right)(t) \\ &= \left(f \star G_{\alpha_1,\dots,\alpha_m}\right)(t), \quad t \in \Omega. \end{aligned}$$

Let y_h be the solution of the equation

$$\sum_{k=1}^{m} A_k {}^C D_{\Delta,t_0}^{\alpha_k} y(t) + A_0 y(t) = 0, \quad t \in \Omega,$$

subject to the initial conditions (9.15). Note that y_h can be found as in the previous section. Then

$$y(t) = y_p(t) + y_h(t), \quad t \in \Omega,$$

is a solution of the problem (9.14), (9.15).

Exercise 9.3. Find the solution of the Cauchy problem

$$^C D^{\frac{28}{9}}_{\Delta, t_0} y(t) - 4y(t) = h_1(t, t_0), \quad t \in \Omega,$$

$$y(t_0) = -1, \quad y^\Delta(t_0) = -3,$$
$$y^{\Delta^2}(t_0) = 2, \quad y^{\Delta^3}(t_0) = -1.$$

Exercise 9.4. Find the solution of the Cauchy problem

$$^C D^{\frac{13}{4}}_{\Delta, t_0} y(t) - 2 {}^C D^{\frac{1}{3}}_{\Delta, t_0} y(t) - y(t) = \sin_1(t, t_0), \quad t \in \Omega,$$

$$y(t_0) = 5, \quad y^\Delta(t_0) = 2,$$
$$y^{\Delta^2}(t_0) = 1, \quad y^{\Delta^3}(t_0) = -3.$$

9.3 Advanced Practical Problems

Problem 9.1. Find the solution of the Cauchy problem

$$^C D^{\frac{22}{7}}_{\Delta, t_0} y(t) - 8y(t) = 0, \quad t \in \Omega,$$

$$y(t_0) = -10, \quad y^\Delta(t_0) = -2,$$
$$y^{\Delta^2}(t_0) = 4, \quad y^{\Delta^3}(t_0) = -1.$$

Problem 9.2. Find the solution of the Cauchy problem

$$^C D^{\frac{17}{4}}_{\Delta, t_0} y(t) - 8y(t) = 0, \quad t \in \Omega,$$

$$y(t_0) = -1, \quad y^\Delta(t_0) = 2,$$
$$y^{\Delta^2}(t_0) = 4, \quad y^{\Delta^3}(t_0) = -1, \quad y^{\Delta^4}(t_0) = 1.$$

Problem 9.3. Find the solution of the Cauchy problem

$$^{C}D^{\frac{21}{5}}_{\Delta,t_0}y(t) + 4^{C}D^{\frac{7}{2}}_{\Delta,t_0}y(t) + {}^{C}D^{\frac{3}{5}}_{\Delta,t_0}y(t) - 8y(t) = 0, \quad t \in \Omega,$$

$$y(t_0) = -1, \quad y^{\Delta}(t_0) = 2,$$
$$y^{\Delta^2}(t_0) = -4, \quad y^{\Delta^3}(t_0) = -1, \quad y^{\Delta^4}(t_0) = 1.$$

Problem 9.4. Find the solution of the Cauchy problem

$$^{C}D^{\frac{25}{6}}_{\Delta,t_0}y(t) + {}^{C}D^{\frac{15}{4}}_{\Delta,t_0}y(t) + {}^{C}D^{\frac{7}{8}}_{\Delta,t_0}y(t) - y(t) = 0, \quad t \in \Omega,$$

$$y(t_0) = -11, \quad y^{\Delta}(t_0) = 2,$$
$$y^{\Delta^2}(t_0) = 1, \quad y^{\Delta^3}(t_0) = -1, \quad y^{\Delta^4}(t_0) = -3.$$

Problem 9.5. Find the solution of the Cauchy problem

$$^{C}D^{\frac{7}{2}}_{\Delta,t_0}y(t) + 2^{C}D^{\frac{1}{2}}y(t) - 8y(t) = e_1(t, t_0), \quad t \in \Omega,$$

$$y(t_0) = 1, \quad y^{\Delta}(t_0) = 2,$$
$$y^{\Delta^2}(t_0) = 3, \quad y^{\Delta^3}(t_0) = -1.$$

Chapter 10
Appendix: The Gamma Function

10.1 Definition of the Gamma Function

Definition 10.1. The gamma function $\Gamma(z)$ is defined by the integral

$$\Gamma(z) = \int_0^\infty e^{-t}t^{z-1}dt. \tag{10.1}$$

Note that $\Gamma(z)$ converges in the right half of the complex plane, $Re(z) > 0$. Indeed, for $z = x + iy$, $x, y \in \mathbb{R}$, we have

$$
\begin{aligned}
\Gamma(x+iy) &= \int_0^\infty e^{-t}t^{x+iy-1}dt \\
&= \int_0^\infty e^{-t}t^{x-1}t^{iy}dt \\
&= \int_0^\infty e^{-t}t^{x-1}e^{iy\log t}dt \\
&= \int_0^\infty e^{-t}t^{x-1}\left(\cos\left(y\log t\right) + i\sin\left(y\log t\right)\right)dt.
\end{aligned}
\tag{10.2}
$$

Since

$$\cos\left(y\log t\right) + i\sin\left(y\log t\right)$$

is bounded for all t, the convergence of (10.2) at infinity is provided by e^{-t}, and for the convergence at $t = 0$, we must have $x = Re(z) > 0$.

10.2 Some Properties of the Gamma Function

Theorem 10.1. *Let $Re(z) > 0$. Then*

$$\Gamma(z+1) = z\Gamma(z).$$

© Springer International Publishing AG, part of Springer Nature 2018
S. G. Georgiev, *Fractional Dynamic Calculus and Fractional Dynamic
Equations on Time Scales*, https://doi.org/10.1007/978-3-319-73954-0_10

Proof. We have

$$\Gamma(z+1) = \int_0^\infty e^{-t} t^z \, dt$$

$$= -\int_0^\infty t^z \, de^{-t}$$

$$= -t^z e^{-t} \Big|_{t=0}^{t=\infty} + z \int_0^\infty e^{-t} t^{z-1} \, dt$$

$$= z\Gamma(z).$$

This completes the proof.

We have

$$\Gamma(1) = \int_0^\infty e^{-t} \, dt$$

$$= -e^{-t} \Big|_{t=0}^{t=\infty}$$

$$= 1,$$

$$\Gamma(2) = \Gamma(1+1)$$

$$= 1\Gamma(1)$$

$$= 1$$

$$= 1!,$$

$$\Gamma(3) = \Gamma(2+1)$$

$$= 2\Gamma(2)$$

$$= 2$$

$$= 2!.$$

Suppose that

$$\Gamma(n) = (n-1)! \tag{10.3}$$

for some $n \in \mathbb{N}$. We will prove that

$$\Gamma(n+1) = n!.$$

Indeed, using Theorem 10.1 and (10.3), we get

$$\Gamma(n+1) = n\Gamma(n)$$

$$= n(n - 1)!$$

$$= n!.$$

Consequently, (10.3) holds for all $n \in \mathbb{N}$.

We recall some useful definitions from complex analysis.

Definition 10.2. Given a complex-valued function f of a single complex variable, the derivative of f at a point z_0 in its domain is defined by the limit

$$f'(z_0) = \lim_{z \to z_0} \frac{f(z) - f(z_0)}{z - z_0}.$$

If f is complex differentiable at every point z_0 in an open set U, we say that f is holomorphic on U. We say that f is holomorphic at the point z_0 if it is holomorphic on some neighborhood of the point z_0. We say that f is holomorphic on some nonopen set A if it is holomorphic in an open set containing A.

Definition 10.3. An entire function, also called an integral function, is a complex-valued function that is holomorphic at every finite point in the complex plane.

Definition 10.4. Suppose U is an open subset of the complex plane $\mathbb{C}$, p is an element of U, $f : U \setminus \{p\} \to \mathbb{C}$ is a function that is holomorphic over its domain. If there exist a holomorphic function $g : U \to \mathbb{C}$ such that $g(p)$ is nonzero and a positive integer n such that for all $z \in U \setminus \{p\}$,

$$f(z) = \frac{g(z)}{(z - p)^n}$$

holds, then p is called a pole of f. The smallest such n is called the order of the pole. A pole of order 1 is called a simple pole.

Theorem 10.2. *The gamma function has simple poles at the points $z = -n, n \in \mathbb{N}_0$.*

Proof. We rewrite (10.1) in the following way:

$$\Gamma(z) = \int_0^1 e^{-t} t^{z-1} dt + \int_1^\infty e^{-t} t^{z-1} dt, \quad Re(z) > 0. \tag{10.4}$$

Let $Re(z) = x > 0$. Then

$$Re(z + n) = x + n > 0$$

and

$$t^{z+n} = t^{x+iy+n} \Big|_{t=0}$$

$$= 0, \quad n \in \mathbb{N}_0.$$

Therefore,

$$\int_0^1 e^{-t} t^{z-1} dt = \int_0^1 \left(\sum_{k=0}^{\infty} \frac{(t)^k}{k!} \right) t^{z-1} dt$$

$$= \sum_{k=0}^{\infty} \frac{(-1)^k}{k!} \int_0^1 t^{k+z-1} dt$$

$$= \sum_{k=0}^{\infty} \frac{(-1)^k}{k!} \frac{1}{k+z} t^{k+z} \Big|_{t=0}^{t=1} \qquad (10.5)$$

$$= \sum_{k=0}^{\infty} \frac{(-1)^k}{k!(k+z)}.$$

We set

$$\phi(z) = \int_1^{\infty} e^{-t} t^{z-1} dt.$$

Then

$$\phi(z) = \int_1^{\infty} e^{-t+(z-1)\log t} dt.$$

Note that $e^{-t+(z-1)\log t}$ is a continuous function of z and t for arbitrary z and $t \geq 1$. Also, if $t \geq 1$, it is an entire function of z. Let D be an arbitrary closed domain in the complex plane and set $x_0 = \max_{z \in D} Re(z)$. Then

$$\left| e^{-t} t^{z-1} \right| = \left| e^{(z-1)\log t - t} \right|$$

$$= \left| e^{(x-1)\log t - t + iy\log t} \right|$$

$$= \left| e^{(x-1)\log t - t} \right| \left| e^{iy\log t} \right|$$

$$= e^{(x-1)\log t - t}$$

$$\leq e^{(x_0-1)\log t - t}$$

$$= e^{-t} t^{x_0-1}.$$

Therefore, $\phi(z)$ converges uniformly in D, and differentiation under the integral sign is allowed. Because D has been arbitrarily chosen, we conclude that the function $\phi(z)$ has the above properties in the whole complex plane. Consequently, $\phi(z)$ is an entire function allowing differentiation under the integral sign. From this and (10.4), (10.5), we conclude that

$$\Gamma(z) = \sum_{k=0}^{\infty} \frac{(-1)^k}{k!(k+z)} + \text{entire   function}.$$

Thus, $\Gamma(z)$ has only simple poles, and these are at the points $z = -n$, $n \in \mathbb{N}_0$. This completes the proof.

10.3 Limit Representation of the Gamma Function

Theorem 10.3. *The gamma function can be represented as follows:*

$$\Gamma(z) = \lim_{n \to \infty} \frac{n! n^z}{z(z+1) \ldots (z+n)}, \quad Re(z) > 0. \tag{10.6}$$

Proof. Suppose that $Re(z) > 0$, $z = x + iy$, x and y are real variables. Let

$$f_n(z) = \int_0^n \left(1 - \frac{t}{n}\right)^n t^{z-1} dt.$$

We substitute $\tau = \frac{t}{n}$, and we get

$$t = n\tau,$$

$$dt = n d\tau,$$

$$f_n(z) = \int_0^1 (1 - \tau)^n n^{z-1} \tau^{z-1} n d\tau \tag{10.7}$$

$$= n^z \int_0^1 (1 - \tau)^n \tau^{z-1} d\tau$$

$$= \frac{n^z}{z} (1 - \tau)^n \tau^z \Big|_{\tau=0}^{\tau=1} + \frac{n^z}{z} n \int_0^1 (1 - \tau)^{n-1} \tau^z d\tau$$

$$= n^z \frac{n}{z} \int_0^1 (1 - \tau)^{n-1} \tau^z d\tau$$

$$= n^z \frac{n}{z} \frac{1}{z+1} \left((1 - \tau)^{n-1} \tau^{z+1} \Big|_{\tau=0}^{\tau=1} + (n-1) \int_0^1 (1 - \tau)^{n-2} \tau^{z+1} d\tau \right)$$

$$= n^z \frac{n}{z} \frac{n-1}{z+1} \int_0^1 (1 - \tau)^{n-2} \tau^{z+1} d\tau$$

$$\vdots$$

$$= \frac{n^z n!}{z(z+1) \ldots (z+n-1)(z+n)}.$$

Note that

$$\lim_{n \to \infty} \left(1 - \frac{t}{n}\right)^n = e^{-t}.$$

We set

$$\psi(z) = \int_0^\infty e^{-t} t^{z-1} dt - f_n(z).$$

Then

$$\psi(z) = \int_0^\infty e^{-t} t^{z-1} dt - \int_0^n \left(1 - \frac{t}{n}\right)^n t^{z-1} dt \tag{10.8}$$
$$= \int_0^n \left(e^{-t} - \left(1 - \frac{t}{n}\right)^n\right) t^{z-1} dt + \int_n^\infty e^{-t} t^{z-1} dt.$$

Let $\epsilon > 0$ be arbitrarily chosen. Then there exists $N = N(\epsilon) \in \mathbb{N}$ such that

$$\int_n^\infty e^{-t} t^{x-1} dt < \frac{\epsilon}{3}, \quad \int_n^\infty \left| e^{-t} - \left(1 - \frac{t}{n}\right)^n \right| t^{x-1} dt < \frac{\epsilon}{3}, \quad \frac{1}{2n(x+2)} N^{x+2} < \frac{\epsilon}{3}$$

for all $n \geq N$. We fix $N \in \mathbb{N}$ and take $n > N$. Then using (10.8), we obtain

$$\psi(z) = \int_0^N \left(e^{-t} - \left(1 - \frac{t}{n}\right)^n \right) t^{z-1} dt$$
$$+ \int_N^n \left(e^{-t} - \left(1 - \frac{t}{n}\right)^n \right) t^{z-1} dt \tag{10.9}$$
$$+ \int_n^\infty e^{-t} t^{z-1} dt.$$

We have

$$\left| \int_N^n \left(e^{-t} - \left(1 - \frac{t}{n}\right)^n \right) t^{z-1} dt \right| \leq \int_N^n \left| \left(e^{-t} - \left(1 - \frac{t}{n}\right)^n \right) t^{z-1} \right| dt$$
$$= \int_N^n \left| e^{-t} - \left(1 - \frac{t}{n}\right)^n \right| t^{x-1} dt \tag{10.10}$$
$$\leq \int_N^\infty \left| e^{-t} - \left(1 - \frac{t}{n}\right)^n \right| t^{x-1} dt$$
$$< \frac{\epsilon}{3},$$

and

$$\left| \int_n^\infty e^{-t} t^{z-1} dt \right| \leq \int_n^\infty e^{-t} \left| t^{z-1} \right| dt$$
$$\leq \int_n^\infty e^{-t} t^{x-1} dt \tag{10.11}$$
$$< \frac{\epsilon}{3}.$$

Note that

$$\int_0^t e^\tau \left(1 - \frac{\tau}{n}\right)^{n-1} \frac{\tau}{n} d\tau = -\int_0^t e^\tau \left(1 - \frac{\tau}{n}\right)^n d\tau + \int_0^t e^\tau \left(1 - \frac{\tau}{n}\right)^{n-1} d\tau$$

$$= -e^\tau \left(1 - \frac{\tau}{n}\right)^n \Big|_{\tau=0}^{\tau=t} - \int_0^t e^\tau \left(1 - \frac{\tau}{n}\right)^{n-1} d\tau + \int_0^t e^\tau \left(1 - \frac{\tau}{n}\right)^{n-1} d\tau$$

$$= 1 - e^t \left(1 - \frac{t}{n}\right)^n, \quad n \in \mathbb{N}.$$

Hence,

$$0 < 1 - e^t \left(1 - \frac{t}{n}\right)^n$$

$$= \int_0^t e^\tau \left(1 - \frac{\tau}{n}\right)^{n-1} \frac{\tau}{n} d\tau$$

$$\leq \int_0^t e^\tau \frac{\tau}{n} d\tau$$

$$< e^t \int_0^t \frac{\tau}{n} d\tau$$

$$= e^t \frac{t^2}{2n}, \quad 0 < t < n,$$

and

$$\left| \int_0^N \left(e^{-t} - \left(1 - \frac{t}{n}\right)^n\right) t^{z-1} dt \right| \leq \int_0^N \left| e^{-t} - \left(1 - \frac{t}{n}\right)^n \right| t^{x-1} dt$$

$$\leq \frac{1}{2n} \int_0^N t^{x+1} dt \tag{10.12}$$

$$= \frac{1}{2n(x+2)} N^{x+2}$$

$$< \frac{\epsilon}{3}.$$

Therefore, using (10.9), (10.10), (10.11), (10.12), we get

$$|\psi(z)| < \frac{\epsilon}{3} + \frac{\epsilon}{3} + \frac{\epsilon}{3}$$

$$= \epsilon$$

and

$$\lim_{n \to \infty} f_n(z) = \lim_{n \to \infty} \int_0^n \left(1 - \frac{t}{n}\right)^n t^{z-1} dt$$

$$= \int_0^\infty e^{-t} t^{z-1} dt$$

$$= \Gamma(z).$$

From this and (10.7), we obtain (10.6). This completes the proof.

Remark 10.1. The condition $Re(z) > 0$ in (10.6) can be weakened for $z \neq 0, -1, -2, \ldots$ in the following manner. If $-m < Re(z) < -m + 1$ for some $m \in \mathbb{N}$, then

$$\Gamma(z) = \frac{\Gamma(z + m)}{z(z + 1) \ldots (z + m - 1)}$$

$$= \frac{1}{z(z + 1) \ldots (z + m - 1)} \lim_{n \to \infty} \frac{n^{z+m} n!}{(z + m) \ldots (z + m + n)}$$

$$= \frac{1}{z(z + 1) \ldots (z + m - 1)} \lim_{n \to \infty} \frac{(n - m)^{z+m}(n - m)!}{(z + m)(z + m + 1) \ldots (z + n)}$$

$$= \lim_{n \to \infty} \frac{n^z n!}{z(z + 1) \ldots (z + n)}.$$

Chapter 11
Appendix: The Beta Function

11.1 Definition of the Beta Function

Definition 11.1. Let $z, w \in \mathbb{C}$, $Re(z) > 0$, $Re(w) > 0$. Define the beta function $B(z, w)$ as follows:

$$B(z, w) = \int_0^1 t^{z-1}(1 - t)^{w-1} dt.$$

11.2 Properties of the Beta Function

Suppose that $z, w \in \mathbb{C}$, $Re(z) > 0$, $Re(w) > 0$. In this section we deduce some of the properties of the beta function.

1. Let

$$\tau = 1 - t.$$

Then

$$d\tau = -dt, \quad t = 1 - \tau.$$

Therefore,

$$B(z, w) = \int_1^0 (1 - \tau)^{z-1} \tau^{w-1}(-d\tau)$$

$$= \int_0^1 \tau^{w-1}(1 - \tau)^{z-1} d\tau$$

$$= B(w, z),$$

© Springer International Publishing AG, part of Springer Nature 2018
S. G. Georgiev, *Fractional Dynamic Calculus and Fractional Dynamic Equations on Time Scales*, https://doi.org/10.1007/978-3-319-73954-0_11

i.e.,

$$B(z, w) = B(w, z). \tag{11.1}$$

2. Let

$$y = (1 - t)^{w-1}, \quad x = \frac{t^z}{z}.$$

Then

$$dy = -(w - 1)(1 - t)^{w-2}dt,$$
$$dx = t^{z-1}dt,$$

and

$$t^z = t^{z-1} - t^{z-1}(1 - t).$$

Then

$$\begin{aligned}
B(z, w) &= \int_0^1 t^{z-1}(1 - t)^{w-1}dt \\
&= \frac{1}{z} \int_0^1 (1 - t)^{w-1}d\left(t^z\right) \\
&= \frac{1}{z}(1 - t)^{w-1}t^z\Big|_{t=0}^{t=1} + \frac{w - 1}{z} \int_0^1 t^z(1 - t)^{w-2}dt \\
&= \frac{w - 1}{z} \int_0^1 t^z(1 - t)^{w-2}dt \\
&= \frac{w - 1}{z} \int_0^1 \left(t^{z-1} - t^{z-1}(1 - t)\right)(1 - t)^{w-2}dt \\
&= \frac{w - 1}{z} \int_0^1 t^{z-1}(1 - t)^{w-2}dt - \frac{w - 1}{z} \int_0^1 t^{z-1}(1 - t)^{w-1}dt \\
&= \frac{w - 1}{z}B(z, w - 1) - \frac{w - 1}{z}B(z, w).
\end{aligned}$$

Hence,

$$B(z, w) + \frac{w - 1}{z}B(z, w) = \frac{w - 1}{z}B(z, w - 1),$$

or

$$\frac{w+z-1}{z}B(z,w) = \frac{w-1}{z}B(z, w-1),$$

or

$$B(z,w) = \frac{w-1}{w+z-1}B(z, w-1).$$

As above, using (11.1), we obtain

$$B(z,w) = B(w,z)$$

$$= \frac{z-1}{w+z-1}B(w, z-1)$$

$$= \frac{z-1}{w+z-1}B(z-1, w).$$

Therefore,

$$\frac{z-1}{w+z-1}B(z-1, w) = \frac{w-1}{w+z-1}B(z, w-1),$$

or

$$B(z-1, w) = \frac{w-1}{z-1}B(z, w-1).$$

If we set

$$p = z-1, \quad q = w-1,$$

we get

$$B(p, q+1) = \frac{q}{p}B(p+1, q).$$

Let $w = n, n \in \mathbb{N}$. Then

$$B(z, n) = \frac{n-1}{z+n-1}B(z, n-1)$$

$$= \frac{n-1}{z+n-1}\frac{n-2}{z+n-2}B(z, n-2)$$

$$\vdots$$

$$= \frac{n-1}{z+n-1}\frac{n-2}{z+n-2}\cdots\frac{1}{z+1}B(z, 1).$$

On the other hand,

$$B(z, 1) = \int_0^1 t^{z-1}dt$$

$$= \frac{t^z}{z}\Big|_{t=0}^{t=1}$$

$$= \frac{1}{z}.$$

Consequently,

$$B(z, n) = \frac{(n-1)!}{z(z+1)\dots(z+n-1)}.$$

If $m \in \mathbb{N}$, then

$$B(m, n) = \frac{(n-1)!}{m(m+1)\dots(m+n-1)}$$

$$= \frac{(n-1)!(m-1)!}{(m+n-1)!}.$$

3. Now consider $B(z, z)$. We have

$$B(z, z) = \int_0^1 t^{z-1}(1-t)^{z-1}dt$$

$$= \int_0^1 \left(\frac{1}{4} - \left(\frac{1}{2} - t\right)^2\right)^{z-1} dt$$

$$= \int_0^{\frac{1}{2}} \left(\frac{1}{4} - \left(\frac{1}{2} - t\right)^2\right)^{z-1} dt + \int_{\frac{1}{2}}^1 \left(\frac{1}{4} - \left(\frac{1}{2} - t\right)^2\right)^{z-1} dt$$

$$= \int_0^{\frac{1}{2}} \left(\frac{1}{4} - \left(\frac{1}{2} - t\right)^2\right)^{z-1} dt + \int_0^{\frac{1}{2}} \left(\frac{1}{4} - \left(\frac{1}{2} - t\right)^2\right)^{z-1} dt$$

$$= 2\int_0^{\frac{1}{2}} \left(\frac{1}{4} - \left(\frac{1}{2} - t\right)^2\right)^{z-1} dt.$$

For $t \in \left[0, \frac{1}{2}\right]$, we set

$$\frac{\sqrt{y}}{2} = \frac{1}{2} - t.$$

Then

$$t = \frac{1 - \sqrt{y}}{2},$$

$$dt = -\frac{1}{4\sqrt{y}}dy, \quad t \in \left[0, \frac{1}{2}\right].$$

Hence,

$$B(z, z) = 2\int_{1}^{0} \left(\frac{1}{4} - \frac{y}{4}\right)^{z-1} \left(-\frac{1}{4\sqrt{y}}\right) dy$$

$$= \frac{2}{2^{2z}} \int_{0}^{1} y^{-\frac{1}{2}}(1 - y)^{z-1} dy$$

$$= \frac{1}{2^{2z-1}} \int_{0}^{1} y^{\frac{1}{2}-1}(1 - y)^{z-1} dy$$

$$= \frac{1}{2^{2z-1}} B\left(\frac{1}{2}, z\right),$$

i.e.,

$$B(z, z) = \frac{1}{2^{2z-1}} B\left(\frac{1}{2}, z\right).$$

4. Let

$$t = \frac{y}{y + 1}, \quad t \in [0, 1].$$

Then

$$1 - t = 1 - \frac{y}{1 + y}$$

$$= \frac{1 + y - y}{1 + y}$$

$$= \frac{1}{1 + y},$$

350 Appendix: The Beta Function

$$dt = \frac{1}{1+y}dy - \frac{y}{(1+y)^2}dy$$

$$= \frac{1+y-y}{(1+y)^2}dy$$

$$= \frac{1}{(1+y)^2}dy.$$

Hence,

$$B(z, w) = \int_0^1 t^{z-1}(1-t)^{w-1}dt$$

$$= \int_0^\infty \frac{y^{z-1}}{(1+y)^{z-1}}\frac{1}{(1+y)^{w-1}}\frac{1}{(1+y)^2}dy$$

$$= \int_0^\infty \frac{y^{z-1}}{(1+y)^{z+w}}dy.$$

5. By (10.1), we have

$$\Gamma(z) = \int_0^\infty t^{z-1}e^{-t}dt.$$

Let

$$t = xy, \quad x > 0.$$

Then

$$dt = xdy$$

and

$$\Gamma(z) = \int_0^\infty x^{z-1}y^{z-1}e^{-xy}xdy$$

$$= \int_0^\infty x^z y^{z-1}e^{-xy}dy,$$

whereupon

$$\frac{\Gamma(z)}{x^z} = \int_0^\infty y^{z-1}e^{-xy}dy.$$

Hence,

$$\frac{\Gamma(z+w)}{(1+x)^{z+w}} = \int_0^\infty y^{z+w-1} e^{-(1+x)y} dy.$$

6. We have

$$\Gamma(z+w)B(z, w) = \Gamma(z+w) \int_0^\infty \frac{t^{z-1}}{(1+t)^{z+w}} dt$$

$$= \int_0^\infty \frac{\Gamma(z+w)}{(1+t)^{z+w}} t^{z-1} dt$$

$$= \int_0^\infty \left(\int_0^\infty y^{z+w-1} e^{-(1+t)y} dy \right) t^{z-1} dt$$

$$= \int_0^\infty \left(\int_0^\infty y^{z+w-1} e^{-y} e^{-ty} dy \right) t^{z-1} dt$$

$$= \int_0^\infty \left(\int_0^\infty t^{z-1} e^{-ty} dt \right) y^{z+w-1} e^{-y} dy$$

$$= \int_0^\infty \left(\int_0^\infty (ty)^{z-1} e^{-ty} dt \right) y^w e^{-y} dy$$

$$= \int_0^\infty \left(\int_0^\infty p^{z-1} e^{-p} dp \right) y^{w-1} e^{-y} dy$$

$$= \Gamma(z) \int_0^\infty y^{w-1} e^{-y} dy$$

$$= \Gamma(z)\Gamma(w).$$

Therefore,

$$B(z, w) = \frac{\Gamma(z)\Gamma(w)}{\Gamma(z+w)}.$$

7. Let

$$e^{-x} = y.$$

Then

$$x = \log \frac{1}{y}$$

$$= \lim_{n\to\infty} \left(n \left(1 - y^{\frac{1}{n}} \right) \right)$$

$$= \lim_{n \to \infty} \frac{1 - y^{\frac{1}{n}}}{\frac{1}{n}}$$

$$= \lim_{n \to \infty} \frac{-y^{\frac{1}{n}} \log y \left(-\frac{1}{n^2} \right)}{-\frac{1}{n^2}}$$

$$= -\log y$$

$$= \log \frac{1}{y}.$$

Therefore,

$$\Gamma(z) = \int_0^\infty x^{z-1} e^{-x} dx$$

$$= -\int_1^0 \lim_{n \to \infty} \left(n \left(1 - y^{\frac{1}{n}} \right) \right)^{z-1} y \frac{dy}{y}$$

$$= \lim_{n \to \infty} \left(n^{z-1} \int_0^1 \left(1 - y^{\frac{1}{n}} \right)^{z-1} dy \right).$$

Let now

$$t^n = y.$$

Then

$$dy = n t^{n-1} dt$$

and

$$\Gamma(z) = \lim_{n \to \infty} \left(n^{z-1} \int_0^1 (1 - t)^{z-1} n t^{n-1} dt \right)$$

$$= \lim_{n \to \infty} \left(n^z \int_0^1 t^{n-1} (1 - t)^{z-1} dt \right)$$

$$= \lim_{n \to \infty} \left(n^z B(n, z) \right)$$

$$= \lim_{n \to \infty} n^z \frac{(n-1)!}{z(z+1) \dots (z+n-1)}$$

$$= \lim_{n \to \infty} n^z \frac{n!}{z(z+1) \dots (z+n)}$$

$$= \lim_{n \to \infty} n^z \frac{1.2 \dots n}{z(z+1) \dots (z+n)}$$

$$= \lim_{n\to\infty} n^z \frac{1.2\ldots n}{z\left(1+\frac{z}{1}\right)2\left(1+\frac{z}{2}\right)\ldots n\left(1+\frac{z}{n}\right)}$$

$$= \frac{1}{z}\lim_{n\to\infty} \frac{n^z}{\left(1+\frac{z}{1}\right)\left(1+\frac{z}{2}\right)\ldots\left(1+\frac{z}{n}\right)}$$

$$= \frac{1}{z}\lim_{n\to\infty} \frac{\left(1+\frac{1}{1}\right)^z \left(1+\frac{1}{2}\right)^z \ldots \left(1+\frac{1}{n-1}\right)^z}{\left(1+\frac{z}{1}\right)\left(1+\frac{z}{2}\right)\ldots\left(1+\frac{z}{n}\right)}$$

$$= \frac{1}{z}\lim_{n\to\infty} \frac{\left(1+\frac{1}{1}\right)^z \left(1+\frac{1}{2}\right)^2 \ldots \left(1+\frac{1}{n}\right)^z}{\left(1+\frac{z}{1}\right)\left(1+\frac{z}{2}\right)\ldots\left(1+\frac{z}{n}\right)}$$

$$= \frac{1}{z}\prod_{n=1}^{\infty} \frac{\left(1+\frac{1}{n}\right)^z}{1+\frac{z}{n}},$$

i.e.,

$$z\Gamma(z) = \prod_{n=1}^{\infty} \frac{\left(1+\frac{1}{n}\right)^z}{1+\frac{z}{n}},$$

whereupon

$$\Gamma(z+1) = \prod_{n=1}^{\infty} \frac{\left(1+\frac{1}{n}\right)^z}{1+\frac{z}{n}}. \tag{11.2}$$

8. By (11.2), we get

$$\Gamma(1-z) = \prod_{n=1}^{\infty} \frac{\left(1+\frac{1}{n}\right)^{-z}}{1-\frac{z}{n}}.$$

Hence,

$$\Gamma(z)\Gamma(1-z) = \left(\frac{1}{z}\prod_{n=1}^{\infty} \frac{\left(1+\frac{1}{n}\right)^z}{1+\frac{z}{n}}\right)\left(\prod_{n=1}^{\infty} \frac{\left(1+\frac{1}{n}\right)^{-z}}{1-\frac{z}{n}}\right)$$

$$= \frac{1}{z\prod_{n=1}^{\infty}\left(1-\frac{z^2}{n^2}\right)}.$$

Now, using that

$$\sin(\pi z) = \pi z \prod_{n=1}^{\infty} \left(1 - \frac{z^2}{n^2}\right),$$

we get

$$\Gamma(z)\Gamma(1-z) = \frac{\pi}{\sin(\pi z)}. \tag{11.3}$$

9. We put $z = \frac{1}{2}$ in (11.3), and we get

$$\left(\Gamma\left(\frac{1}{2}\right)\right)^2 = \frac{\pi}{\sin\frac{\pi}{2}}$$

$$= \pi.$$

Therefore,

$$\Gamma\left(\frac{1}{2}\right) = \sqrt{\pi}.$$

10. We have

$$\Gamma\left(z+\frac{1}{2}\right) = \lim_{n\to\infty} n^{z+\frac{1}{2}} \frac{1.2\ldots n}{\left(z+\frac{1}{2}\right)\left(z+\frac{1}{2}+1\right)\ldots\left(z+\frac{1}{2}+n\right)},$$

$$\Gamma(2z) = \lim_{n\to\infty} (2n)^{2z} \frac{1.2\ldots(2n)}{2z(2z+1)(2z+2)\ldots(2z+2n)}.$$

Hence,

$$2^{2z-1} \frac{\Gamma(z)\Gamma\left(z+\frac{1}{2}\right)}{\Gamma(2z)} = 2^{2z-1} \lim_{n\to\infty} \frac{n^{2z+\frac{1}{2}}(n!)^2}{(2n)!(2n)^{2z}}$$

$$\times \frac{2z(2z+1)\ldots(2z+2n)}{z(z+1)\ldots(z+n)\left(z+\frac{1}{2}\right)\left(z+\frac{1}{2}+1\right)\ldots\left(z+\frac{1}{2}+n\right)}$$

$$= \frac{1}{2} \lim_{n\to\infty} \frac{(n!)^2 n^{\frac{1}{2}}}{(2n)!}$$

$$\times \frac{2^{2n+2} z(2z+1)(z+1)(2z+3)\ldots(z+n)}{z(z+1)\ldots(z+n)(2z+1)(2z+3)\ldots(2z+2n+1)}$$

$$= \lim_{n \to \infty} \frac{(n!)^2 n^{\frac{1}{2}} 2^{2n+1}}{(2n)!(2z + 2n + 1)}$$

$$= \lim_{n \to \infty} \frac{(n!)^2 (2n) 2^{2n}}{(2n)!(2z + 2n + 1)n^{\frac{1}{2}}}$$

$$= \lim_{n \to \infty} \frac{2^{2n}(n!)^2}{(2n)!n^{\frac{1}{2}}},$$

i.e.,

$$2^{2z-1} \frac{\Gamma(z)\Gamma\left(z + \frac{1}{2}\right)}{\Gamma(2z)} = \lim_{n \to \infty} \frac{2^{2n}(n!)^2}{(2n)!n^{\frac{1}{2}}}. \tag{11.4}$$

11. We put $z = \frac{1}{2}$ in (11.4), and we get

$$\lim_{n \to \infty} \frac{2^{2n}(n!)^2}{(2n)!n^{\frac{1}{2}}} = \sqrt{\pi}.$$

11.3   An Application

Consider the integral

$$I = \int_0^{\frac{\pi}{2}} \sin^n x \cos^m x dx.$$

We set

$$y = \sin^2 x, \quad x \in \left[0, \frac{\pi}{2}\right].$$

Then

$$\sin^n x = \left(\sin^2 x\right)^{\frac{n}{2}}$$

$$= y^{\frac{n}{2}},$$

$$\cos^m x = \left(\cos^2 x\right)^{\frac{m}{2}}$$

$$= \left(1 - \sin^2 x\right)^{\frac{m}{2}}$$

$$= (1 - y)^{\frac{m}{2}},$$

$$dy = 2 \sin x \cos x dx$$

$$= 2 \left(\sin^2 x \right)^{\frac{1}{2}} \left(\cos^2 x \right) dx$$

$$= 2 y^{\frac{1}{2}} (1 - y)^{\frac{1}{2}} dx,$$

$$dx = \frac{1}{2} y^{-\frac{1}{2}} (1 - y)^{-\frac{1}{2}} dy.$$

Hence,

$$I = \frac{1}{2} \int_0^1 y^{\frac{n}{2}} (1 - y)^{\frac{m}{2}} y^{-\frac{1}{2}} (1 - y)^{-\frac{1}{2}} dy$$

$$= \frac{1}{2} \int_0^1 y^{\frac{n+1}{2} - 1} (1 - y)^{\frac{m+1}{2} - 1} dy$$

$$= \frac{1}{2} B \left(\frac{n+1}{2}, \frac{m+1}{2} \right)$$

$$= \frac{1}{2} \frac{\Gamma \left(\frac{n+1}{2} \right) \Gamma \left(\frac{m+1}{2} \right)}{\Gamma \left(\frac{m+n}{2} + 1 \right)}.$$

References

1. Atici, F., & Eloe, P. (2007). Fractional q-calculus on a time scale. *Journal of Nonlinear Mathematical Physics, 14*(3), 341–352.
2. Bohner, M., & Georgiev, S. (2017). *Multivariable dynamic calculus on time scales.* Cham: Springer.
3. Bohner, M., & Guseinov, G. Sh. (2007). The convolution on time scales. *Abstract and Applied Analysis, 2007.*
4. Bohner, M., & Guseinov, G. Sh. (2010). The Laplace transform on isolated time scales. *Computers and Mathematics with Applications, 60,* 1536–1547.
5. Bohner, M., & Peterson, A. (2003). *Dynamic equations on time scales: An introduction with applications.* Boston: Birkhäuser.
6. Hilger, S. (1988). *Ein Maßkettenkalkül mit Anwendung auf Zentrumsmannigfaltigkeiten.* PhD Thesis, Universität Würzburg.
7. Hilger, S. (1999). Special functions, Laplace and Fourier transform on measure chains. *Dynamic Systems and Applications, 8*(3–4), 471–488 [Special Issue on "Discrete and Continuous Hamiltonian Systems," edited by R. P. Agarwal and M. Bohner].
8. Jury, E. (1964). *Theory and applications of the z-transform.* New York: John Wiley & Sons, Inc.
9. Marks, R., Gravagne, I., & Davis, J. (2008). A generalized Fourier transform and convolution on time scales. *Journal of Mathematical Analysis and Applications, 340,* 901–919.
10. Thomas, A. *Transforms on time scales.* Master's Thesis. University of Georgia, 2003.
11. Zhu, J., & Wu, L. (2015). Fractional Cauchy problem with caputo nabla derivative on time scales. *Abstract and Applied Analysis, 2015.*
12. Zhu, J., & Zhu, Y. (2013). Fractional Cauchy problem with Riemann–Liouville fractional delta derivative on time scales. *Abstract and Applied Analysis, 2013.*

Index

© Springer International Publishing AG, part of Springer Nature 2018
S. G. Georgiev, *Fractional Dynamic Calculus and Fractional Dynamic
Equations on Time Scales*, https://doi.org/10.1007/978-3-319-73954-0

Printed in the United States
By Bookmasters